Stress, Trauma and Synaptic Plasticity

Maxwell Bennett • Jim Lagopoulos

Stress, Trauma and Synaptic Plasticity

Springer

Maxwell Bennett
The University of Sydney
Brain and Mind Centre
Camperdown NSW, NSW, Australia

Jim Lagopoulos
Sunshine Coast Mind and Neuroscience
Thompson Institute
University of Sunshine Coast
Birtinya, QLD, Australia

ISBN 978-3-319-91115-1 ISBN 978-3-319-91116-8 (eBook)
https://doi.org/10.1007/978-3-319-91116-8

Library of Congress Control Number: 2018962162

Cover illustration: Robert Klein-Boonschate, Born, Deveter, The Netherlands. Currently, Walla Walla, NSW, Australia.

This Springer imprint is published by the registered company Springer Nature Switzerland AG
The registered company address is: Gewerbestrasse 11, 6330 Cham, Switzerland

Preface

Stress and trauma bring untold suffering into the lives of individuals, whether that of a genius with depression as described in Virginia Wolf and neuropsychiatry by one of us or in retired ex-service men and women diagnosed with post-traumatic stress disorder. It is for great literary figures to illuminate the humanistic dimensions to this suffering and for neuropsychiatrists to determine what has gone awry in the brain that gives rise to it. This work is concerned with the neuroscientific perspective. Some object to the description of animal studies on stress and trauma occurring between the same covers as a description of human suffering, believing that the former denigrates those experiencing the latter. But evolution has conserved many of the neural networks that support our normal behaviour pertaining to fear and pain; hence it would be very foolish to abandon the study of what has gone awry in the animal brain that gives rise to, for example, anxiety behaviour.

This approach is vindicated here in Chap. 1 which shows that the pattern of grey matter loss in the rodent brain is similar to that found in humans following stress and trauma, as revealed by magnetic resonance imaging. This supports the idea that similar cellular mechanisms might be in play in the two different species. Subsequent research on the cellular basis of this gray matter loss shows that there are changes in the loss of synapses leading to the retraction of neuronal dendrites that quantitatively accounts for the loss of grey matter (Chap. 2).

The discovery that synapse loss is a major accompaniment of stress and trauma, giving rise to the particular pattern of grey matter loss observed in the brain, implicates these regions as constituting a core neural network. This comprises the amygdala, the medial prefrontal cortex and the hippocampus. The synaptic connections forming this network are then considered in Chap. 3, including its modulation by neurons in the basal ganglia of the brain. Given the identification of synapse loss in the core synaptic network as a major phenomenon in stress and trauma, the next two sections (Chaps. 4 and 5) are concerned with molecular mechanisms involved in synaptic plasticity. In particular, two principal molecules implicated in synapse function and elimination are considered, namely brain-derived neurotrophic factor and endocannabinoids.

Emphasis is also given to the importance in synaptic plasticity of long-term potentiation and the role of *N*-methyl-D-aspartate receptors in this form of potentiation. This is because such potentiation is strongly implicated in the sustained memory of the emotional concomitants accompanying stressful and traumatic events. Here two appendices deal with these mechanisms in detail, particularly in relation to synapse regression.

The field of research concerned with the identification of the core synaptic network and its control at the molecular level in stress and trauma is growing apace. This is because of its obvious importance in offering insights into how to ameliorate human suffering. But also we have good animal models involving stress and trauma that provide the opportunity for elucidating the mechanisms involved, which is not the case for other psychiatric disturbances. Given the rate of change of this field, it is likely that this monograph will fall short of the latest developments in the field over the last 18 months or so, 2016–2017. Nevertheless, we hope that this work will at least give senior graduate research students in the subject some insights into the synaptic phenomena involved and plausible hypotheses for how they go awry in stress and trauma.

Camperdown, NSW — Maxwell Bennett
Birtinya, QLD — Jim Lagopoulos

Acknowledgements

We are most grateful to the following for their very considerable contributions, discussions and criticisms of different sections of the present work.

Arnold JC, Chitty KM, Chohan TW, Farnell L, Gibson WG, Hatton SN, Kassem MS, O'Doherty DC, Paquola C, Price WS, Saddiqui S and Stait-Gardner T.

Contents

List of Abbreviations

2-AG	2-arachidonoylglycerol
ACC	Anterior cingulate cortex
ADP	Adenosine diphosphate
AEA	Anandamide
AKAP	A-kinase anchoring protein
AMPAR	Alpha-amino-3-hydroxy-5-methyl-4-isoxazolepropionic acid receptor
AR	Androgen receptor
ATP	Adenosine triphosphate
BA	Basal amygdala
BAe	BA extinction neurons
BAf	BA fear neurons
BLA	Basolateral amygdala
BNST	Bed nucleus of the stria terminalis
CA	*Cornu Ammonis*
CaM	Calmodulin
CaMKII	Calcium calmodulin kinase II
cAMP	Cyclic adenosine monophosphate
CB	Cannabinoid
CCK	Cholecystokinin
CD	Catalytic domain
CEA	Central amygdala
CeA	Central nucleus of the amygdala
CEl	Central lateral
CeL	Lateral central nucleus
CEm	Central medial
CMRglc(ox)	Cortical metabolic rate of oxidized glucose
CR	Conditioned response
CREB	cAMP response element-binding
CRF	Corticotropin-releasing factor
CS	Conditioned stimulus
DGL	Diacylglycerol
DHPG	Dihydroxyphenylglycine
DMN	Default mode network

DSM-IV	Diagnostic and Statistical Manual of Mental Disorders, 4th Edition
DTI	Diffusion tensor imaging
E	Extinction
EBP	Enhancer-binding protein
eCBR	Endocannabinoid receptor
eCBs	Endocannabinoids
EGF	Epidermal growth factor
EGFR	Epidermal growth factor receptor
ERK	Extracellular signal-regulated kinase
F	Fear
FA	Fractional anisotropy
FAAH	Fatty acid amide hydrolase
F-actin	Filamentous actin
fMRI	Functional magnetic resonance imaging
GABA	Gamma-Aminobutyric acid
G-actin	Globular actin
GILZ	Glucocorticoid-induced leucine zipper
Gl	Glucocorticoid
GLP	G9a-like protein
GlR	Glucocorticoid receptor
GMV	Grey matter volume
GPC	G-Protein-Coupled
GR	Glucocorticoid receptor
GRE	Glucocorticoid response elements
HAT	Histone acetyltransferase
HDAC	Histone deacetylases
HMT	Histone methyltransferases
HPA	Hypothalamic–pituitary–adrenocortical
HPC	Hippocampus
iGluRs	Ionotropic glutamate receptors
IL	Infralimbic
IPSC	Inhibitory postsynaptic currents
ITC	Inhibitory intercalated region
ITCd	Dorsal ITC
ITCv	Ventral ITC
LA	Lateral amygdala
LAd	Dorsal lateral nucleus
LTD	Long-term depression
LTP	Long-term potentiation
MAPK	Mitogen-activated protein kinase
MC	Monte Carlo
MeAd	Anterior dorsal
MeCP2	Methyl-CpG binding protein 2
mGluRs	Metabotropic glutamate receptors

MLC	Myosin regulatory light chain
MLCP	Myosin light chain phosphatase
mPFC	Medial prefrontal cortex
MR	Mineralocorticoid receptors
MRI	Magnetic resonance imaging
NAc	Nucleus accumbens
NAPE	*N*-arachidonoyl-phosphatidylethanolamine
NAT	*N*-acetyltransferase
NGF	Nerve growth factor
NMDA	*N*-methyl-D-aspartate
NMDAR	*N*-methyl-D-aspartate receptor
NRG	Neuregulin
N-WASP	Neuronal Wiskott–Aldrich syndrome protein
PAG	Periaqueductal grey
PB	Peribrachial nuclei
PCR	Polymerase chain reaction
PET	Positron emission tomography
PFL	Prefrontal cortex
PKA	Protein kinase A
PL	Prelimbic
PLD	Phospholipase D
PP1	Protein phosphatase 1
PTSD	Post-traumatic stress disorder
PV	Parvalbumin
rACC	Rostral anterior cingulate cortex
rCBF	Regional cerebral blood flow
SEM	Standard error of mean
SERT	Serotonin transporter
SN	Salience network
SNP	Single-nucleotide polymorphism
SOM	Somatostatin
TARP	Transmembrane AMPAR regulatory protein
THC	Tetrahydrocannabinol
tPA	Tissue plasminogen activator
TRCP3	Transient receptor potential canonical 3
TrkB	Tropomyosin receptor kinase B
UR	Unconditioned response
US	Unconditioned stimulus
VASP	Vasodilator-stimulated phosphoprotein
VBM	Voxel-based morphometry
VGCC	Voltage-gated calcium (Ca2+) channels
VIP	Vasoactive intestinal peptide
vmPFC	Ventral-medial prefrontal cortex
VTA	Ventral tegmental area

List of Figures

distinct CpG sites, which can be methylated and interact selectively with MeCp2 to form complexes that repress gene transcription (see also Fig. 4.1). Histone methyltransferases (HMT) are responsible for adding methyl groups at histone tails [panel (**a**)], whereas histone deacetylases (HDAC) remove acetylation at histone tails [panel (**b**)], both processes that repress gene transcription. Moreover, low levels of nicotinamine adenine dinucleotide (NAD) promote DNA methylation at the BDNF locus. BDNF gene activation is associated with increased histone H3 and H4 acetylation, which is mediated by histone acetyl transferase (HAT) activity. DNA demethylation might be facilitated by growth arrest and DNA damage proteins such as Gadd45b. An increased binding of CREB to its specific binding protein, CREB binding protein (CBP), is also associated with an increase in BDNF gene transcription. See Boulle et al. (2012) 93

Fig. 4.4 Schematic representation of the signaling and epigenetic pathways in granule neurons of the dentate gyrus thought to be involved in the consolidation process of memory formation after a psychologically stressful challenge. Activation of NMDAR results in stimulation of the MAPK/ERK signaling cascade, the AC•/PKA cascade and the CaMKII cascade. In conjunction with activated GR, these signaling cascades result in the activation of MSK and ERK leading to the formation of dual histone acetylation marks along the c-Fos promoter and subsequently induction of gene transcription. Signaling via CREB also leads to the same outcome. The induction of gene transcription is thought to be instrumental in the consolidation of memory formation in various stressful learning events. See Trollope et al. (2012) 95

Fig. 4.5 Model for G9a/GLP complex transcriptional activity in the hippocampus during fear memory consolidation. Shown [panels (**a**) and (**b**)] is the role of G9a/GLP in the regulation of chromatin remodelling during long-term memory consolidation. Regulation of histone lysine methylation mediates active and repressive transcriptional regulation of genes in the hippocampus. The changes in chromatin structure result in transcriptional gene silencing in the hippocampus. H3K9me2 dimethylation is associated with transcriptional silencing (not shown). The G9a/GLP complex methyltransferase is specific for producing this modification. Abbreviations: *Ac* acetylation, *M* methylation, *MLLI* histone H3 lysine 4 methyltransferase (which regulates memory formation), *H3K9me2* histone H3 lysine 9 dimethylation, *HAT* histone acetyltransferase, *G9a/GLP* G9a/G9a-like protein (GLP) complex methyltransferase (from Bennett and Lagopoulos 2014) 97

Fig. 5.5 Schematic diagram depicting the regional heterogeneity in the expression of the molecular components of EC signalling between different amygdala subnuclei. The localization of $DAGL_a$, CB1 receptors, FAAH and MGL are depicted based on data summarized in the text. Left top inset: diagram of the amygdala with simplified cellular organization. BLA projection neuron in black, GABAergic interneuron in yellow and glutamatergic inputs in grey. Boxed area is expanded in the main figure. Synaptic localization of EC signalling components in a schematized BLA neuron receiving synaptic inputs from a GABAergic interneuron and glutamatergic input from outside the amygdala, e.g. infralimbic cortex. Known forms of EC-mediated synaptic signalling at these synapses are overlaid, and specific EC ligands noted where data is available (from Ramikie and Patel 2012). (For interpretation of the references to colour in this figure legend, the reader is referred to the web version of this article.) 138

Fig. 6.1 Neural networks subserving the expression of conditioned fear (**a**) and of its extinction (**b**). 'The networks include known pathways and hypothetical links (marked by asterisk) that collectively account for most of the available evidence reviewed in Chaps. 2–4. Solid and dashed lines represent connections that are more or less active, respectively. (**a**) The increased CS responsiveness of CeM output neurons after conditioning likely results from two parallel mechanisms: excitation by glutamatergic BA neurons plus disinhibition from CeL and ICM_{mv} inputs. CeM excitation: CS-induced LA activation causes a BA neuron subtype ('fear neurons', F) to fire and excite CeM cells, whereas another type of BA neuron ('extinction cells', E) is inhibited, possibly by CCK^+ interneurons. Although LA neurons respond transiently to the CS, BA fear neurons, through excitatory interactions with each other and/or with prelimbic (PL) (lower left), would prolong the transient tone signal emanating from LA into persistent CS responses. CeM disinhibition: the excitation of LA cells also leads to the recruitment of ICM_{MD} neurons and of a subgroup of CeL cells, likely PKCδ− (CeL-On) cells. ICM_{MD} neurons would then inhibit ICM_{MV} cells, disinhibiting CeM neurons. In addition, ICM_{MD} cells would inhibit subgroups of CeL neurons, possibly PKCδ+ (CeL-Off) cells. The recruitment of PKCδ− (CeL-On) cells by LAd neurons would cause a further inhibition of PKCδ+ neurons and disinhibition of CeM cells. (**b**) The decreased CS responsiveness of CeM output neurons after extinction likely depends on two parallel mechanisms: disfacilitation of CeM cells and increased feedforward inhibition of CeM neurons. CeM disfacilitation: the rapid extinction of LAd responses to the CS

results in a diminished recruitment of BA fear neurons, disfacilitation of CeM neurons, and, possibly, of CCKδ+ interneurons. As a result, BA extinction cells are disinhibited. Reciprocal excitatory interactions between IL (lower left) and BA might also contribute to enhance the excitability of extinction cells. The disinhibition of extinction cells causes increased excitation of a different set of BA interneurons, possibly PV^+ interneurons, controlling fear cells. CeM inhibition: the reduced CS responsiveness of LAd neurons would cause a disfacilitation of ICM_{MD} neurons and consequent disinhibition of ICM_{MV} neurons. This effect would coincide with an increased excitation of ICM_{MV} cells by inputs from BA extinction cells, thus resulting in an increased feedforward inhibition of CeM cells. The disfacilitation of ICM_{MD} neurons would also cause a disinhibition of subsets of CeL cells, possibly corresponding to PKCδ+ neurons. This effect would be reinforced by the reduced activation of PKC^- cells secondary to reduced LAd inputs. Hypothetical connections (marked by asterisks): (*1 and *2) While it was shown that BA fear and extinction cells differentially project to PL and IL, respectively, whether return mPFC projections are similarly segregated is unknown. (*3 and *4) Currently, there is no data available on the connections of fear and extinction neurons with other amygdala neurons. The differential connections shown are hypotheses based on the available literature. (*5) Little data is available on the connectivity of CCK cells with different types of BA neurons. Trouche et al. (2013) reported that they contact fear (not shown here) and extinction-resistant neurons. CCK synapses to extinction-resistant (but nor fear) neurons showed an upregulation of CB1 receptors after extinction training. The input and output connections of CCK^+ interneurons shown in the figure are all hypothetical. It is possible that other subtypes of interneurons are differentially connected to fear and extinction neurons.' [from Duvarci and Pare (2014); legend to their Fig. 1] 165

Fig. A.1 The actin nucleation model of growth and regression of synaptic spines. There are two pools of F-actin: a stable pool that is retained at the base of the spine head and a dynamic pool, where growth or regression can occur. Processes occurring in the dynamic pool are as follows: *1*, glutamate binds to NMDA receptors and activates a signalling pathway that leads via RhoA and ROCK to the release of profilin which then binds ADP-actin to give ATP-actin-profilin. *2*, activated Arp2/3 complexes form new actin filaments by attachment to an existing filament to form a new side branch. *3*, filaments grow by the attachment of profilin-ATP-actin to their barbed ends, thus pushing the

List of Tables

1 Grey Matter Changes in the Brain Following Stress and Trauma

1.1 During Maturity

1.1.1 Previous Studies

The pathophysiology of post-traumatic stress disorder (PTSD) is characterized by a heightened sensitivity to certain stimuli perceived to be threatening, followed by inability to extinguish the resulting fear (Garfinkel and Liberzon 2009). Elucidating the neurobiology and neurocircuitry of PTSD constitutes an essential step in informing the detection and treatment of the disorder. To this end, significant efforts have focussed on determining why only a proportion of trauma-affected individuals go on to develop PTSD symptoms. Early studies have hypothesized an association between PTSD and atrophy in the hippocampus, amygdala and the prefrontal cortex (Karl et al. 2006). The link between PTSD and stress responses (Davis 1992; McEwen 1995; LeDoux 2000) has underscored the role of the amygdala and hippocampus in particular, although all three of these brain regions are critical for normal fear extinction and the regulation of emotions (Chen et al. 2012a, b, c).

Studies investigating grey matter loss in the amygdala and hippocampus in association with other psychiatric disorders including depression (Siegle et al. 2003; Campbell et al. 2004) and anxiety (Milham et al. 2005) have reported decreased grey matter volumes (GMV). In relation to PTSD, attenuated recruitment of these structures (Vogt et al. 2003) and reduced volumes have also been reported; however, these findings have not been conclusive.

Reprinted from Psychiatry Research. O'Doherty DC, Chitty KM, Saddiqui S, Bennett MR, Lagopoulos J. A systematic review and meta-analysis of magnetic resonance imaging measurement of structural volumes in posttraumatic stress disorder. Vol 232(1):1–33. Copyright (2015). With permission from Elsevier.
Reprinted from Neuroscien Biobehav Rev. Paquola C, Bennett MR, Lagopoulos J. Understanding heterogeneity in grey matter research of adults with childhood maltreatment-A meta-analysis and review. Vol. 2016:299–312. Copyright (2016). With permission from Elsevier.

M. Bennett, J. Lagopoulos, *Stress, Trauma and Synaptic Plasticity*,
https://doi.org/10.1007/978-3-319-91116-8_1

Numerous studies undertaken within the past decade have attempted to identify associations between specific brain regions linked to PTSD and whether indeed volumetric differences are present in people with PTSD. Functional magnetic resonance imaging (fMRI) studies have reported increased activation of the hippocampus, amygdala and the anterior cingulate cortex (ACC) when placing subjects with PTSD under stress. All three regions have been said to play important roles in normal fear extinction as well as the regulation of memories (Herry and Garcia 2002; Myers and Davis 2007). As such, studies have sought to identify morphological characteristics of these specific regions and elucidate their association to core PTSD symptomatology. In this regard one of the most consistent associations is that of hypervigilance and changes within the amygdala, which has been suggested to be resultant from attenuation of top-down inhibition by the pregenual ACC and the hippocampus (Bremner et al. 1999; Liberzon et al. 1999; De Bellis et al. 2000; Rauch et al. 2000, 2006).

Even amongst the research that focuses on a particular brain region, the literature invariably consists of studies that use different imaging modalities, different techniques for measurement, different tools for diagnosing PTSD and heterogeneous groups of subjects with different histories of trauma (Smith 2005). To date, numerous studies have looked at a variety of brain regions and their association with PTSD symptomatology with the majority investigating the hippocampus (Kitayama et al. 2005; Smith 2005; Woon and Hedges 2008, 2011; Hedges and Woon 2010; Woon et al. 2010; Rodrigues et al. 2011), although several others have specifically investigated the amygdala and ACC (de-Almeida et al. 2012; Woon and Hedges 2008, 2009).

1.1.2 Hippocampal Volume

As memory impairment and emotion regulation are core features of PTSD, a significant corpus of work has targeted the hippocampus given its involvement in memory processing (Squire et al. 2004; Brewin et al. 2007; Apfel et al. 2011) as well as the regulation of the hypothalamic–pituitary–adrenocortical (HPA) axis (Buchanan et al. 2004). In human studies, it remains inconclusive as to whether hippocampal grey matter volume loss occurs as a result of PTSD (Bremner 1999) or whether premorbid hippocampal volume loss merely places an individual at a higher risk of developing PTSD (Gilbertson et al. 2002).

Since Bremner et al. 1995 finding of decreased hippocampal volume in veterans suffering from PTSD, numerous studies have produced similar results (Gurvits et al. 1996; Bremner et al. 1997, 2003; Villarreal et al. 2002; Hedges et al. 2003; Lindauer et al. 2004; Shin et al. 2004b; Vythilingam et al. 2005; Emdad et al. 2006; Li et al. 2006; Bonne et al. 2008; Bossini et al. 2008; Felmingham et al. 2009; Apfel et al. 2011; Zhang et al. 2011a, b). Moreover, several meta-analyses have confirmed these observations reporting volume reduction in the hippocampus (Kitayama et al. 2005; Smith 2005).

More recently, there has been an interest in subjects that have been exposed to trauma but that do not develop PTSD. In this regard it has been found that sufferers of PTSD have reduced volumes in the right and left hippocampi as compared to people who have not been exposed to trauma (Karl et al. 2006). Interestingly in these studies, the volume of the left hippocampus has consistently been reported as being smaller in the PTSD patients when compared with trauma-exposed subjects without PTSD. As such it has been suggested that exposure to trauma may be associated with diminished hippocampal volume in both people who develop PTSD and those who do not (Winter and Irle 2004). In addition hippocampal volumes have been found to be related to the length of time since trauma (Villarreal et al. 2002) as well as the severity of the underlying trauma (Gurvits et al. 1996; Bremner et al. 1997; Winter and Irle 2004).

1.1.3 Amygdala Volume

In contrast to the hippocampus, fewer studies have examined the role of the amygdala and its relation to PTSD. Findings from fMRI studies into amygdala activation in PTSD (Rauch et al. 2000; Bremner 2003; Hendler et al. 2003; Shin et al. 2004a, 2005; Lanius et al. 2006; Williams et al. 2006; Etkin and Wager 2007; Morey et al. 2009) have provided theoretical support for further investigation into amygdala volumetric changes. The amygdala has been postulated to play a role in the pathophysiology of PTSD (Bremner et al. 1997; Bonne et al. 2001; Wignall et al. 2004; Pavlisa et al. 2006) given its putative role in the regulation of memory of stressful and traumatic events, behaviour and emotion (Dolan 2007), as well as in fear conditioning and generalisation (LaBar et al. 1998; Pape and Pare 2010; Dunsmoor et al. 2011). Successful learned fear extinction is dependent upon *N*-methyl-D-aspartate (NMDA) receptors in the amygdala. Lower levels of NMDA are thought to impair the function of the amygdala in extinguishing fear (Myers and Davis 2007).

Early volumetric studies of the amygdala report both larger (Bremner et al. 1997) and smaller volumes (Gurvits et al. 1996). More recently, similar, inconclusive findings have been reported. Several studies have reported smaller left amygdala volumes in patients with PTSD (Karl et al. 2006; Woon and Hedges 2009), while others have reported larger left amygdala volume (Pavlisa et al. 2006; Kuo et al. 2012). Interestingly, yet other studies have found no significant differences in amygdala volume (Gurvits et al. 1996; Bonne et al. 2001; Lindauer et al. 2004; Wignall et al. 2004).

Even where volumetric differences have been reported in different subregions of the amygdala in patients with PTSD (Pavlisa et al. 2006; Rogers et al. 2009), it is perhaps prudent to consider the potential limitations of past studies. Some studies have combined data across age groups and used a wider sample of patients (Karl et al. 2006) that have included adolescent data from a number of other studies (De Bellis et al. 1999, 2001b, 2002) in order to examine the moderating effects of age on volumes. This practice of including paediatric data in neural morphological studies is potentially problematic given the disparity in the growth and development of children and adolescents (Sisk and Foster 2004; Hedges et al. 2007; Whitford et al. 2007). Thus the inconsistency across findings for the amygdala volume may in part be attributed to a range of methodological differences rather than the core pathophysiology of the disorder.

1.1.4 Anterior Cingulate Cortex Volume

The third area implicated in the neurocircuitry and pathophysiology of PTSD is the prefrontal cortex and in particular the ACC which includes the dorsal, rostral and ventral subdivisions (Vogt et al. 1992; Bush et al. 1998; Whalen et al. 1998). Investigation into the role the ACC assumes in the anatomy of fear conditioning and PTSD is especially interesting given the numerous neuroimaging studies that have demonstrated associations between activation within this structure and volume in patients with PTSD.

Positron emission tomography (PET) and fMRI studies have indicated how the ACC is involved in the regulation of negative feedback of the HPA during emotional distress (Kitayama et al. 2006), behavioural inhibition (Menon et al. 2001; Kerns et al. 2004) and the regulation of emotion (Ochsner et al. 2002). One theory with regard to the role of the ACC in PTSD is that symptoms occur as the result of amygdala hyperactivity and a corresponding failure of the ACC (Yamasue et al. 2003) to inhibit this response (Shin et al. 2001; Murphy et al. 2003; Damsa et al. 2005; Rauch et al. 2006; Kasai et al. 2008). Animal stress studies have provided further support for this putative role, demonstrating changes to ACC dendritic architecture and the implication of impaired or altered ACC functioning in rodents (Radley et al. 2004, 2006; Cerqueira et al. 2005a, b) as well as in primates (Mathew et al. 2003).

As with the hippocampus and amygdala, previous studies of the ACC in PTSD (using PET, fMRI and magnetic resonance spectroscopy) have found anomalous ACC engagement in PTSD subjects compared with subjects without PTSD. Typically, the ACC demonstrates hypoactivity when compared with subjects without PTSD (Bremner et al. 1999, 2004; Shin et al. 2001; Lanius et al. 2003; Yang et al. 2004).

Similar to findings from neuroimaging studies examining ACC activation and function, different volumetric measurement techniques (i.e. manual tracing, voxel-based analysis and automated segmentation) have been used to investigate ACC morphometry in PTSD, and the various techniques have yielded similar findings of reductions in grey matter, lending support for the ACC's function and effect in emotional regulation and fear conditioning (Cohen et al. 2006).

One study that employed automated segmentation to measure ACC volumes in subjects who reported early life stressors found reduced grey matter volume. In terms of architectural anomalies, changes in ACC dendritic growth have also been observed in these patients following stress tests (Wellman 2001; Radley et al. 2004), and a meta-analysis conducted by Karl et al. (2006) reported significantly reduced ACC volumes in PTSD compared with trauma-exposed controls. Similarly, a study that used a manual tracing method (Woodward et al. 2006) reported similarly reduced ACC volumes in PTSD compared with controls as has Chen et al. (2006a, b), using a morphological Voxel-based morphometry (VBM) analysis, as has Rauch et al. (2003). In the case of the latter study, volumetric differences were detected in the pregenual ACC, while the volumetric difference reported by Yamasue et al. (2003) was localized to the dorsal ACC. Interestingly, voxel-based methods, as employed by Kasai et al. (2008) found smaller ACC volume in combat-exposed twins with PTSD when compared with twins that had not been exposed to combat, as well as when compared with combat-exposed twins with and without PTSD. These findings suggest that ACC volume loss may be a morphological characteristic resulting from PTSD, rather than a pre-existing risk factor for PTSD.

1.1.5 Discussion

Reduced hippocampal and ACC volumes in subjects with PTSD have been consistently reported in the literature. This is in contrast to mixed results for amygdala volumes, which have not shown significant differences between PTSD cohorts and trauma-exposed controls but instead have a medium effect size reduction in bilateral volume when compared with healthy controls.

Whilst the cellular mechanisms resulting in PTSD-related grey matter volume changes have not been fully explicated in adults, the following three mechanisms have been postulated: (1) dendritic remodelling caused by fluctuations in synaptic spine density on different-order dendrites, (2) reduction in brain-derived neurotrophic factor (BDNF) resulting in impaired neurogenesis and (3) glial dysfunctions resulting in an impaired glutamate-glutamine regulation.

Fluctuations in synaptic spine density within cortical regions associated with PTSD have been examined using animal models. A recent study examining stress-induced grey matter loss in rodents (Kassem et al. 2013) found a linear relationship between changes in ACC and hippocampal grey matter and cumulative changes of dendritic volume. This finding suggests that GMV reductions observed in humans afflicted by PTSD may have a similar dendritic basis.

BDNF has been shown to be one of the most active neurotrophins involved in neurogenesis and has a neuroprotective effect against glutamate excitotoxicity (Nibuya et al. 1995, 1996; Vaidya et al. 1999; Almeida et al. 2005). Reduced levels of BDNF in structures implicated in PTSD have been found to correlate to impairments of neurogenesis and subsequent grey matter atrophy (Sapolsky 2000; Sala et al. 2004).

Glial cell dysfunction or reduction is another suspected mechanism of grey matter atrophy in PTSD. Glial cells are necessary for the regulation of glutamate throughout the brain—especially the synaptic cleft via conversion of glutamate into glutamine using the enzyme glutamine synthetase. Any dysfunction and/or loss of glial cells results in an impaired glutamate-glutamine cycle potentially subjecting any afflicted brain structure to glutamate excitotoxicity (Sheline 2000; Reul and Nutt 2008).

The role of a sustained increased level of glucocorticoids involved in stress-related excitotoxicity has been well documented in relation to hippocampal and ACC grey matter atrophy (Arborelius et al. 1999; Tata and Anderson 2010), but differences in the plasticity of these structures compared with the amygdala in the context of PTSD recovery is less understood. The neuropil of the hippocampus, in particular, has been found to be more capable of recovery from atrophy, compared with the ability of the amygdala to recover from stress-related dendritic hypertrophy (Vyas et al. 2002, 2004; Adamec et al. 2012).

Some studies investigating volumetric structural changes raise the issue of genetic influences being a predetermined risk factor for grey matter changes associated with PTSD. A recent study that used combat-exposed monozygotic twins with and without PTSD to investigate ACC volumes (Kasai et al. 2008) concluded that grey matter reductions in limbic and paralimbic structures were acquired due to stress-induced PTSD-related loss, rather than any genetic or familial risk factors.

The intrusive recollection of traumatic events associated with PTSD commonly known as "flashbacks" during waking hours and nightmares when sleeping has been associated with volumetric reductions in the hippocampus and related disruptions in memory (Squire 1992; Eichenbaum 2000; Brewin et al. 2009, 2010). A model developed by Brewin et al. (1996, 2010) and Brewin (2001) explains these flashbacks in terms of two types of hippocampal dependant memory: (1) contextual representational memory, which provides context to memories during retrieval, and (2) sensation-based representational memory, which is involved in the emotional recollections of an event. The model suggests that flashbacks occur when sensation-based representational memory is activated by either an environmental cue or internal thought, without the corresponding contextual representational memory—causing PTSD sufferers to experience the emotional memory out of context and propelling them into a hypervigilant state.

A similar model explored in animals (Rudy et al. 2002, 2004; Rudy and O'Reilly 1999) also suggests that failures in hippocampal encoding of traumatic events result in flashbacks. In this model the hippocampus is responsible in the creation of memories during a traumatic event and correctly encoding multiple conjunctive cues (e.g. visual, aural, odour) during a conditioned fear response. Flashbacks are the result of one or more of these conjunctive cues being recalled outside the context of the original conditioned fear event. Both these models involve reduced hippocampal function associated with reduced neuronal and dendritic-related atrophy found in PTSD patients following traumatic events. Each model can explain the flashbacks experienced in PTSD in terms of hippocampal dysfunction either in memory encoding or retrieval as triggered by external environmental cues or internal thoughts.

Both the amygdala and hippocampus are identified as regions of interest in PTSD due to their role in the processing and coordination of traumatic memories and perception (Rolls 1996; Kensinger and Corkin 2004) and in salience network (SN) processes. SN function within the brain is in the assessment of salient external stimuli and pertinent sensory data (Seeley et al. 2007; Menon and Uddin 2010).

Impaired functioning in the SN can result in dysfunctional processing of stimuli, autonomic arousal, cognition and behaviour regulation.

Reductions in ACC volume are consistent with the SN model and, due to its role in fear extinction and its moderating effect on the amygdala, suggest that dysfunction of the ACC may assume a larger role in PTSD symptoms and severity than previously suspected. As disruption of the SN in PTSD is associated with volume reduction and hypoactivation of the ACC (Rauch et al. 2006), the current and previous mixed findings in relation to amygdala volumes may be explained in relation to an impairment of the top-down processing of a correctly functioning SN.

The top-down processing of the SN may explain why PTSD sufferers are more prone to being affected by certain types of triggers (Cohen and Servanschreiber 1992), e.g. trauma type that may result in a selective dominance of traumatic memory networks (McFarlane et al. 2002). These memory networks are primed to respond in a particular way due to a neural network bias in the way a trigger is perceived and processed (Hoffman and McGlashan 1993).

The nature of the trauma may not be as critical as the way it is processed by the neural network and the particular effect this processing has on an individual's memory and perception (McFarlane et al. 2002). Further to this, impairment within the default mode network (DMN) has also been suggested as a contributing factor to PTSD symptoms, in conjunction with dysfunctional SN (Lanius et al. 2011; Patel et al. 2012). This cascading, inter- and intra-network dysfunction may account for the range of PTSD symptoms experienced by sufferers, along with the discrepancies in the results of studies examining amygdala volumes. Rather than amygdala volume itself being a biomarker and predictor of PTSD susceptibility and severity, recent results suggest the pre- and post-morbid ratio between the ACC and the amygdala might be indicative of SN dysfunction within PTSD.

An additional region involved in SN function is the anterior insula. The anterior insula has been found to co-activate with the ACC and is involved in engaging higher-order control processes, such as attention and working memory, while disengaging other systems not important to a current task (Menon and Uddin 2010). Enhanced activity in the anterior insula has been reported in PTSD fMRI studies (Paulus and Stein 2006; Shin and Liberzon 2009; Lanius et al. 2011) and is thought to be involved in valuing a stimulus in terms of how it might affect the body state. In a dysfunctional SN (postulated to occur in PTSD), the anterior insula is primed to overvalue stimuli as potential threats (Paulus and Stein 2006), contributing to classical PTSD hypervigilance symptoms.

1.2 During Childhood

Childhood trauma is associated with poor social, academic, mental and physical health outcomes (McLeod et al. 2014; Rapoza et al. 2014; Romano et al. 2015). Exposure to severe stress in childhood, through interpersonal, socioeconomic or natural trauma, more than doubles the risk of developing a mental illness by adulthood (Chapman et al. 2004; Varese et al. 2012). The prefrontal-limbic system encompassing the ACC, hippocampus and amygdala is essential for emotion and stress regulation. It has been suggested that childhood trauma-induced structural abnormalities in this prefrontal-limbic system underpin functional deficits and confer enhanced psychiatric risk subsequent to childhood trauma (Teicher and Samson 2013).

The hippocampus was one of the first brain regions investigated in childhood trauma research given its putative involvement in the formation, organization and storage of memory. Early hypotheses on the involvement of the hippocampus were further bolstered by evidence of reduced hippocampal volumes in adults with PTSD (Bremner et al. 1995) as well as preclinical evidence of stress-induced impairments to hippocampal dendrites and cell organization (Sapolsky et al. 1990; Uno et al. 1989). Together these observations formulated the notion that hippocampal development could be altered by early life stress.

Childhood trauma has also been associated with abnormalities of the amygdala, a subcortical structure associated with memory, decision-making and emotional reactions (Janak and Tye 2015), though findings have been mixed thus far. The first MRI study in this field used the amygdala as a comparison region for the hippocampus (Bremner et al. 1997). Later, the amygdala was investigated in its own right due to its close neuroanatomical association with the hippocampus and its involvement in memory and fear. While some studies have shown amygdala volume to be significantly decreased in adults with a history of childhood trauma (Aas et al. 2012; Hoy et al. 2012), others have reported increased volume (Baldacara et al. 2014) or no link between amygdala volume and childhood trauma (Brambilla et al. 2004; Bremner et al. 1997; Cohen et al. 2006; Driessen et al. 2000; Schmahl et al. 2003).

The most common finding to arise from whole brain analyses has been decreased prefrontal grey matter in relation to childhood trauma (Chaney et al. 2014; Kumari et al. 2013; Sheffield et al. 2013; Thomaes et al. 2010; Tomoda et al. 2009b; van Harmelen et al. 2010). The prefrontal cortex (functionally important for coordinated neural responses and executive functions) (Funahashi and Andreau 2013) is sensitive to childhood trauma due to its protracted development, as well as its connections to the HPA axis, hippocampus and amygdala. The prefrontal subregions linked to childhood trauma have varied between studies, however, and not all whole brain analyses have reported a relationship between prefrontal grey matter and childhood trauma (Benedetti et al. 2012; Labudda et al. 2013; Lu et al. 2013a, b; Tomoda et al. 2009b; Van Dam et al. 2014).

Inconsistencies in neuroimaging findings may be due to variability in parameters such as cohort demographics, definition of childhood trauma and the choice between investigating healthy and psychiatric cohorts. For the latter, investigating healthy individuals provides a basis for understanding the effects of childhood maltreatment independent of psychiatric illness and highlights grey matter alterations that underpin psychiatric resilience following childhood trauma (Baker et al. 2013; Carballedo et al. 2012; Cohen et al. 2006; Dannlowski et al. 2012; Lu et al. 2013a, b; Walsh et al. 2014).

To address the association of childhood trauma in conjunction with the development of a psychiatric illness, some studies have investigated individuals with mental illnesses and a history of childhood maltreatment. To reduce the effect of disorder-specific pathologies, high and low trauma groups have been matched for psychiatric health (Chaney et al. 2014; Liao et al. 2013; Sheffield et al. 2013) as well as investigating heterogeneous psychiatric cohorts (Tomoda et al. 2009a, 2011, 2012). These transdiagnostic investigations aid understanding of the general effects of childhood maltreatment in vulnerable individuals, yet they require additional control of confounding factors related to psychiatric health. The combination of these three approaches (healthy, psychiatrically matched and mixed psychiatric cohorts) has led to a more comprehensive understanding of how childhood maltreatment confers psychiatric risk and the different brain regions involved.

Childhood trauma encompasses any severely stressful event occurring before the age of 16, including neglect, abuse, natural disasters and major family disturbances. Studies have used psychiatrist-led interviews and retrospective questionnaires to measure childhood trauma. Depending on the clinical instruments implemented, childhood trauma has been variably restricted to experiences of abuse or neglect, only abuse, only non-emotional abuse (physical or sexual) and only sexual abuse. Differential patterns of behavioural problems and neuroendocrine activity have been reported depending on type of childhood trauma (Cicchetti and Rogosch 2001; van der Put et al. 2015). In juvenile offenders, sexual abuse has been linked to internalizing problems, whereas physical maltreatment has been linked to externalizing problems and violent crimes (van der Put et al. 2015). Physically and sexually abused children have also been shown to have greater morning cortisol than non-maltreated, emotionally maltreated or neglected children (Cicchetti and Rogosch 2001).

In regards to brain structure, it has been postulated that childhood trauma has a general effect on stress-related systems and trauma-specific effects on sensory systems (Tomoda et al. 2012). This proposition is based on evidence of reduced visual cortex volume in adults exposed to domestic violence in childhood and increased superior temporal gyrus grey matter in adults that experienced severe parental verbal abuse (Tomoda et al. 2011, 2012). Additionally, in a study looking at the differential effects of childhood trauma types on the cortex, childhood sexual abuse was linked to thinning of the genital region of the somatosensory cortex, whereas childhood emotional abuse was linked to thinning of the face region (Heim et al. 2013).

Three meta-analyses have previously been conducted on grey matter abnormalities in individuals with a history of childhood trauma. Woon and Hedges (2008) focused on hippocampal and amygdala studies of PTSD subsequent to childhood abuse. Due to strict inclusion parameters, only four adult studies were used in the meta-analysis. They reported that adult-abused cohorts exhibited bilateral reductions in hippocampus. A study by Riem et al. (2015) presented evidence for peak hippocampal sensitivity to childhood maltreatment occurring in middle childhood. Furthermore, they reported the combined effect size of childhood maltreatment on reduction of hippocampal volume was significant in studies measuring multiple types of maltreatment, but not in studies measuring solely sexual abuse, emotional abuse or deprivation.

A recent meta-analysis of whole-brain structural MRI studies reported distributed frontal and temporal grey matter reductions in individuals with a history of childhood maltreatment, as well as increased grey matter in right superior frontal and left middle temporal gyri (Lim et al. 2014). The latter two meta-analyses included a study of PTSD with an unknown number of traumas occurring in childhood or young adulthood (Landre et al. 2010). All three meta-analyses showed the exclusion of child studies altered their main findings, which aligns with the hypothesis of childhood trauma altering the developmental trajectory of brain volumes rather than causing a deleterious insult.

Systematic and critical reviews of childhood maltreatment have been more plentiful (Frodl and O'Keane 2013; Grassi-Oliveira et al. 2008; Hart and Rubia 2012; McCrory et al. 2011; Teicher et al. 2003). These reviews have highlighted the prefrontal-limbic system as most likely to be affected by childhood trauma; both structurally and functionally emphasis has been placed on the complexity of the relationship between childhood trauma and structural abnormalities.

1.2.1 Hippocampal Volume

The hippocampus has been the central focus of grey matter childhood trauma studies with several meta-analyses reporting the hippocampus to be consistently smaller in childhood trauma cohorts. The reported effect size of hippocampal differences between individuals with PTSD subsequent to childhood trauma and healthy individuals without childhood trauma is similar to the comparison of healthy traumatized individuals and healthy no-trauma individuals. Reduced hippocampal volume has been repeatedly reported in adult PTSD, but conjecture still exists as to whether a smaller hippocampus represents a risk factor or acquired feature of the illness (Pitman et al. 2012). Without studies implementing PTSD-matched controls, it is difficult to discern whether reduced hippocampal volume amongst adults with PTSD secondary to childhood abuse is linked to diagnosis or childhood trauma.

Studies composed of three adult groups—PTSD secondary to childhood abuse, history of childhood abuse without PTSD and neither PTSD nor childhood abuse—have presented conflicting results (Bremner et al. 2003; Pederson et al. 2004; Weniger et al. 2009). All three studies have reported both abused groups to have smaller mean hippocampal volumes than the healthy-non abused group. Only two studies reported the PTSD group to exhibit smaller hippocampi than the non PTSD abused group. Furthermore, high levels of early life trauma have been linked to greater hippocampal volumes in individuals with PTSD from urban violence (Baldacara et al. 2014), suggesting an interaction of childhood trauma and adulthood PTSD on hippocampal volume.

Reduced hippocampal volume in depressed individuals with childhood trauma compared to depressed non-maltreated counterparts has been consistently detected (Chaney et al. 2014; Opel et al. 2014; Vythilingam et al. 2002); however, this was not replicated in a study of emotional and physical neglect (Frodl et al. 2010). The interaction of early life adversity and hippocampal volume has been shown to be predictive of the number of depressive episodes later in life (Rao et al. 2010) as well as cumulative illness duration (Frodl et al. 2010). Regression analyses have revealed the difference in hippocampal volume between healthy and depressed individuals can be accounted for by childhood trauma (Opel et al. 2014).

It has been suggested that reduced hippocampal volumes evident in major depression (Schmaal et al. 2016) may stem from the combination of childhood maltreatment and a genetic susceptibility to hippocampal decline, with further hippocampal volume loss occurring throughout disorder course (Frodl et al. 2010). In line with this notion, a negative correlation of the number of stressful early life events with hippocampal volume was detected amongst depressed Val66Met carriers of the BDNF gene, but not depressed Val/Val carriers (Gatt et al. 2009). This gene-environment interaction on hippocampal volume was also found in a psychosis cohort (Aas et al. 2014), but was not evident when healthy individuals were included (Gerritsen et al. 2015; Molendijk et al. 2012).

The definition of childhood trauma and/or the type of experimental design implemented have an effect on the findings of bipolar disorder studies. Whereas hippocampal volume is consistently reduced in abused groups compared to non-abused groups (Brambilla et al. 2004; Weniger et al. 2009), associative studies have not detected a relationship of child abuse/neglect severity with hippocampal volume (Boen et al. 2014; Driessen et al. 2000; Kuhlmann et al. 2013). Hippocampal differences have only been observed in severe or abusive cases of childhood maltreatment amongst individuals who develop bipolar disorder.

Finally, no studies of first episode psychosis have detected a main effect of childhood trauma on hippocampal volume (Aas et al. 2012, 2014; Hoy et al. 2012; Sheffield et al. 2013). However, after accounting for BDNF met carriers or age of psychosis onset, regression analyses including severity of childhood trauma have been able to account for the variation in hippocampal volume (Aas et al. 2014; Hoy et al. 2012). The commonality of hippocampal reduction in psychosis cohorts may overshadow the effect of childhood trauma (van Erp et al. 2016).

1.2.2 Amygdala Volume

It was originally hypothesized that amygdala volume would be greater in individuals with a history of childhood trauma, due to evidence of greater amygdala volumes in PTSD. However, only three studies have reported increased amygdala volume in adults with a history of childhood trauma (Baldacara et al. 2014). Rather, the presence of childhood trauma appears to direct the effect of adulthood traumatic stress on amygdala volume. Kuo et al. (2012) reported a negative correlation between combat exposure and amygdala volume in war veterans with a history of childhood trauma, but a positive correlation was exhibited amongst war veterans without a history of childhood trauma.

In support of this notion, compared to healthy controls, individuals with PTSD exhibit right amygdala enlargement only after controlling for severity of childhood trauma (Baldacara et al. 2014). Additionally, a recent meta-analysis showed no significant difference in amygdala volume of individuals with PTSD compared to trauma-exposed controls (O'Doherty et al. 2015), suggesting adulthood trauma rather than diagnosis determines amygdala volume. Therefore, the interaction of childhood trauma with high levels of adulthood stress (prevalent in psychiatric cohorts (Eckert et al. 2006; Navarro-Mateu et al. 2015) appears to lead to reduced amygdala volume. On the other hand, in healthy cohorts where adulthood trauma is less common, amygdala volume does not appear to be determined by childhood trauma.

1.2.3 Prefrontal Cortex Volume

The relationship of childhood trauma to prefrontal grey matter, as well as the subregions affected, appears to vary widely depending on experimental design. The most consistent report of prefrontal abnormalities in healthy maltreated cohorts has been of grey matter reductions in the ventromedial prefrontal cortex/ACC (Carballedo et al. 2012; Cohen et al. 2006; Dannlowski et al. 2012; Frodl et al. 2010; Gorka et al. 2014). However, this finding has not been replicated in studies defining childhood maltreatment by one trauma or an adverse family environment (Gerritsen et al. 2012; Lu et al. 2013a, b; Walsh et al. 2014).

In terms of psychiatric cohorts, no prefrontal differences related to childhood trauma have been observed amongst individuals with bipolar disorder (Kuhlmann et al. 2013; Labudda et al. 2013) or in the majority of mixed mood and anxiety cohorts (Tomoda et al. 2009a, 2011, 2012) [the exception being (Tomoda et al. 2009b)]. Individuals with PTSD subsequent to childhood trauma appear to have decreased grey matter in dorsal prefrontal regions (Fonzo et al. 2013; Thomaes et al. 2010).

Prefrontal grey matter reductions are present in psychosis cohorts with childhood maltreatment. The regional specificity of this reduction is unclear, however, as different prefrontal regions have been utilized across studies (Kumari et al. 2013; Sheffield et al. 2013).

Finally, studies of individuals with depression have been mixed. Evidence of decreased, increased or no difference in prefrontal grey matter in depressed adults with a history of childhood trauma has been reported (Chaney et al. 2014; Frodl et al. 2010; Treadway et al. 2009).

1.2.4 Discussion

Reduced hippocampal, amygdala and prefrontal grey matter volumes have been reported in adults with a history of childhood trauma. However, overall there is significant variability in the literature on childhood trauma-induced structural abnormalities. In this regard the following moderating factors have been suggested to contribute to this variability.

1.2.4.1 Age

Child studies of maltreatment have frequently reported greater amygdala volumes in neglected children (Lupien et al. 2011; Mehta et al. 2009; Tottenham et al. 2010); however, no difference in amygdala volume has been evident in abused children with PTSD (De Bellis et al. 1999, 2001a, b, 2002). Typically, the amygdala increases in volume throughout adolescence (Wierenga et al. 2014). One longitudinal study has shown reduced amygdala growth throughout mid-adolescence in individuals with high childhood maltreatment (Whittle et al. 2014). This finding supports the notion that abnormal amygdala development results in smaller amygdala volumes in adults with childhood trauma. Reduced grey matter in the prefrontal cortex and right postcentral gyrus also appeared to be driven by older cohorts, as they were no longer significant following regression of mean age. Although no studies have focused on the longitudinal changes in these regions following childhood trauma, it is possible that alterations to the developmental trajectory of grey matter only become apparent later in life.

1.2.4.2 Gender

Gender has been shown to moderate the effect of childhood trauma on depressive symptoms in adulthood (Khan et al. 2015), internalization problems (Godinet et al. 2014), affective processing (Crozier et al. 2014) and cerebral, corpus callosum and ventricular development (De Bellis and Keshavan 2003). Gender differences may arise from the synergistic actions of stress and sex hormones on grey matter plasticity (McEwen 2010). Alternatively, differences in the prevalence of maltreatment types for boys and girls could lead to different effects on grey matter development (Christoffersen et al. 2013; Radford et al. 2013).

1.2.4.3 Diagnosis and Type of Maltreatment

The type of maltreatment appears to be an important factor in determining the structural consequences of childhood trauma. The differential effect of childhood trauma on grey matter volume depending on psychiatric grouping and maltreatment type stems from a multitude of neurobiological and genetic factors. An individual's genetic susceptibility to a certain mental illness plays an important role in determining the effect of childhood trauma. Genes involved in monoaminergic, neurotrophic and stress systems are the primary candidates of where early life environment influences grey matter development. For example, expression of BDNF genes has been shown to mediate the effect of stress hormones on dendritic spines (Bennett and Lagopoulos 2014).

Preclinical studies have shown the neurobiological effects of stress differ depending on stressor type. In adult mice, chronic immobilization stress enhances arborization of excitatory neurons in the basal amygdala and leads to atrophy of basal dendrites in CA3 of the hippocampus. On the other hand, chronic unpredictable stress results in atrophy of amygdala interneurons and no change in CA3 pyramidal neurons (Vyas et al. 2002). Furthermore, only chronic immobilization stress significantly increases anxiety behaviour, suggesting the type of stress experienced can affect psychiatric outcomes and should be thoroughly examined in future studies.

1.3 Animal Models

1.3.1 Changes in Grey Matter Volume Following Stress

Animal studies on rodents (McEwen et al. 1992; Sunanda et al. 1995; Magarinos et al. 1996; Sapolsky 1996; Gould et al. 1998; McEwen 2000b; Stewart et al. 2005) and primates (Sapolsky et al. 1990) suggest an association between stress and structural changes observed in the hippocampus. Damage to the hippocampus is postulated to occur as the result of prolonged, stress-related glucocorticoid and glutamate neurotoxicity, or due to brain-derived neurotrophic factor (BDNF) changes (Magarinos and McEwen 1995; Smith et al. 1995; Sapolsky 2000; Moghaddam 2002; Abrous et al. 2005; Bennett and Lagopoulos 2014), that eventually causes atrophy, cell death and inhibits neurogenesis (McEwen 2000a; Sapolsky 2000), with higher levels of stress hormone leading to greater levels of neurotoxicity.

The outcome of animal stress models and human studies into neuronal architecture in the amygdala has provided further support for the amygdala's association with PTSD. Animal models involving fear conditioning have demonstrated volumetric changes in the basolateral complex of the amygdala (BLA) (Yang et al. 2008), increased arborization and the growth of dendritic spines in the extended amygdala (Vyas et al. 2002, 2003, 2004; Roozendaal et al. 2009). Altered fear learning and stress behaviours have also been observed in the BLA of rats following lesions induced to this area. Finally, human studies have demonstrated heightened fear response and fear conditioning in PTSD patients (Schmahl et al. 2002; Shin et al. 2006), as well as increased glucocorticoid response to stress (Hartley et al. 2011). These findings support both heightened activation and morphological changes of the amygdala (e.g. decreased volume) being associated with PTSD (Ganzel et al. 2008).

References

Aas M, Navari S, Gibbs A, Mondelli V, Fisher HL, Morgan C et al (2012) Is there a link between childhood trauma, cognition, and amygdala and hippocampus volume in first-episode psychosis? Schizophr Res 137(1–3):73–79. https://doi.org/10.1016/j.schres.2012.01.035

Aas M, Haukvik UK, Djurovic S, Tesli M, Athanasiu L, Bjella T et al (2014) Interplay between childhood trauma and BDNF val66met variants on blood BDNF mRNA levels and on hippocampus subfields volumes in schizophrenia spectrum and bipolar disorders. J Psychiatr Res 59:14–21. https://doi.org/10.1016/j.jpsychires.2014.08.011

Abrous DN, Koehl M, Le Moal M (2005) Adult neurogenesis: from precursors to network and physiology. Physiol Rev 85(2):523–569. https://doi.org/10.1152/physrev.00055.2003

Adamec R, Hebert M, Blundell J, Mervis RF (2012) Dendritic morphology of amygdala and hippocampal neurons in more and less predator stress responsive rats and more and less spontaneously anxious handled controls. Behav Brain Res 226(1):133–146. https://doi.org/10.1016/j.bbr.2011.09.009

Almeida RD, Manadas BJ, Melo CV, Gomes JR, Mendes CS, Graos MM et al (2005) Neuroprotection by BDNF against glutamate-induced apoptotic cell death is mediated by ERK and PI3-kinase pathways. Cell Death Differ 12(10):1329–1343. https://doi.org/10.1038/sj.cdd.4401662

Apfel BA, Ross J, Hlavin J, Meyerhoff DJ, Metzler TJ, Marmar CR et al (2011) Hippocampal volume differences in Gulf War veterans with current versus lifetime posttraumatic stress disorder symptoms. Biol Psychiatry 69(6):541–548. https://doi.org/10.1016/j.biopsych.2010.09.044

Arborelius L, Owens MJ, Plotsky PM, Nemeroff CB (1999) The role of corticotropin-releasing factor in depression and anxiety disorders. J Endocrinol 160(1):1–12

Baker LM, Williams LM, Korgaonkar MS, Cohen RA, Heaps JM, Paul RH (2013) Impact of early vs. late childhood early life stress on brain morphometrics. Brain Imaging Behav 7(2):196–203. https://doi.org/10.1007/s11682-012-9215-y

Baldacara L, Zugman A, Araujo C, Cogo-Moreira H, Lacerda AL, Schoedl A et al (2014) Reduction of anterior cingulate in adults with urban violence-related PTSD. J Affect Disord 168:13–20. https://doi.org/10.1016/j.jad.2014.06.036

Benedetti F, Poletti S, Radaelli D, Pozzi E, Giacosa C, Ruffini C et al (2012) Caudate gray matter volume in obsessive-compulsive disorder is influenced by adverse childhood experiences and ongoing drug treatment. J Clin Psychopharmacol 32(4):544–547. https://doi.org/10.1097/JCP.0b013e31825cce05

Bennett MR, Lagopoulos J (2014) Stress and trauma: BDNF control of dendritic-spine formation and regression. Prog Neurobiol 112:80–99. https://doi.org/10.1016/j.pneurobio.2013.10.005

Boen E, Westlye LT, Elvsashagen T, Hummelen B, Hol PK, Boye B et al (2014) Smaller stress-sensitive hippocampal subfields in women with borderline personality disorder without post-traumatic stress disorder. J Psychiatry Neurosci 39(2):127–134

Bonne O, Brandes D, Gilboa A, Gomori JM, Shenton ME, Pitman RK, Shalev AY (2001) Longitudinal MRI study of hippocampal volume in trauma survivors with PTSD. Am J Psychiatry 158(8):1248–1251. https://doi.org/10.1176/appi.ajp.158.8.1248

Bonne O, Vythilingam M, Inagaki M, Wood S, Neumeister A, Nugent AC et al (2008) Reduced posterior hippocampal volume in posttraumatic stress disorder. J Clin Psychiatry 69 (7):1087–1091

Bossini L, Tavanti M, Calossi S, Lombardelli A, Polizzotto NR, Galli R et al (2008) Magnetic resonance imaging volumes of the hippocampus in drug-naive patients with post-traumatic stress disorder without comorbidity conditions. J Psychiatr Res 42(9):752–762. https://doi.org/10.1016/j.jpsychires.2007.08.004

Brambilla P, Soloff PH, Sala M, Nicoletti MA, Keshavan MS, Soares JC (2004) Anatomical MRI study of borderline personality disorder patients. Psychiatry Res 131(2):125–133. https://doi.org/10.1016/j.pscychresns.2004.04.003

Bremner JD (1999) Does stress damage the brain? Biol Psychiatry 45(7):797–805. https://doi.org/10.1016/S0006-3223(99)00009-8

Bremner JD (2003) Functional neuroanatomical correlates of traumatic stress revisited 7 years later, this time with data. Psychopharmacol Bull 37(2):6–25

Bremner JD, Randall P, Scott TM, Bronen RA, Seibyl JP, Southwick SM et al (1995) MRI-based measurement of hippocampal volume in patients with combat-related posttraumatic stress disorder. Am J Psychiatry 152(7):973–981. https://doi.org/10.1176/ajp.152.7.973

Bremner JD, Randall P, Vermetten E, Staib L, Bronen RA, Mazure C et al (1997) Magnetic resonance imaging-based measurement of hippocampal volume in posttraumatic stress disorder related to childhood physical and sexual abuse—a preliminary report. Biol Psychiatry 41 (1):23–32

Bremner JD, Narayan M, Staib LH, Southwick SM, McGlashan T, Charney DS (1999) Neural correlates of memories of childhood sexual abuse in women with and without posttraumatic stress disorder. Am J Psychiatry 156(11):1787–1795. https://doi.org/10.1176/ajp.156.11.1787
Bremner JD, Vythilingam M, Vermetten E, Southwick SM, McGlashan T, Nazeer A et al (2003) MRI and PET study of deficits in hippocampal structure and function in women with childhood sexual abuse and posttraumatic stress disorder. Am J Psychiatry 160(5):924–932. https://doi.org/10.1176/appi.ajp.160.5.924
Bremner JD, Vermetten E, Vythilingam M, Afzal N, Schmahl C, Elzinga B, Charney DS (2004) Neural correlates of the classic color and emotional stroop in women with abuse-related posttraumatic stress disorder. Biol Psychiatry 55(6):612–620. https://doi.org/10.1016/j.biopsych.2003.10.001
Brewin CR (2001) A cognitive neuroscience account of posttraumatic stress disorder and its treatment. Behav Res Ther 39(4):373–393
Brewin CR, Dalgleish T, Joseph S (1996) A dual representation theory of posttraumatic stress disorder. Psychol Rev 103(4):670–686
Brewin CR, Kleiner JS, Vasterling JJ, Field AP (2007) Memory for emotionally neutral information in posttraumatic stress disorder: a meta-analytic investigation. J Abnorm Psychol 116 (3):448–463. https://doi.org/10.1037/0021-843X.116.3.448
Brewin CR, Lanius RA, Novac A, Schnyder U, Galea S (2009) Reformulating PTSD for DSM-V: life after Criterion A. J Trauma Stress 22(5):366–373. https://doi.org/10.1002/jts.20443
Brewin CR, Gregory JD, Lipton M, Burgess N (2010) Intrusive images in psychological disorders: characteristics, neural mechanisms, and treatment implications. Psychol Rev 117(1):210–232. https://doi.org/10.1037/a0018113
Buchanan TW, Kern S, Allen JS, Tranel D, Kirschbaum C (2004) Circadian regulation of cortisol after hippocampal damage in humans. Biol Psychiatry 56(9):651–656. https://doi.org/10.1016/j.biopsych.2004.08.014
Bush G, Whalen PJ, Rosen BR, Jenike MA, McInerney SC, Rauch SL (1998) The counting Stroop: an interference task specialized for functional neuroimaging—validation study with functional MRI. Hum Brain Mapp 6(4):270–282
Campbell S, Marriott M, Nahmias C, MacQueen GM (2004) Lower hippocampal volume in patients suffering from depression: a meta-analysis. Am J Psychiatry 161(4):598–607. https://doi.org/10.1176/appi.ajp.161.4.598
Carballedo A, Lisiecka D, Fagan A, Saleh K, Ferguson Y, Connolly G et al (2012) Early life adversity is associated with brain changes in subjects at family risk for depression. World J Biol Psychiatry 13(8):569–578. https://doi.org/10.3109/15622975.2012.661079
Cerqueira JJ, Catania C, Sotiropoulos I, Schubert M, Kalisch R, Almeida OF et al (2005a) Corticosteroid status influences the volume of the rat cingulate cortex—a magnetic resonance imaging study. J Psychiatr Res 39(5):451–460. https://doi.org/10.1016/j.jpsychires.2005.01.003
Cerqueira JJ, Pego JM, Taipa R, Bessa JM, Almeida OF, Sousa N (2005b) Morphological correlates of corticosteroid-induced changes in prefrontal cortex-dependent behaviors. J Neurosci 25 (34):7792–7800. https://doi.org/10.1523/JNEUROSCI.1598-05.2005
Chaney A, Carballedo A, Amico F, Fagan A, Skokauskas N, Meaney J, Frodl T (2014) Effect of childhood maltreatment on brain structure in adult patients with major depressive disorder and healthy participants. J Psychiatry Neurosci 39(1):50–59. https://doi.org/10.1503/jpn.120208
Chapman DP, Whitfield CL, Felitti VJ, Dube SR, Edwards VJ, Anda RF (2004) Adverse childhood experiences and the risk of depressive disorders in adulthood. J Affect Disord 82(2):217–225. https://doi.org/10.1016/j.jad.2003.12.013
Chen S, Xia W, Li L, Liu J, He Z, Zhang Z et al (2006a) Gray matter density reduction in the insula in fire survivors with posttraumatic stress disorder: a voxel-based morphometric study. Psychiatry Res 146(1):65–72. https://doi.org/10.1016/j.pscychresns.2005.09.006
Chen TJ, Gehler S, Shaw AE, Bamburg JR, Letourneau PC (2006b) Cdc42 participates in the regulation of ADF/cofilin and retinal growth cone filopodia by brain derived neurotrophic factor. J Neurobiol 66(2):103–114. https://doi.org/10.1002/neu.20204

Chen DY, Bambah-Mukku D, Pollonini G, Alberini CM (2012a) Glucocorticoid receptors recruit the CaMKIIalpha-BDNF-CREB pathways to mediate memory consolidation. Nat Neurosci 15 (12):1707–1714. https://doi.org/10.1038/nn.3266

Chen X, Li Y, Li S, Kirouac GJ (2012b) Early fear as a predictor of avoidance in a rat model of post-traumatic stress disorder. Behav Brain Res 226(1):112–117. https://doi.org/10.1016/j.bbr.2011.09.004

Chen Y, Andres AL, Frotscher M, Baram TZ (2012c) Tuning synaptic transmission in the hippocampus by stress: the CRH system. Front Cell Neurosci 6:13. https://doi.org/10.3389/fncel.2012.00013

Christoffersen MN, Armour C, Lasgaard M, Andersen TE, Elklit A (2013) The prevalence of four types of childhood maltreatment in denmark. Clin Pract Epidemiol Ment Health 9:149–156. https://doi.org/10.2174/1745017901309010149

Cicchetti D, Rogosch FA (2001) Diverse patterns of neuroendocrine activity in maltreated children. Dev Psychopathol 13(3):677–693

Cohen JD, Servanschreiber D (1992) A neural network model of disturbances in the processing of context in schizophrenia. Psychiatr Ann 22(3):131–136

Cohen RA, Grieve S, Hoth KF, Paul RH, Sweet L, Tate D et al (2006) Early life stress and morphometry of the adult anterior cingulate cortex and caudate nuclei. Biol Psychiatry 59 (10):975–982. https://doi.org/10.1016/j.biopsych.2005.12.016

Crozier JC, Wang L, Huettel SA, De Bellis MD (2014) Neural correlates of cognitive and affective processing in maltreated youth with posttraumatic stress symptoms: does gender matter? Dev Psychopathol 26(2):491–513. https://doi.org/10.1017/S095457941400008X

Damsa C, Maris S, Pull CB (2005) New fields of research in posttraumatic stress disorder: brain imaging. Curr Opin Psychiatry 18(1):55–64

Dannlowski U, Stuhrmann A, Beutelmann V, Zwanzger P, Lenzen T, Grotegerd D et al (2012) Limbic scars: long-term consequences of childhood maltreatment revealed by functional and structural magnetic resonance imaging. Biol Psychiatry 71(4):286–293. https://doi.org/10.1016/j.biopsych.2011.10.021

Davis M (1992) The role of the amygdala in fear and anxiety. Annu Rev Neurosci 15:353–375. https://doi.org/10.1146/annurev.neuro.15.1.353

De Bellis MD, Keshavan MS (2003) Sex differences in brain maturation in maltreatment-related pediatric posttraumatic stress disorder. Neurosci Biobehav Rev 27(1–2):103–117

De Bellis MD, Keshavan MS, Clark DB, Casey BJ, Giedd JN, Boring AM et al (1999) Developmental traumatology part II: brain development. Biol Psychiatry 45(10):1271–1284. https://doi.org/10.1016/S0006-3223(99)00045-1

De Bellis MD, Keshavan MS, Spencer S, Hall J (2000) N-acetylaspartate concentration in the anterior cingulate of maltreated children and adolescents with PTSD. Am J Psychiatr 157 (7):1175–1177. https://doi.org/10.1176/appi.ajp.157.7.1175

De Bellis MD, Hall J, Boring AM, Frustaci K, Moritz G (2001a) A pilot longitudinal study of hippocampal volumes in pediatric maltreatment-related posttraumatic stress disorder. Biol Psychiatry 50(4):305–309

De Bellis MD, Keshavan MS, Harenski KA (2001b) Anterior cingulate N-acetylaspartate/creatine ratios during clonidine treatment in a maltreated child with posttraumatic stress disorder. J Child Adolesc Psychopharmacol 11(3):311–316. https://doi.org/10.1089/10445460152595649

De Bellis MD, Keshavan MS, Shifflett H, Iyengar S, Beers SR, Hall J, Moritz G (2002) Brain structures in pediatric maltreatment-related posttraumatic stress disorder: a sociodemographically matched study. Biol Psychiatry 52(11):1066–1078

de-Almeida CP, Wenzel A, de-Carvalho CS, Powell VB, Araujo-Neto C, Quarantini LC, de-Oliveira IR (2012) Amygdalar volume in borderline personality disorder with and without comorbid post-traumatic stress disorder: a meta-analysis. CNS Spectr 17(2):70–75. https://doi.org/10.1017/S1092852912000466

Dolan RJ (2007) The human amygdala and orbital prefrontal cortex in behavioural regulation. Philos Trans R Soc Lond Ser B Biol Sci 362(1481):787–799. https://doi.org/10.1098/rstb.2007.2088

Driessen M, Herrmann J, Stahl K, Zwaan M, Meier S, Hill A et al (2000) Magnetic resonance imaging volumes of the hippocampus and the amygdala in women with borderline personality disorder and early traumatization. Arch Gen Psychiatry 57(12):1115–1122

Dunsmoor JE, Prince SE, Murty VP, Kragel PA, LaBar KS (2011) Neurobehavioral mechanisms of human fear generalization. NeuroImage 55(4):1878–1888. https://doi.org/10.1016/j.neuroimage.2011.01.041

Eckert KA, Wilkinson D, Taylor AW, Stewart S, Tucker GR (2006) A population view of mental illness in South Australia: broader issues than location. Rural Remote Health 6(2):541

Eichenbaum H (2000) A cortical-hippocampal system for declarative memory. Nat Rev Neurosci 1 (1):41–50. https://doi.org/10.1038/35036213

Emdad R, Bonekamp D, Sondergaard HP, Bjorklund TB, Agartz I, Ingvar M, Theorell TR (2006) Morphometric and psychometric comparisons between non-substance- abusing patients with posttraumatic stress disorder and normal controls. Psychother Psychosom 75(2):122–132. https://doi.org/10.1159/000090897

Etkin A, Wager TD (2007) Functional neuroimaging of anxiety: a meta-analysis of emotional processing in PTSD, social anxiety disorder, and specific phobia. Am J Psychiatr 164 (10):1476–1488. https://doi.org/10.1176/appi.ajp.2007.07030504

Felmingham K, Williams LM, Whitford TJ, Falconer E, Kemp AH, Peduto A, Bryant RA (2009) Duration of posttraumatic stress disorder predicts hippocampal grey matter loss. Neuroreport 20 (16):1402–1406. https://doi.org/10.1097/WNR.0b013e3283300fbc

Fonzo GA, Flagan TM, Sullivan S, Allard CB, Grimes EM, Simmons AN et al (2013) Neural functional and structural correlates of childhood maltreatment in women with intimate-partner violence-related posttraumatic stress disorder. Psychiatry Res 211(2):93–103. https://doi.org/10.1016/j.pscychresns.2012.08.006

Frodl T, O'Keane V (2013) How does the brain deal with cumulative stress? A review with focus on developmental stress, HPA axis function and hippocampal structure in humans. Neurobiol Dis 52:24–37. https://doi.org/10.1016/j.nbd.2012.03.012

Frodl T, Reinhold E, Koutsouleris N, Reiser M, Meisenzahl EM (2010) Interaction of childhood stress with hippocampus and prefrontal cortex volume reduction in major depression. J Psychiatr Res 44(13):799–807. https://doi.org/10.1016/j.jpsychires.2010.01.006

Funahashi S, Andreau JM (2013) Prefrontal cortex and neural mechanisms of executive function. J Physiol Paris 107(6):471–482. https://doi.org/10.1016/j.jphysparis.2013.05.001

Ganzel BL, Kim P, Glover GH, Temple E (2008) Resilience after 9/11: multimodal neuroimaging evidence for stress-related change in the healthy adult brain. NeuroImage 40(2):788–795. https://doi.org/10.1016/j.neuroimage.2007.12.010

Garfinkel SN, Liberzon I (2009) Neurobiology of PTSD: a review of neuroimaging findings. Psychiatr Ann 39(6):370–372, 376–381. https://doi.org/10.3928/00485713-20090527-01

Gatt JM, Nemeroff CB, Dobson-Stone C, Paul RH, Bryant RA, Schofield PR et al (2009) Interactions between BDNF Val66Met polymorphism and early life stress predict brain and arousal pathways to syndromal depression and anxiety. Mol Psychiatry 14(7):681–695. https://doi.org/10.1038/mp.2008.143

Gerritsen L, Tendolkar I, Franke B, Vasquez AA, Kooijman S, Buitelaar J et al (2012) BDNF Val66Met genotype modulates the effect of childhood adversity on subgenual anterior cingulate cortex volume in healthy subjects. Mol Psychiatry 17(6):597–603. https://doi.org/10.1038/mp.2011.51

Gerritsen L, Kalpouzos G, Westman E, Simmons A, Wahlund LO, Backman L et al (2015) The influence of negative life events on hippocampal and amygdala volumes in old age: a life-course perspective. Psychol Med 45(6):1219–1228. https://doi.org/10.1017/S0033291714002293

Gilbertson MW, Shenton ME, Ciszewski A, Kasai K, Lasko NB, Orr SP, Pitman RK (2002) Smaller hippocampal volume predicts pathologic vulnerability to psychological trauma. Nat Neurosci 5(11):1242–1247. https://doi.org/10.1038/nn958

Godinet MT, Li F, Berg T (2014) Early childhood maltreatment and trajectories of behavioral problems: exploring gender and racial differences. Child Abuse Negl 38(3):544–556. https://doi.org/10.1016/j.chiabu.2013.07.018

Gorka AX, Hanson JL, Radtke SR, Hariri AR (2014) Reduced hippocampal and medial prefrontal gray matter mediate the association between reported childhood maltreatment and trait anxiety in adulthood and predict sensitivity to future life stress. Biol Mood Anxiety Disord 4:12. https://doi.org/10.1186/2045-5380-4-12

Gould E, Tanapat P, McEwen BS, Flugge G, Fuchs E (1998) Proliferation of granule cell precursors in the dentate gyrus of adult monkeys is diminished by stress. Proc Natl Acad Sci USA 95 (6):3168–3171. https://doi.org/10.1073/pnas.95.6.3168

Grassi-Oliveira R, Ashy M, Stein LM (2008) Psychobiology of childhood maltreatment: effects of allostatic load? Rev Bras Psiquiatr 30(1):60–68

Gurvits TV, Shenton ME, Hokama H, Ohta H, Lasko NB, Gilbertson MW et al (1996) Magnetic resonance imaging study of hippocampal volume in chronic, combat-related posttraumatic stress disorder. Biol Psychiatry 40(11):1091–1099. https://doi.org/10.1016/S0006-3223(96)00229-6

Hart H, Rubia K (2012) Neuroimaging of child abuse: a critical review. Front Hum Neurosci 6:52. https://doi.org/10.3389/fnhum.2012.00052

Hartley CA, Fischl B, Phelps EA (2011) Brain structure correlates of individual differences in the acquisition and inhibition of conditioned fear. Cereb Cortex 21(9):1954–1962. https://doi.org/10.1093/cercor/bhq253

Hedges DW, Woon FL (2010) Alcohol use and hippocampal volume deficits in adults with posttraumatic stress disorder: a meta-analysis. Biol Psychol 84(2):163–168. https://doi.org/10.1016/j.biopsycho.2010.03.002

Hedges DW, Allen S, Tate DF, Thatcher GW, Miller MJ, Rice SA et al (2003) Reduced hippocampal volume in alcohol and substance naive Vietnam combat veterans with posttraumatic stress disorder. Cogn Behav Neurol 16(4):219–224

Hedges DW, Thatcher GW, Bennett PJ, Sood S, Paulson D, Creem-Regehr S et al (2007) Brain integrity and cerebral atrophy in Vietnam combat veterans with and without posttraumatic stress disorder. Neurocase 13(5):402–410. https://doi.org/10.1080/13554790701851551

Heim CM, Mayberg HS, Mletzko T, Nemeroff CB, Pruessner JC (2013) Decreased cortical representation of genital somatosensory field after childhood sexual abuse. Am J Psychiatry 170(6):616–623. https://doi.org/10.1176/appi.ajp.2013.12070950

Hendler T, Rotshtein P, Yeshurun Y, Weizmann T, Kahn I, Ben-Bashat D et al (2003) Sensing the invisible: differential sensitivity of visual cortex and amygdala to traumatic context. NeuroImage 19(3):587–600

Herry C, Garcia R (2002) Prefrontal cortex long-term potentiation, but not long-term depression, is associated with the maintenance of extinction of learned fear in mice. J Neurosci 22(2):577–583

Hoffman RE, Mcglashan TH (1993) Parallel distributed-processing and the emergence of schizophrenic symptoms. Schizophr Bull 19(1):119–140

Hoy K, Barrett S, Shannon C, Campbell C, Watson D, Rushe T et al (2012) Childhood trauma and hippocampal and amygdalar volumes in first-episode psychosis. Schizophr Bull 38 (6):1162–1169. https://doi.org/10.1093/schbul/sbr085

Janak PH, Tye KM (2015) From circuits to behaviour in the amygdala. Nature 517(7534):284–292. https://doi.org/10.1038/nature14188

Karl A, Schaefer M, Malta LS, Dorfel D, Rohleder N, Werner A (2006) A meta-analysis of structural brain abnormalities in PTSD. Neurosci Biobehav Rev 30(7):1004–1031. https://doi.org/10.1016/j.neubiorev.2006.03.004

Kasai K, Yamasue H, Gilbertson MW, Shenton ME, Rauch SL, Pitman RK (2008) Evidence for acquired pregenual anterior cingulate gray matter loss from a twin study of combat-related posttraumatic stress disorder. Biol Psychiatry 63(6):550–556. https://doi.org/10.1016/j.biopsych.2007.06.022

Kassem MS, Lagopoulos J, Stait-Gardner T, Price WS, Chohan TW, Arnold JC et al (2013) Stress-induced grey matter loss determined by MRI Is primarily due to loss of dendrites and their synapses. Mol Neurobiol 47(2):645–661. https://doi.org/10.1007/s12035-012-8365-7

Kensinger EA, Corkin S (2004) Two routes to emotional memory: distinct neural processes for valence and arousal. Proc Natl Acad Sci USA 101(9):3310–3315. https://doi.org/10.1073/pnas.0306408101

Kerns JG, Cohen JD, MacDonald AW, Cho RY, Stenger VA, Carter CS (2004) Anterior Cingulate conflict monitoring and adjustments in control. Science 303(5660):1023–1026. https://doi.org/10.1126/science.1089910

Khan A, McCormack HC, Bolger EA, McGreenery CE, Vitaliano G, Polcari A, Teicher MH (2015) Childhood maltreatment, depression, and suicidal ideation: critical importance of parental and peer emotional abuse during developmental sensitive periods in males and females. Front Psychiatry 6:42. https://doi.org/10.3389/fpsyt.2015.00042

Kitayama N, Vaccarino V, Kutner M, Weiss P, Bremner JD (2005) Magnetic resonance imaging (MRI) measurement of hippocampal volume in posttraumatic stress disorder: a meta-analysis. J Affect Disord 88(1):79–86. https://doi.org/10.1016/j.jad.2005.05.014

Kitayama N, Quinn S, Bremner JD (2006) Smaller volume of anterior cingulate cortex in abuse-related posttraumatic stress disorder. J Affect Disord 90(2–3):171–174

Kuhlmann A, Bertsch K, Schmidinger I, Thomann PA, Herpertz SC (2013) Morphometric differences in central stress-regulating structures between women with and without borderline personality disorder. J Psychiatry Neurosci 38(2):129–137. https://doi.org/10.1503/jpn.120039

Kumari V, Gudjonsson GH, Raghuvanshi S, Barkataki I, Taylor P, Sumich A et al (2013) Reduced thalamic volume in men with antisocial personality disorder or schizophrenia and a history of serious violence and childhood abuse. Eur Psychiatry 28(4):225–234. https://doi.org/10.1016/j.eurpsy.2012.03.002

Kuo JR, Kaloupek DG, Woodward SH (2012) Amygdala volume in combat-exposed veterans with and without posttraumatic stress disorder a cross-sectional study. Arch Gen Psychiatry 69 (10):1080–1086

LaBar KS, Gatenby JC, Gore JC, LeDoux JE, Phelps EA (1998) Human amygdala activation during conditioned fear acquisition and extinction: a mixed-trial fMRI study. Neuron 20(5):937–945. https://doi.org/10.1016/S0896-6273(00)80475-4

Labudda K, Kreisel S, Beblo T, Mertens M, Kurlandchikov O, Bien CG et al (2013) Mesiotemporal volume loss associated with disorder severity: a VBM study in borderline personality disorder. PLoS One 8(12):e83677. https://doi.org/10.1371/journal.pone.0083677

Landre L, Destrieux C, Baudry M, Barantin L, Cottier JP, Martineau J et al (2010) Preserved subcortical volumes and cortical thickness in women with sexual abuse-related PTSD. Psychiatry Res 183(3):181–186. https://doi.org/10.1016/j.pscychresns.2010.01.015

Lanius RA, Williamson PC, Hopper J, Densmore M, Boksman K, Gupta MA et al (2003) Recall of emotional states in posttraumatic stress disorder: an fMRI investigation. Biol Psychiatry 53 (3):204–210

Lanius RA, Bluhm R, Lanius U, Pain C (2006) A review of neuroimaging studies in PTSD: heterogeneity of response to symptom provocation. J Psychiatr Res 40(8):709–729. https://doi.org/10.1016/j.jpyschires.2005.07.007

Lanius RA, Bluhm RL, Frewen PA (2011) How understanding the neurobiology of complex post-traumatic stress disorder can inform clinical practice: a social cognitive and affective neuroscience approach. Acta Psychiatr Scand 124(5):331–348. https://doi.org/10.1111/j.1600-0447.2011.01755.x

LeDoux JE (2000) Emotion circuits in the brain. Annu Rev Neurosci 23:155–184. https://doi.org/10.1146/annurev.neuro.23.1.155

Li L, Chen S, Liu J, Zhang J, He Z, Lin X (2006) Magnetic resonance imaging and magnetic resonance spectroscopy study of deficits in hippocampal structure in fire victims with recent-onset posttraumatic stress disorder. Can J Psychiatr 51(7):431–437. https://doi.org/10.1177/070674370605100704

Liao M, Yang F, Zhang Y, He Z, Song M, Jiang T et al (2013) Childhood maltreatment is associated with larger left thalamic gray matter volume in adolescents with generalized anxiety disorder. PLoS One 8(8):e71898. https://doi.org/10.1371/journal.pone.0071898

Liberzon I, Taylor SF, Amdur R, Jung TD, Chamberlain KR, Minoshima S et al (1999) Brain activation in PTSD in response to trauma-related stimuli. Biol Psychiatry 45(7):817–826

Lim L, Radua J, Rubia K (2014) Gray matter abnormalities in childhood maltreatment: a voxel-wise meta-analysis. Am J Psychiatry 171(8):854–863. https://doi.org/10.1176/appi.ajp.2014.13101427

Lindauer RJ, Vlieger EJ, Jalink M, Olff M, Carlier IV, Majoie CB et al (2004) Smaller hippocampal volume in Dutch police officers with posttraumatic stress disorder. Biol Psychiatry 56(5):356–363. https://doi.org/10.1016/j.biopsych.2004.05.021

Lu B, Nagappan G, Guan X, Nathan PJ, Wren P (2013a) BDNF-based synaptic repair as a disease-modifying strategy for neurodegenerative diseases. Nat Rev Neurosci 14(6):401–416. https://doi.org/10.1038/nrn3505

Lu S, Gao W, Wei Z, Wu W, Liao M, Ding Y et al (2013b) Reduced cingulate gyrus volume associated with enhanced cortisol awakening response in young healthy adults reporting childhood trauma. PLoS One 8(7):e69350. https://doi.org/10.1371/journal.pone.0069350

Lupien SJ, Parent S, Evans AC, Tremblay RE, Zelazo PD, Corbo V et al (2011) Larger amygdala but no change in hippocampal volume in 10-year-old children exposed to maternal depressive symptomatology since birth. Proc Natl Acad Sci USA 108(34):14324–14329. https://doi.org/10.1073/pnas.1105371108

Magarinos AM, McEwen BS (1995) Stress-induced atrophy of apical dendrites of hippocampal CA3c neurons: involvement of glucocorticoid secretion and excitatory amino acid receptors. Neuroscience 69(1):89–98

Magarinos AM, McEwen BS, Flugge G, Fuchs E (1996) Chronic psychosocial stress causes apical dendritic atrophy of hippocampal CA3 pyramidal neurons in subordinate tree shrews. J Neurosci 16(10):3534–3540

Mathew SJ, Shungu DC, Mao X, Smith EL, Perera GM, Kegeles LS et al (2003) A magnetic resonance spectroscopic imaging study of adult nonhuman primates exposed to early-life stressors. Biol Psychiatry 54(7):727–735

McCrory E, De Brito SA, Viding E (2011) The impact of childhood maltreatment: a review of neurobiological and genetic factors. Front Psychiatry 2:48. https://doi.org/10.3389/fpsyt.2011.00048

McEwen BS (1995) Adrenal steroid actions on brain: dissecting the fine line between protection and damage. In: Neurobiological and clinical consequences of stress: from normal adaptation to post-traumatic stress disorder. Lippincott Williams & Wilkins Publishers, Philadelphia, PA, pp 135–147

McEwen BS (2000a) Effects of adverse experiences for brain structure and function. Biol Psychiatry 48(8):721–731

McEwen BS (2000b) The neurobiology of stress: from serendipity to clinical relevance. Brain Res 886(1–2):172–189

McEwen BS (2010) Stress, sex, and neural adaptation to a changing environment: mechanisms of neuronal remodeling. Ann N Y Acad Sci 1204(Suppl):E38–E59. https://doi.org/10.1111/j.1749-6632.2010.05568.x

McEwen BS, Angulo J, Cameron H, Chao HM, Daniels D, Gannon MN et al (1992) Paradoxical effects of adrenal steroids on the brain: protection versus degeneration. Biol Psychiatry 31(2):177–199

McFarlane AC, Yehuda R, Clark CR (2002) Biologic models of traumatic memories and post-traumatic stress disorder. The role of neural networks. Psychiatr Clin North Am 25(2):253–270

McLeod GF, Fergusson DM, Horwood LJ (2014) Childhood physical punishment or maltreatment and partnership outcomes at age 30. Am J Orthopsychiatry 84(3):307–315. https://doi.org/10.1037/h0099807

Mehta MA, Golembo NI, Nosarti C, Colvert E, Mota A, Williams SC et al (2009) Amygdala, hippocampal and corpus callosum size following severe early institutional deprivation: the English and Romanian Adoptees study pilot. J Child Psychol Psychiatry 50(8):943–951. https://doi.org/10.1111/j.1469-7610.2009.02084.x

Menon V, Uddin LQ (2010) Saliency, switching, attention and control: a network model of insula function. Brain Struct Funct 214(5–6):655–667. https://doi.org/10.1007/s00429-010-0262-0

Menon V, Adleman NE, White CD, Glover GH, Reiss AL (2001) Error-related brain activation during a Go/NoGo response inhibition task. Hum Brain Mapp 12(3):131–143

Milham MP, Nugent AC, Drevets WC, Dickstein DP, Leibenluft E, Ernst M et al (2005) Selective reduction in amygdala volume in pediatric anxiety disorders: a voxel-based morphometry investigation. Biol Psychiatry 57(9):961–966. https://doi.org/10.1016/j.biopsych.2005.01.038

Moghaddam B (2002) Stress activation of glutamate neurotransmission in the prefrontal cortex: implications for dopamine-associated psychiatric disorders. Biol Psychiatry 51(10):775–787

Molendijk ML, van Tol MJ, Penninx BW, van der Wee NJ, Aleman A, Veltman DJ et al (2012) BDNF val66met affects hippocampal volume and emotion-related hippocampal memory activity. Transl Psychiatry 2:e74. https://doi.org/10.1038/tp.2011.72

Morey RA, Dolcos F, Petty CM, Cooper DA, Hayes JP, LaBar KS, McCarthy G (2009) The role of trauma-related distractors on neural systems for working memory and emotion processing in posttraumatic stress disorder. J Psychiatr Res 43(8):809–817. https://doi.org/10.1016/j.jpsychires.2008.10.014

Murphy FC, Nimmo-Smith I, Lawrence AD (2003) Functional neuroanatomy of emotions: a meta-analysis. Cogn Affect Behav Neurosci 3(3):207–233

Myers KM, Davis M (2007) Mechanisms of fear extinction. Mol Psychiatry 12(2):120–150. https://doi.org/10.1038/sj.mp.4001939

Navarro-Mateu F, Tormo MJ, Salmeron D, Vilagut G, Navarro C, Ruiz-Merino G et al (2015) Prevalence of mental disorders in the South-East of Spain, one of the European regions most affected by the economic crisis: the cross-sectional PEGASUS-murcia project. PLoS One 10(9): e0137293. https://doi.org/10.1371/journal.pone.0137293

Nibuya M, Morinobu S, Duman RS (1995) Regulation of BDNF and trkB mRNA in rat brain by chronic electroconvulsive seizure and antidepressant drug treatments. J Neurosci 15 (11):7539–7547

Nibuya M, Nestler EJ, Duman RS (1996) Chronic antidepressant administration increases the expression of cAMP response element binding protein (CREB) in rat hippocampus. J Neurosci 16(7):2365–2372

O'Doherty DC, Chitty KM, Saddiqui S, Bennett MR, Lagopoulos J (2015) A systematic review and meta-analysis of magnetic resonance imaging measurement of structural volumes in posttraumatic stress disorder. Psychiatry Res 232(1):1–33. https://doi.org/10.1016/j.pscychresns.2015.01.002

Ochsner KN, Bunge SA, Gross JJ, Gabrieli JD (2002) Rethinking feelings: an FMRI study of the cognitive regulation of emotion. J Cogn Neurosci 14(8):1215–1229. https://doi.org/10.1162/089892902760807212

Opel N, Redlich R, Zwanzger P, Grotegerd D, Arolt V, Heindel W et al (2014) Hippocampal atrophy in major depression: a function of childhood maltreatment rather than diagnosis? Neuropsychopharmacology 39(12):2723–2731. https://doi.org/10.1038/npp.2014.145

Pape HC, Pare D (2010) Plastic synaptic networks of the amygdala for the acquisition, expression, and extinction of conditioned fear. Physiol Rev 90(2):419–463. https://doi.org/10.1152/physrev.00037.2009

Patel R, Spreng RN, Shin LM, Girard TA (2012) Neurocircuitry models of posttraumatic stress disorder and beyond: a meta-analysis of functional neuroimaging studies. Neurosci Biobehav Rev 36(9):2130–2142. https://doi.org/10.1016/j.neubiorev.2012.06.003

Paulus MP, Stein MB (2006) An insular view of anxiety. Biol Psychiatry 60(4):383–387. https://doi.org/10.1016/j.biopsych.2006.03.042

Pavlisa G, Papa J, Pavic L, Pavlisa G (2006) Bilateral MR volumetry of the amygdala in chronic PTSD patients. Coll Antropol 30(3):565–568

Pederson CL, Maurer SH, Kaminski PL, Zander KA, Peters CM, Stokes-Crowe LA, Osborn RE (2004) Hippocampal volume and memory performance in a community-based sample of women with posttraumatic stress disorder secondary to child abuse. J Trauma Stress 17 (1):37–40. https://doi.org/10.1023/B:JOTS.0000014674.84517.46

Pitman RK, Rasmusson AM, Koenen KC, Shin LM, Orr SP, Gilbertson MW et al (2012) Biological studies of post-traumatic stress disorder. Nat Rev Neurosci 13(11):769–787. https://doi.org/10.1038/nrn3339

Radford L, Corral S, Bradley C, Fisher HL (2013) The prevalence and impact of child maltreatment and other types of victimization in the UK: findings from a population survey of caregivers, children and young people and young adults. Child Abuse Negl 37(10):801–813. https://doi.org/10.1016/j.chiabu.2013.02.004

Radley JJ, Sisti HM, Hao J, Rocher AB, McCall T, Hof PR et al (2004) Chronic behavioral stress induces apical dendritic reorganization in pyramidal neurons of the medial prefrontal cortex. Neuroscience 125(1):1–6. https://doi.org/10.1016/j.neuroscience.2004.01.006

Radley JJ, Rocher AB, Miller M, Janssen WG, Liston C, Hof PR et al (2006) Repeated stress induces dendritic spine loss in the rat medial prefrontal cortex. Cereb Cortex 16(3):313–320. https://doi.org/10.1093/cercor/bhi104

Rao U, Chen LA, Bidesi AS, Shad MU, Thomas MA, Hammen CL (2010) Hippocampal changes associated with early-life adversity and vulnerability to depression. Biol Psychiatry 67 (4):357–364. https://doi.org/10.1016/j.biopsych.2009.10.017

Rapoza KA, Wilson DT, Widmann WA, Riley MA, Robertson TW, Maiello E et al (2014) The relationship between adult health and childhood maltreatment, as moderated by anger and ethnic background. Child Abuse Negl 38(3):445–456. https://doi.org/10.1016/j.chiabu.2014.01.009

Rauch SL, Whalen PJ, Shin LM, McInerney SC, Macklin ML, Lasko NB et al (2000) Exaggerated amygdala response to masked facial stimuli in posttraumatic stress disorder: a functional MRI study. Biol Psychiatry 47(9):769–776

Rauch SL, Shin LM, Segal E, Pitman RK, Carson MA, McMullin K et al (2003) Selectively reduced regional cortical volumes in post-traumatic stress disorder. Neuroreport 14(7):913–916. https://doi.org/10.1097/01.wnr.0000071767.24455.10

Rauch SL, Shin LM, Phelps EA (2006) Neurocircuitry models of posttraumatic stress disorder and extinction: human neuroimaging research—past, present, and future. Biol Psychiatry 60 (4):376–382. https://doi.org/10.1016/j.biopsych.2006.06.004

Reul JM, Nutt DJ (2008) Glutamate and cortisol—a critical confluence in PTSD? J Psychopharmacol 22(5):469–472. https://doi.org/10.1177/0269881108094617

Riem MM, Alink LR, Out D, Van Ijzendoorn MH, Bakermans-Kranenburg MJ (2015) Beating the brain about abuse: empirical and meta-analytic studies of the association between maltreatment and hippocampal volume across childhood and adolescence. Dev Psychopathol 27(2):507–520. https://doi.org/10.1017/S0954579415000127

Rodrigues E, Wenzel A, Ribeiro MP, Quarantini LC, Miranda-Scippa A, de Sena EP, de Oliveira IR (2011) Hippocampal volume in borderline personality disorder with and without comorbid posttraumatic stress disorder: a meta-analysis. Eur Psychiatry 26(7):452–456. https://doi.org/10.1016/j.eurpsy.2010.07.005

Rogers MA, Yamasue H, Abe O, Yamada H, Ohtani T, Iwanami A et al (2009) Smaller amygdala volume and reduced anterior cingulate gray matter density associated with history of post-traumatic stress disorder. Psychiatry Res 174(3):210–216. https://doi.org/10.1016/j.pscychresns.2009.06.001

Rolls ET (1996) A theory of hippocampal function in memory. Hippocampus 6(6):601–620. https://doi.org/10.1002/(SICI)1098-1063(1996)6:6<601::AID-HIPO5>3.0.CO;2-J

Romano E, Babchishin L, Marquis R, Frechette S (2015) Childhood maltreatment and educational outcomes. Trauma Violence Abuse 16(4):418–437. https://doi.org/10.1177/1524838014537908
Roozendaal B, McEwen BS, Chattarji S (2009) Stress, memory and the amygdala. Nat Rev Neurosci 10(6):423–433. https://doi.org/10.1038/nrn2651
Rudy JW, O'Reilly RC (1999) Contextual fear conditioning, conjunctive representations, pattern completion, and the hippocampus. Behav Neurosci 113(5):867–880
Rudy JW, Barrientos RM, O'Reilly RC (2002) Hippocampal formation supports conditioning to memory of a context. Behav Neurosci 116(4):530–538
Rudy JW, Huff NC, Matus-Amat P (2004) Understanding contextual fear conditioning: insights from a two-process model. Neurosci Biobehav Rev 28(7):675–685. https://doi.org/10.1016/j.neubiorev.2004.09.004
Sala M, Perez J, Soloff P, Ucelli di Nemi S, Caverzasi E, Soares JC, Brambilla P (2004) Stress and hippocampal abnormalities in psychiatric disorders. Eur Neuropsychopharmacol 14 (5):393–405. https://doi.org/10.1016/j.euroneuro.2003.12.005
Sapolsky RM (1996) Why stress is bad for your brain. Science 273(5276):749–750
Sapolsky RM (2000) Glucocorticoids and hippocampal atrophy in neuropsychiatric disorders. Arch Gen Psychiatry 57(10):925–935
Sapolsky RM, Uno H, Rebert CS, Finch CE (1990) Hippocampal damage associated with prolonged glucocorticoid exposure in primates. J Neurosci 10(9):2897–2902
Schmaal L, Veltman DJ, van Erp TG, Samann PG, Frodl T, Jahanshad N et al (2016) Subcortical brain alterations in major depressive disorder: findings from the ENIGMA Major Depressive Disorder working group. Mol Psychiatry 21(6):806–812. https://doi.org/10.1038/mp.2015.69
Schmahl CG, McGlashan TH, Bremner JD (2002) Neurobiological correlates of borderline personality disorder. Psychopharmacol Bull 36(2):69–87
Schmahl CG, Vermetten E, Elzinga BM, Douglas Bremner J (2003) Magnetic resonance imaging of hippocampal and amygdala volume in women with childhood abuse and borderline personality disorder. Psychiatry Res 122(3):193–198
Seeley WW, Menon V, Schatzberg AF, Keller J, Glover GH, Kenna H et al (2007) Dissociable intrinsic connectivity networks for salience processing and executive control. J Neurosci 27 (9):2349–2356. https://doi.org/10.1523/JNEUROSCI.5587-06.2007
Sheffield JM, Williams LE, Woodward ND, Heckers S (2013) Reduced gray matter volume in psychotic disorder patients with a history of childhood sexual abuse. Schizophr Res 143:185–191
Sheline YI (2000) 3D MRI studies of neuroanatomic changes in unipolar major depression: the role of stress and medical comorbidity. Biol Psychiatry 48(8):791–800
Shin LM, Liberzon I (2009) The neurocircuitry of fear, stress, and anxiety disorders. Neuropsychopharmacology 35:169–191
Shin LM, Whalen PJ, Pitman RK, Bush G, Macklin ML, Lasko NB et al (2001) An fMRI study of anterior cingulate function in posttraumatic stress disorder. Biol Psychiatry 50(12):932–942
Shin LM, Orr SP, Carson MA, Rauch SL, Macklin ML, Lasko NB et al (2004a) Regional cerebral blood flow in the amygdala and medial prefrontal cortex during traumatic imagery in male and female Vietnam veterans with PTSD. Arch Gen Psychiatry 61(2):168–176. https://doi.org/10.1001/archpsyc.61.2.168
Shin LM, Shin PS, Heckers S, Krangel TS, Macklin ML, Orr SP et al (2004b) Hippocampal function in posttraumatic stress disorder. Hippocampus 14(3):292–300. https://doi.org/10.1002/hipo.10183
Shin LM, Wright CI, Cannistraro PA, Wedig MM, McMullin K, Martis B et al (2005) A functional magnetic resonance imaging study of amygdala and medial prefrontal cortex responses to overtly presented fearful faces in posttraumatic stress disorder. Arch Gen Psychiatry 62 (3):273–281. https://doi.org/10.1001/archpsyc.62.3.273
Shin LM, Rauch SL, Pitman RK (2006) Amygdala, medial prefrontal cortex, and hippocampal function in PTSD. Ann N Y Acad Sci 1071:67–79. https://doi.org/10.1196/annals.1364.007

Siegle GJ, Konecky RO, Thase ME, Carter CS (2003) Relationships between amygdala volume and activity during emotional information processing tasks in depressed and never-depressed individuals: an fMRI investigation. Ann N Y Acad Sci 985:481–484

Sisk CL, Foster DL (2004) The neural basis of puberty and adolescence. Nat Neurosci 7 (10):1040–1047. https://doi.org/10.1038/nn1326

Smith ME (2005) Bilateral hippocampal volume reduction in adults with post-traumatic stress disorder: a meta-analysis of structural MRI studies. Hippocampus 15(6):798–807. https://doi.org/10.1002/hipo.20102

Smith MA, Makino S, Kvetnansky R, Post RM (1995) Stress and glucocorticoids affect the expression of brain-derived neurotrophic factor and neurotrophin-3 mRNAs in the hippocampus. J Neurosci 15(3 Pt 1):1768–1777

Squire LR (1992) Memory and the hippocampus: a synthesis from findings with rats, monkeys, and humans. Psychol Rev 99(2):195–231

Squire LR, Stark CE, Clark RE (2004) The medial temporal lobe. Annu Rev Neurosci 27:279–306. https://doi.org/10.1146/annurev.neuro.27.070203.144130

Stewart MG, Davies HA, Sandi C, Kraev IV, Rogachevsky VV, Peddie CJ et al (2005) Stress suppresses and learning induces plasticity in CA3 of rat hippocampus: a three-dimensional ultrastructural study of thorny excrescences and their postsynaptic densities. Neuroscience 131 (1):43–54. https://doi.org/10.1016/j.neuroscience.2004.10.031

Sunanda, Rao MS, Raju TR (1995) Effect of chronic restraint stress on dendritic spines and excrescences of hippocampal CA3 pyramidal neurons—a quantitative study. Brain Res 694 (1–2):312–317

Tata DA, Anderson BJ (2010) The effects of chronic glucocorticoid exposure on dendritic length, synapse numbers and glial volume in animal models: implications for hippocampal volume reductions in depression. Physiol Behav 99(2):186–193. https://doi.org/10.1016/j.physbeh.2009.09.008

Teicher MH, Samson JA (2013) Childhood maltreatment and psychopathology: a case for ecophenotypic variants as clinically and neurobiologically distinct subtypes. Am J Psychiatry 170(10):1114–1133. https://doi.org/10.1176/appi.ajp.2013.12070957

Teicher MH, Andersen SL, Polcari A, Anderson CM, Navalta CP, Kim DM (2003) The neurobiological consequences of early stress and childhood maltreatment. Neurosci Biobehav Rev 27 (1–2):33–44

Thomaes K, Dorrepaal E, Draijer N, de Ruiter MB, van Balkom AJ, Smit JH, Veltman DJ (2010) Reduced anterior cingulate and orbitofrontal volumes in child abuse-related complex PTSD. J Clin Psychiatry 71(12):1636–1644. https://doi.org/10.4088/JCP.08m04754blu

Tomoda A, Navalta CP, Polcari A, Sadato N, Teicher MH (2009a) Childhood sexual abuse is associated with reduced gray matter volume in visual cortex of young women. Biol Psychiatry 66(7):642–648. https://doi.org/10.1016/j.biopsych.2009.04.021

Tomoda A, Suzuki H, Rabi K, Sheu YS, Polcari A, Teicher MH (2009b) Reduced prefrontal cortical gray matter volume in young adults exposed to harsh corporal punishment. NeuroImage 47(Suppl 2):T66–T71. https://doi.org/10.1016/j.neuroimage.2009.03.005

Tomoda A, Sheu YS, Rabi K, Suzuki H, Navalta CP, Polcari A, Teicher MH (2011) Exposure to parental verbal abuse is associated with increased gray matter volume in superior temporal gyrus. NeuroImage 54(Suppl 1):S280–S286. https://doi.org/10.1016/j.neuroimage.2010.05.027

Tomoda A, Polcari A, Anderson CM, Teicher MH (2012) Reduced visual cortex gray matter volume and thickness in young adults who witnessed domestic violence during childhood. PLoS One 7(12):e52528. https://doi.org/10.1371/journal.pone.0052528

Tottenham N, Hare TA, Quinn BT, McCarry TW, Nurse M, Gilhooly T et al (2010) Prolonged institutional rearing is associated with atypically large amygdala volume and difficulties in emotion regulation. Dev Sci 13(1):46–61. https://doi.org/10.1111/j.1467-7687.2009.00852.x

Treadway MT, Grant MM, Ding Z, Hollon SD, Gore JC, Shelton RC (2009) Early adverse events, HPA activity and rostral anterior cingulate volume in MDD. PLoS One 4(3):e4887. https://doi.org/10.1371/journal.pone.0004887

Uno H, Tarara R, Else JG, Suleman MA, Sapolsky RM (1989) Hippocampal damage associated with prolonged and fatal stress in primates. J Neurosci 9(5):1705–1711
Vaidya VA, Terwilliger RM, Duman RS (1999) Role of 5-HT2A receptors in the stress-induced down-regulation of brain-derived neurotrophic factor expression in rat hippocampus. Neurosci Lett 262(1):1–4
van Erp TG, Hibar DP, Rasmussen JM, Glahn DC, Pearlson GD, Andreassen OA, Agartz I, Westlye LT, Haukvik UK, Dale AM, Melle I, Hartberg CB, Gruber O, Kraemer B, Zilles D, Donohoe G, Kelly S, McDonald C, Morris DW, Cannon DM, Corvin A, Machielsen MW, Koenders L, de Haan L, Veltman DJ, Satterthwaite TD, Wolf DH, Gur RC, Gur RE, Potkin SG, Mathalon DH, Mueller BA, Preda A, Macciardi F, Ehrlich S, Walton E, Hass J, Calhoun VD, Bockholt HJ, Sponheim SR, Shoemaker JM, van Haren NE, Hulshoff Pol HE, Ophoff RA, Kahn RS, Roiz-Santiañez R, Crespo-Facorro B, Wang L, Alpert KI, Jönsson EG, Dimitrova R, Bois C, Whalley HC, McIntosh AM, Lawrie SM, Hashimoto R, Thompson PM, Turner JA (2016) Subcortical brain volume abnormalities in 2028 individuals with schizophrenia and 2540 healthy controls via the ENIGMA consortium. Mol Psychiatry 21(4):547–553
Van Dam NT, Rando K, Potenza MN, Tuit K, Sinha R (2014) Childhood maltreatment, altered limbic neurobiology, and substance use relapse severity via trauma-specific reductions in limbic gray matter volume. JAMA Psychiatry 71(8):917–925. https://doi.org/10.1001/jamapsychiatry.2014.680
van der Put CE, Lanctot N, de Ruiter C, van Vugt E (2015) Child maltreatment among boy and girl probationers: does type of maltreatment make a difference in offending behavior and psychosocial problems? Child Abuse Negl 46:142–151. https://doi.org/10.1016/j.chiabu.2015.05.012
van Harmelen AL, van Tol MJ, van der Wee NJ, Veltman DJ, Aleman A, Spinhoven P et al (2010) Reduced medial prefrontal cortex volume in adults reporting childhood emotional maltreatment. Biol Psychiatry 68(9):832–838. https://doi.org/10.1016/j.biopsych.2010.06.011
Varese F, Smeets F, Drukker M, Lieverse R, Lataster T, Viechtbauer W et al (2012) Childhood adversities increase the risk of psychosis: a meta-analysis of patient-control, prospective- and cross-sectional cohort studies. Schizophr Bull 38(4):661–671. https://doi.org/10.1093/schbul/sbs050
Villarreal G, Hamilton DA, Petropoulos H, Driscoll I, Rowland LM, Griego JA et al (2002) Reduced hippocampal volume and total white matter volume in posttraumatic stress disorder. Biol Psychiatry 52(2):119–125
Vogt BA, Finch DM, Olson CR (1992) Functional heterogeneity in cingulate cortex: the anterior executive and posterior evaluative regions. Cereb Cortex 2(6):435–443
Vogt BA, Berger GR, Derbyshire SW (2003) Structural and functional dichotomy of human midcingulate cortex. Eur J Neurosci 18(11):3134–3144
Vyas A, Mitra R, Shankaranarayana Rao BS, Chattarji S (2002) Chronic stress induces contrasting patterns of dendritic remodeling in hippocampal and amygdaloid neurons. J Neurosci 22(15):6810–6818. https://doi.org/10.1523/JNEUROSCI.22-15-06810.2002
Vyas A, Bernal S, Chattarji S (2003) Effects of chronic stress on dendritic arborization in the central and extended amygdala. Brain Res 965(1–2):290–294
Vyas A, Pillai AG, Chattarji S (2004) Recovery after chronic stress fails to reverse amygdaloid neuronal hypertrophy and enhanced anxiety-like behavior. Neuroscience 128(4):667–673. https://doi.org/10.1016/j.neuroscience.2004.07.013
Vythilingam M, Heim C, Newport J, Miller AH, Anderson E, Bronen R et al (2002) Childhood trauma associated with smaller hippocampal volume in women with major depression. Am J Psychiatry 159(12):2072–2080. https://doi.org/10.1176/appi.ajp.159.12.2072
Vythilingam M, Luckenbaugh DA, Lam T, Morgan CA 3rd, Lipschitz D, Charney DS et al (2005) Smaller head of the hippocampus in Gulf War-related posttraumatic stress disorder. Psychiatry Res 139(2):89–99. https://doi.org/10.1016/j.pscychresns.2005.04.003
Walsh ND, Dalgleish T, Lombardo MV, Dunn VJ, Van Harmelen AL, Ban M, Goodyer IM (2014) General and specific effects of early-life psychosocial adversities on adolescent grey matter volume. Neuroimage Clin 4:308–318. https://doi.org/10.1016/j.nicl.2014.01.001

Wellman CL (2001) Dendritic reorganization in pyramidal neurons in medial prefrontal cortex after chronic corticosterone administration. J Neurobiol 49(3):245–253

Weniger G, Lange C, Sachsse U, Irle E (2009) Reduced amygdala and hippocampus size in trauma-exposed women with borderline personality disorder and without posttraumatic stress disorder. J Psychiatry Neurosci 34(5):383–388

Whalen PJ, Bush G, McNally RJ, Wilhelm S, McInerney SC, Jenike MA, Rauch SL (1998) The emotional counting Stroop paradigm: a functional magnetic resonance imaging probe of the anterior cingulate affective division. Biol Psychiatry 44(12):1219–1228

Whitford TJ, Rennie CJ, Grieve SM, Clark CR, Gordon E, Williams LM (2007) Brain maturation in adolescence: concurrent changes in neuroanatomy and neurophysiology. Hum Brain Mapp 28 (3):228–237. https://doi.org/10.1002/hbm.20273

Whittle S, Simmons JG, Dennison M, Vijayakumar N, Schwartz O, Yap MB et al (2014) Positive parenting predicts the development of adolescent brain structure: a longitudinal study. Dev Cogn Neurosci 8:7–17. https://doi.org/10.1016/j.dcn.2013.10.006

Wierenga L, Langen M, Ambrosino S, van Dijk S, Oranje B, Durston S (2014) Typical development of basal ganglia, hippocampus, amygdala and cerebellum from age 7 to 24. NeuroImage 96:67–72. https://doi.org/10.1016/j.neuroimage.2014.03.072

Wignall EL, Dickson JM, Vaughan P, Farrow TF, Wilkinson ID, Hunter MD, Woodruff PW (2004) Smaller hippocampal volume in patients with recent-onset posttraumatic stress disorder. Biol Psychiatry 56(11):832–836. https://doi.org/10.1016/j.biopsych.2004.09.015

Williams LM, Kemp AH, Felmingham K, Barton M, Olivieri G, Peduto A et al (2006) Trauma modulates amygdala and medial prefrontal responses to consciously attended fear. NeuroImage 29(2):347–357. https://doi.org/10.1016/j.neuroimage.2005.03.047

Winter H, Irle E (2004) Hippocampal volume in adult burn patients with and without posttraumatic stress disorder. Am J Psychiatry 161(12):2194–2200. https://doi.org/10.1176/appi.ajp.161.12.2194

Woodward SH, Kaloupek DG, Streeter CC, Kimble MO, Reiss AL, Eliez S et al (2006) Hippocampal volume, PTSD, and alcoholism in combat veterans. Am J Psychiatry 163(4):674–681. https://doi.org/10.1176/appi.ajp.163.4.674

Woon FL, Hedges DW (2008) Hippocampal and amygdala volumes in children and adults with childhood maltreatment-related posttraumatic stress disorder: a meta-analysis. Hippocampus 18 (8):729–736. https://doi.org/10.1002/hipo.20437

Woon FL, Hedges DW (2009) Amygdala volume in adults with posttraumatic stress disorder: a meta-analysis. J Neuropsychiatry Clin Neurosci 21(1):5–12. https://doi.org/10.1176/appi.neuropsych.21.1.5, https://doi.org/10.1176/jnp.2009.21.1.5

Woon F, Hedges DW (2011) Gender does not moderate hippocampal volume deficits in adults with posttraumatic stress disorder: a meta-analysis. Hippocampus 21(3):243–252. https://doi.org/10.1002/hipo.20746

Woon FL, Sood S, Hedges DW (2010) Hippocampal volume deficits associated with exposure to psychological trauma and posttraumatic stress disorder in adults: a meta-analysis. Prog Neuro-Psychopharmacol Biol Psychiatry 34(7):1181–1188. https://doi.org/10.1016/j.pnpbp.2010.06.016

Yamasue H, Kasai K, Iwanami A, Ohtani T, Yamada H, Abe O et al (2003) Voxel-based analysis of MRI reveals anterior cingulate gray-matter volume reduction in posttraumatic stress disorder due to terrorism. Proc Natl Acad Sci USA 100(15):9039–9043. https://doi.org/10.1073/pnas.1530467100

Yang P, Wu MT, Hsu CC, Ker JH (2004) Evidence of early neurobiological alternations in adolescents with posttraumatic stress disorder: a functional MRI study. Neurosci Lett 370 (1):13–18. https://doi.org/10.1016/j.neulet.2004.07.033

Yang RJ, Mozhui K, Karlsson RM, Cameron HA, Williams RW, Holmes A (2008) Variation in mouse basolateral amygdala volume is associated with differences in stress reactivity and fear learning. Neuropsychopharmacology 33(11):2595–2604. https://doi.org/10.1038/sj.npp.1301665

Zhang J, Tan Q, Yin H, Zhang X, Huan Y, Tang L et al (2011a) Decreased gray matter volume in the left hippocampus and bilateral calcarine cortex in coal mine flood disaster survivors with recent onset PTSD. Psychiatry Res 192(2):84–90. https://doi.org/10.1016/j.pscychresns.2010.09.001

Zhang L, Zhang Y, Li L, Li Z, Li W, Ma N et al (2011b) Different white matter abnormalities between the first-episode, treatment-naive patients with posttraumatic stress disorder and generalized anxiety disorder without comorbid conditions. J Affect Disord 133(1–2):294–299. https://doi.org/10.1016/j.jad.2011.03.040

2 Synaptic Changes Responsible for Grey Matter Changes in the Brain of Animal Models Following Stress

2.1 Changes in MRI-Determined GMV Following Stress

Following stress, GMV has been determined using high-resolution MRI for the ACC, hippocampus, amygdala and RSG (Fig. 2.1a–d). Significant volume changes are observed in the ACC of hippocampus but not in the amygdala nor the RSG (Table 2.1).

2.2 Changes in the Number of Neurons, Astrocytes and Oligodendrocytes Following Stress

No significant change in the number of neurons, astrocytes or oligodendrocytes following stress, according to Nissl stains, in any of the anatomical brain regions have been observed (see Fig. 2.2; Table 2.1). Moreover, no change in the surface area of the somas of the aforementioned cells occurs following stress.

The cumulative lengths of the apical dendrites of individual neurons were measured in the ACC, CA1, amygdala and RSG of control animals as well as following stress (Fig. 2.3a, b). Significant decreases in the lengths of apical dendrites are found for dendrites in the ACC and CA1 but not in the amygdala nor in the RSG. Percentage changes of over 45% have been determined for the ACC and CA1. The cumulative apical dendritic volume per neuron across the different brain regions amounts to decreases of 56% in the ACC, 45% in CA1, 5% in the amygdala and an increase of 2% in the RSG. These volumes are determined from the length of the dendrites given in Fig. 2.3 and an average dendritic diameter of 0.9 μm as previously determined in the mouse (Braitenberg and Schuz 1998).

M. Bennett, J. Lagopoulos, *Stress, Trauma and Synaptic Plasticity*,
https://doi.org/10.1007/978-3-319-91116-8_2

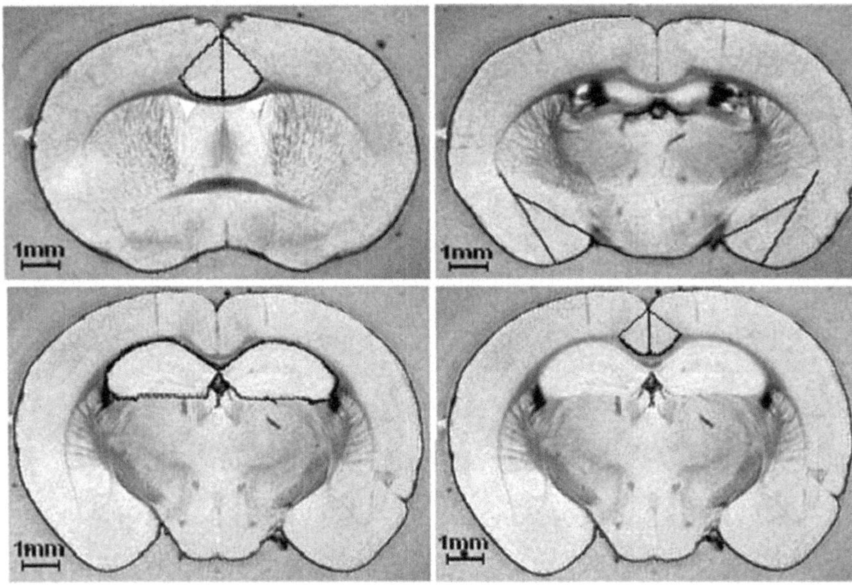

Fig. 2.1 GMV in the cortex of mice, as determined with high-resolution MRI, following 120 cumulative hours of restraint stress. (**a**) Coronal image showing the ACC, delineated by the dorsal triangular regions, used in its volume estimates. (**b**) Coronal image showing the amygdala, delineated by the ventral triangular regions, used in its volume estimates. (**c**) Coronal image showing the dorsal region in the hippocampus that is clearly delineated for the purpose of volume measurements. (**d**) Coronal image showing the RSG, delineated by the dorsal triangular regions, used in its volume estimates (from Kassem et al. 2013)

Table 2.1 Changes in the number of neurons, astrocytes and oligodendrocytes following stress

ACC		Hippocampus		Amygdala		RSG	
F(1, 8)	*p*	*F*(1, 8)	*p*	*F*(1, 8)	*P*	*F*(1, 8)	*p*
Neurons							
1.296	<0.288	0.968	<0.354	2.173	<0.179	0.008	<0.930
Astrocytes							
1.296	<0.288	0.039	<0.848	0.103	<0.757	0.183	<0.680
Oligodendrocytes							
7.111	<0.029	0.003	<0.957	0.521	<0.491	0.089	<0.773

The predicted volume change of each brain region (in cubic millimetres) has been determined, transforming the above values into the cumulative volume of all dendrites within each region (in cubic millimetres per cubic millimetres; Table 2.2). Comparison has been made of the before and after stress values to find the change in dendrite volume in cubic millimetres per cubic millimetres. Multiplying this difference by the measured volume of the brain region (in cubic millimetres) predicts the volume change in cubic millimetres (Table 2.2).

The volumes of cells and their processes in 1 mm^3 of rodent cortical grey matter can be calculated based on a review of existing literature (Table 2.3; Fig. 2.4).

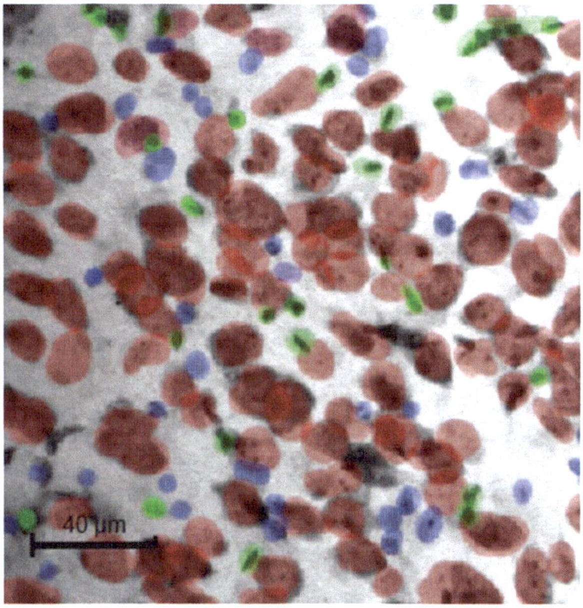

Fig. 2.2 No loss of neurons, astrocytes or oligodendrocytes were detected in stressed animals. Nissl-stained section through the ACC showing neurons (N, red), astrocytes (A, blue) and oligodendrocytes (O, green) (from Kassem et al. 2013)

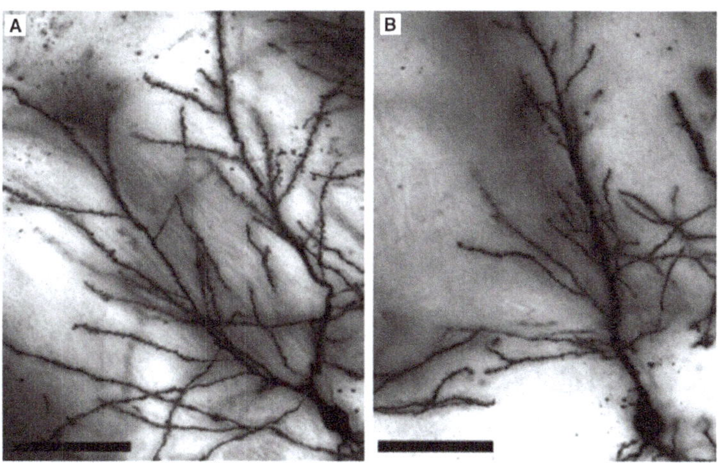

Fig. 2.3 Significant decreases occur in stressed animals in the cumulative length of apical dendrites of neurons, that is, the sum of all different-order dendrites in the ACC and CA1. (**a**) Photomicrograph of Golgi-stained pyramidal CA1 neuron of a non-stressed mouse. (**b**) Dendrites [at the same magnification as in (**a**)] of a pyramidal CA1 neuron of a stressed mouse with shorter cumulative dendritic length. Note that there was no significant change in dendritic length or volume in the amygdala or RSG, and these regions showed no change in GMV loss (from Kassem et al. 2013)

Table 2.2 Changes in volume of grey matter due to changes in volume of dendrites following stress, compared with the changes in volume of grey matter determined with high-resolution MRI

Calculation	Hippocampus	Hippocampus (stress)	ACC	ACC (stress)	Amygdala	Amygdala (stress)	RSG	RSG (stress)
Apical dendrite length/neuron (μm)	1753	964	2025	902	884	837	650	675
Total dendrite length/neuron[a] (μm)	3506	1928	4050	1804	1768	1674	1300	1350
Dendrite diameter[b] (μm)	0.9	0.9	0.9	0.9	0.9	0.9	0.9	0.9
Volume of dendrite/neuron (μm^3)	2271	1249	2624	1160	1145	1084	842	874
Number of neurons (mm^3)	9.2×10^4	9.2×10^4	7.7×10^4	7.7×10^4	6.57×10^4	6.57×10^4	7.0×10^4	7.0×10^4
Volume of dendrites (mm^3/mm^3)	0.21	0.115	0.20	0.09	0.074	0.0714	0.0589	0.0612
Dendrite volume change (mm^3/mm^3)		−0.095		−0.11		−0.0026		+0.0023
Volume of region of interest (mm^3)	27.0	24.0	5.2	4.4	9.7	9.9	5.6	5.6
Predicted volume change of brain region[c] (mm^3)		−2.565 (−3.222)		−0.572 (−0.675)		−0.025 (−0.026)		+0.012 (+0.002)
Observed volume change of brain region (mm^3)		−3.01		−0.85		+0.21		+0.08

[a]Based on the total length of basal dendrites being equivalent to that of the apical dendrites (see Bennett 2011, Table 4), this can vary in some instances of up to 20% in mouse hippocampus (see Pillai et al. 2012, Table 1). It follows that the percentage changes in apical dendrites determined here under stress has been taken to be the same for basal dendrites, which is approximately the case in CA3 subregion of rat hippocampus (see Magarinos et al. 2011, Fig. 1)

[b]This 'average' dendritic diameter, over the different-order dendritic branches in the mouse, has been determined by Braitenberg and Schuz (1998) (see also, Bennett 2011, Table 4)

[c]These values are calculated from the dendritic volume changes. In parenthesis are shown the calculations that include dendritic and spine volume changes

Table 2.3 Volumes of cells and their processes in 1 mm^3 of rodent cortical grey matter

Neuron vol. = (soma volume)[a] × (neuron density)[b] = $(1400 \times 10^{-9}) \times (8 \times 10^4) = 0.11\ mm^3$
Dendritic vol. = (dendritic volume of single neuron)[c, d] × (neuron density)[b] = ((length of dendrite/neuron) (l) × (cross-sectional area of dendrite, diameter d)) × (neuron density) = $\left(l \times \frac{\pi d^2}{4}\right)$ × (neurodensity) = $\left(4819 \times \frac{\pi d^2}{4}\right) \times \left(8 \times 10^4\right) = \left(4819 \times 0.79 \times 0.9^2\right) \times (8 \times 10^4) = 0.25\ mm^3$
Dendritic vol. = (total dendritic length per mm^3)[e] × (cross-sectional area of dendrite)[d] = $(0.4 \times 10^6) \times \left(\frac{\pi d^2}{4} \times 10^{-6}\right) = 0.259\ mm^3$
Axon collateral vol. = (collateral length per mm^3)[f] × (collateral cross-sectional area)[g] = $(0.4 \times 10^6) \times \left(\frac{\pi 0.3^2}{4} \times 10^{-6}\right) = 0.29\ mm^3$
Astrocyte cell body = (astrocyte soma volume)[h] × (astrocyte density)[i] = $(355 \times 10^{-9}) \times (2.5 \times 10^4) = 0.009\ mm^3$
Astrocyte process = (astrocyte process volume)[j] × (astrocyte density)[i] = $(3300 \times 10^{-9}) \times (2.5 \times 10^4) = 0.08\ mm^3$
Oligodendrocyte cell body = (oligodendrocyte soma volume)[k] × (oligodendrocyte density)[l] = $(285 \times 10^{-9}) \times (2.5 \times 10^4) = 0.007\ mm^3$
Oligodendrocyte process = (oligodendrocyte process volume)[m] × (oligodendrocyte density)[l] = $(624 \times 10^{-9}) \times (2.5 \times 10^4) = 0.016\ mm^3$
Volume fraction of extracellular space[n] = $0.18\ mm^3$
Capillaries vol. = (average length per mm^3)[o] × (Average cross sectional area, diameter d)[o] = $(378) \times \left(\frac{\pi d^2}{4} \times 10^{-6}\right) = 0.006\ mm^3$
Microglial vol. = (microglial soma and process volume)[p] × (microglial density)[q] = $(0.69 \times 10^{-6}) \times (6.5 \times 10^3) = 0.005\ mm^3$
Synaptic (presynaptic bouton) vol. = (single synapse volume)[r] × (synapse density)[s] = $(0.03 \times 10^{-9}) \times (7 \times 10^8) = 0.021\ mm^3$
Synaptic spine end-bulb vol. = (single end-bulb volume)[t] × (end-blub density)[u] = $(0{:}15 \times 10^{-9}) \times (2.4 \times 10^8) = 0.036\ mm^3$
Total = $1.016\ mm^3$

Neuron/glial ratio = 1.43 (from footnotes b, i, l and q above). For humans, this is 1.49 (Pelvig et al. 2008). Oligodendrocyte/astrocyte/microglia number ratios are 1:1:0.6. For humans, these are 1:0.8:0.2 (Blinkov and Glezer 1968) and 1:0.27:0.07 (Pelvig et al. 2008)

[a]Soma volume, 1400 μm^3 (Tata et al. 2006)

[b]Neuron density, $8 \times 10^4\ mm^{-3}$ (Tata et al. 2006) and $9 \times 10^4\ mm^{-3}$ (Schuz and Palm 1989)

[c]Value of l, 4819 μm (Braitenberg and Schuz 1998; Alpar et al. 2006; Englisch et al. 1974; Lloyd et al. 2003; Minkwitz and Holz 1975; Pokorny and Yamamoto 1981; Trommald et al. 1995)

[d]Value of d, 0.9 μm (Braitenberg and Schuz 1998; Trommald et al. 1995)

[e]Dendritic length/mm^3, 0.4 km/mm^3 (Stepanyants et al. 2004)

[f]Collateral length/mm^3, 4.1 km/mm^3 (Stepanyants et al. 2004; Foh et al. 1973)

[g]Collateral diameter, 0.3 μm (Braitenberg and Schuz 1998; Bolstad et al. 2007)

[h]Astrocyte soma volume, d 0 8.6 μm (Hama et al. 2004; Oberheim et al. 2009)

[i]Astrocyte density, $2.5 \times 10^4\ mm^{-3}$ (Ogata and Kosaka 2002)

[j]Astrocyte process volume, $3300 \times 10^{-9}\ mm^3$ (Tata et al. 2006; Wenzel et al. 1991)

[k]Oligodendrocyte soma volume, $285 \times 10^{-9}\ mm^3$ (D'Ambrosio et al. 1998)

[l]Oligodendrocyte density, $2.5 \times 10\ mm^{-3}$(Hamidi et al. 2004; Hof et al. 2003); these references are to humans

[m]Oligodendrocyte process volume, $624 \times 10^{-9}\ mm^3$ (D'Ambrosio et al. 1998)

[n]Volume fraction of extracellular space, 0.18 ± 0.02 (Lehmenkuhler et al. 1993; Nicholson and Phillips 1981; Sykova et al. 2005a, b; Yao et al. 2008)

(continued)

Table 2.3 (continued)

[o]Capillary volume, average length of capillary = 378 mm/mm^3; diameter = 4.5 × 10^{-3} mm; so volume = 0.006 mm^3/mm^3. Respective values for arterioles are 18 mm/mm^3, 13 × 10^{-3} mm and 0.003 mm^3/mm^3. Respective values for venules are about the same as for arterioles, so volume = 0.003 mm^3/mm^3 (Bell and Ball 1985, Table 4; Knox and Oliveira 1980)
[p]Microglial soma and process volume, soma volume with diameter 7.7 × 10^{-3} mm is 0.24 × 10^{-6} mm^3. If d = diameter of process, 0.8 μm; l = total length of processes, 417 μm (Nimmerjahn et al. 2005). So process volume is 0:45 × 10^{-6} mm^3 = total process length per cell $(l)\frac{\pi d^2}{4}$
[q]Microglial density, 6.5 × 10^3 mm^{-3} (Nimmerjahn et al. 2005)
[r]Single synapse bouton volume from diameter of bouton, d = −0.64 μm (Cragg 1967, this is for cat)
[s]Synapse density, 0.7 × 10^9 mm^{-3} (Braitenberg and Schuz 1998; Schuz and Palm 1989)
[t]Single end-bulb plus stalk volume of spine, 0.155 × 10^{-9} mm^3 (Spacek and Hartmann 1983)
[u]End-bulb spine density, average number of spines over all cortical layers in the cat = 2.4 × 10^8/mm^3 (Stepanyants et al. 2009)

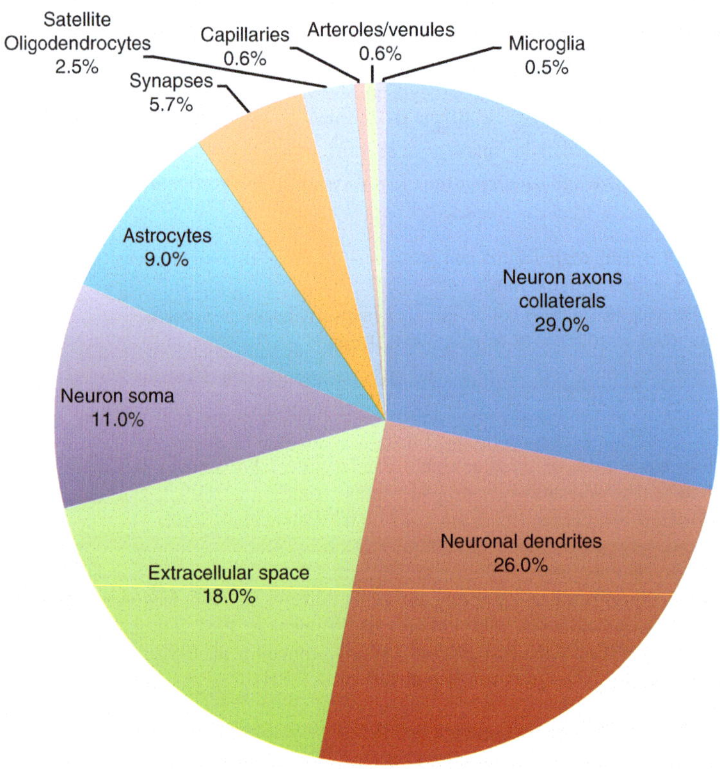

Fig. 2.4 Volume of grey matter taken up by different volumes of its cellular constituents and their processes. Percentages are determined from the review of existing literature calculated in Table 2.3 (from Kassem et al. 2013)

Plotting the absolute changes of GMV against the loss of dendritic volume (both in cubic millimetres) for each brain region produces a line of best fit as $y = 1.18x + 0.03$ (Fig. 2.5a). Plotting the percentage change in GMV following stress plotted against the percentage change in volume of dendrites indicates a line of best fit as $y = 0.30x + 1.9$ (Fig. 2.5b).

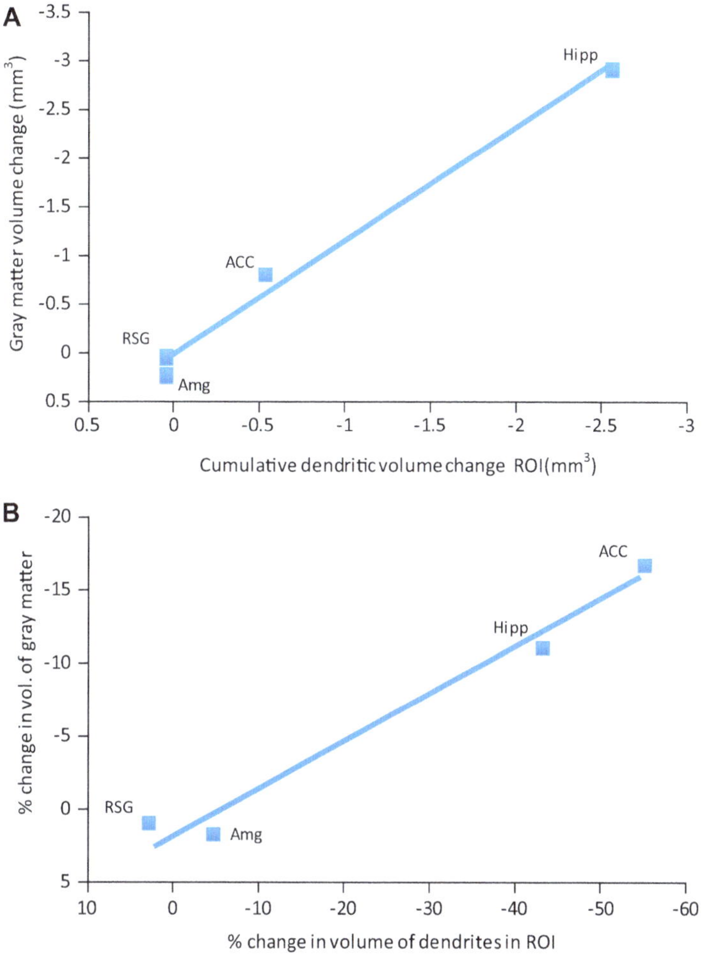

Fig. 2.5 In stressed animals, there is a linear relationship between the loss of GMV determined with high-resolution MRI and the cumulative loss of dendritic volume such that the former can mostly be accounted for by the latter. (**a**) The absolute changes of GMV are plotted against that of the loss of dendritic volume (both in cubic millimetre), for the different ROI. The line of best fit is y 0 $1.18x + 0.03$ (R^2 0 0.99), indicating approximately 80% of the volume of grey matter decrease is accounted for by the loss of dendritic volume. (**b**) The percentage changes in GMV following stress plotted against the percentage change in volume of dendrites, showing that these percentages are similar with y 0 $0.30x + 1.9$ (R^2 0 0.97), indicating that if there was a 100% loss of dendrites, the GMV should decrease by 32%. Figure 2.4, determined from the literature evaluations given in Table 2.3, predicts that this value should be 26%. If allowance is made for the loss of dendritic spines when dendrites are lost, as well as the decrease in spine density on the remaining dendrites following stress, the lines in (**a**) and (**b**) are y 0 $1.01x - 0.07$ (R^2 0 0.99) and y 0 $0.30x + 1.84$ (R^2 0 0.98), respectively (from Kassem et al. 2013)

2.3 Discussion

2.3.1 Density of Cell Types in the Cortex Before and After Stress

Given that neither the amygdala nor the RSG show significant GMV changes following stress, these regions can act as useful controls. It might be anticipated that no changes in cells occurs within the amygdala or RSG as compared with the ACC and hippocampus where substantial GMV loss was observed. Table 2.3 sets out the density of different cell types found in the grey matter and averaged over different areas in the rodent brain as determined by the authors referenced at the end of the table. These are 8×10^4, 2.5×10^4 and $2.5 \times 10^4/\text{mm}^3$ for neurons, astrocytes and oligodendrocytes, respectively. These values approximate the same values as determined in Nissl-stained regions of interest (see Fig. 2.2). No significant changes in these densities in the ACC, CA1, amygdala and RSG are determined for the stressed mice (Fig. 2.2), regardless of whether the aforementioned brain regions include those with GMV loss or not. Although there have been a large number of studies on the effects of different kinds of stress regimes on the density of different cell types in the cortex of rodents, they do not allow any definite conclusions.

It is of interest to consider if the claims for decreases in a particular cell type in the cortex following stress, even if inconsistent from study to study, could explain the observed changes in GMV. For example, the claim for a 35% decrease in the number of oligodendrocytes in the prefrontal cortex following a period of chronic stress (Banasr et al. 2007) would, according to Fig. 2.4, give rise to a decrease in GMV of 0.8%, a change that on its own would not be detectable with MRI. Indeed, none of the changes in the numbers of glial cells reported, either up or down, give rise to significant changes in GMV. However, changes in neuron numbers could have a large effect on GMV because of the very large percentage of GMV taken up by dendrites and axon collaterals, cumulatively 55% (Fig. 2.4). Quantitative evaluation of the loss of neurons following stress has been mostly carried out for granule cells of the dentate gyrus (see, e.g. Veena et al. 2009), which cannot be used to estimate changes in hippocampal CA1 and CA2 GMV. On the other hand, the claim of a 25% loss of CA1 neurons following predatory stress, an extreme stressor known to produce neuron damage and loss (McEwen 2000a, b), would, according to Fig. 2.4, give rise to a 16% decrease in GMV of the CA1 region of the hippocampus. This is considerably in excess of the GMV loss of 11% stated above (see Fig. 2.1). However, under the conditions of restraint stress, there is no loss of cells in any of the brain regions investigated, so that changes in cellular processes that significantly contribute to GMV must be considered.

2.3.2 Dendritic Lengths and Spine Densities Before and After Stress

There are a number of studies on the length of dendrites and of their spine densities before and after restraint stress of rats. Here, consideration is given for each brain region and of observations on rats for comparison with studies on mice. For rat CA1 hippocampus, the length of apical dendrites is on average, over reported values, 2023 ± 183 μm, and the density of dendritic spines is 1.7 ± 0.23/μm with these reduced by 23 ± 2.5 and 26 ± 6%, respectively, following between 1 and 6 h of restraint stress/day over 21 days (Bannister and Larkman 1995; Chen et al. 2009, 2010; Conrad et al. 1999; Konur et al. 2003; McLaughlin et al. 2007; Sousa et al. 2000). The decrease in spine density is about the same for apical and basal dendrites in CA1 hippocampus (Dalla et al. 2009). The largest decreases in dendrites and spines reported in single studies were 41.8 (prefrontal cortex) and 22% (CA1), respectively. Values represented above for mouse CA1 hippocampal apical dendrites were of length 1750 ± 250 μm and the density of spines 0.6 ± 0.04/μm. Following restraint stress for 6 h/day for 21 days, these were reduced by 40 ± 10 and 20 ± 4%, respectively. These lengths for CA1 dendrites and for the density of dendritic spines are less than that reported on average for the rat, whereas the extent of reduction of the dendrites in the mouse is greater on average following stress as it is for dendritic spines (Pawlak et al. 2005). This difference is unlikely to be due to differences in the period of restraint stress used in the rat studies, for the percentage differences in reduction of dendrites following cumulative periods of restraint in the rats from 20 h to over 125 h is not very different whether considering neurons in CA1/CA3 hippocampus, prefrontal cortex or ACC (Conrad et al. 1999; McLaughlin et al. 2007; Sousa et al. 2000; Liston et al. 2006; Radley et al. 2004, 2005; Shansky et al. 2009; Watanabe et al. 1992; Cook and Wellman 2004; Perez-Cruz et al. 2007).

For the rat ACC, medial prefrontal and motor-cortex apical dendrite length is 1710 ± 196 μm and the density of dendritic spines 1.44 ± 0.49/μm, over the reported values, and these are reduced by 23 ± 4 and 16 ± 2.8%, respectively, following 2 to 6 h stress between 10 and 21 days (Chen et al. 2009; Balu et al. 2012; Liston et al. 2006; Radley et al. 2004, 2005, 2008; Shansky et al. 2009; Cook and Wellman 2004; Perez-Cruz et al. 2007). Comparison can be made with the mouse ACC apical dendrites reported with a length of 2000 ± 300 μm and the density of spines at 0.7 ± 0.04/μm, reduced following stress for 6 h/day for 21 days by 55 ± 10 and 50 ± 7%, respectively. These values for the length of ACC dendrites are therefore greater and that for spine density less than that reported for the rat, whereas the extent of reduction of both dendrites and spines in the mouse is greater following restraint stress.

For the rat amygdala, the length of apical dendrites on average across reports is 1124 ± 128 μm, and the density of dendritic spines is 0.72 ± 0.19/μm with these increasing by 12 ± 4 and 1.0 ± 4%, respectively, following 0.5 to 6 h of restraint stress for 10 to 21 days (Pego et al. 2008; Bennur et al. 2007; Marcuzzo et al. 2007; Qin et al. 2011; Vyas et al. 2004). The increase in length of dendrites in some reports is very small (Pego et al. 2008). However, consideration has to be given to the type of neuron and the specific nuclei of the amygdala in which they occur as the effects of stress on dendrites depend on these factors. For example, chronic restraint stress has been reported to significantly increase the length of dendrites of both pyramidal and stellate neurons in the basolateral complex of the amygdala but to decrease dendrites of bipolar and bitufted neurons in this complex (Vyas et al. 2002) and to have no effect on dendrites of neurons in the central nuclei of the amygdala (Vyas et al. 2003). Dendritic lengths of pyramidal neurons are significantly increased in the mouse basolateral complex following restraint stress (Vyas et al. 2003; Hill et al. 2011; Johnson et al. 2009). Furthermore, the density of spines increases significantly on neurons in the medial amygdala but decreases significantly in the basolateral amygdala (Bennur et al. 2007). In the present considerations, pyramidal apical dendrites in the amygdala, over all nuclei, possessed lengths of 880 ± 100 μm and spine densities of 0.6 ± 0.04/μm and these were not significantly altered following restraint stress. It may then be that the heterogeneity in changes in dendritic length and spine density observed over neurons in different nuclei is the basis for the lack of significant changes being recorded following stress as reported above, for which dendritic length and spine densities were noted over all nuclei.

In summary, there are significant changes in the length of dendrites and the density of spines on the remaining dendrites in the ACC and hippocampus following restraint stress but not in the amygdala or RSG. Given that the ACC and hippocampus show significant decreases in GMV whereas the amygdala and RSG do not, it is likely that the loss of GMV is principally due to the loss of neuronal dendrites. This immediately raises the question as to why dendrites and their spines are lost in some brain regions and not in others following stress? This is clearly an important topic for future research that likely involves study of the molecular mechanisms that govern spine formation and regression.

2.4 The Relation Between Stress-Induced Decreases in Synapses/Dendrites and Grey Matter Loss

If the changes in GMV occurring in a brain region following stress are due to changes in the volume occupied by dendrites and their synapses, what are the quantitative relations between the two? Figure 2.5a, based on the calculations given in Table 2.2, shows that there is a linear relationship between the GMV changes observed using MRI and the changes in volume occupied by the dendrites in the ACC, hippocampus, amygdala and the RSG. If this is converted to percentage changes, as shown in Fig. 2.5b, then the regression line ($R^2 = 0.97$) through the data points is $y = 0.30 + 1.9$ so that the percentage change in GMV = 0.30 × percentage change in dendritic volume + 1.9. It follows that if all the dendrites were removed, then there would be a 31.9% change in GMV. Recently, estimates have been made of the volume composition of the GMV in human neocortex, with this dominated by dendrites (30%), axon collaterals (29%), neuronal somas (7.8%) and synapses (6%) (Bennett 2011). Using the same approach to determine these values for rodents, the present considerations show that these are not much different to those in humans, namely, dendrites (26%), axon collaterals (29%), neuronal somas (11%) and synapses (5.7%) (Fig. 2.4). This figure indicates that a GMV change of 31.7% is expected for a loss of dendrites and their synapses, the same change (31.9%) as that determined independently from the observations in this paper. The approximate agreement between the two independent estimates of the GMV occupied by dendrites and their synapses is interesting and lends support to the suggestion that the major changes in GMV in the ACC and the hippocampus following stress are due to the loss of dendrites and their synapses.

2.5 The Relation Between Decreases in Synapses, Dendrites and Grey Matter

Given the robust evidence for a substantial loss of synapses in some cortical areas following stress, the questions arise as to whether, as claimed above, such synapse loss is accompanied by a similar percentage loss of dendrites, and if so whether such loss of synapses and dendrites is likely to give rise to a decrease in GMV. Such a decrease implies that the grey matter compacts. An alternative is that any additional extracellular space made available within the grey matter is compensated for by the growth of other cellular processes such as those of astrocytes. Deafferentation of the cat's lateral cervical nucleus leads to similar percentage decreases in synapses and dendrites on neurons that have lost innervation (Benes et al. 1977; Griph and Westman 1977). Deafferentation of specific dendrites on neurons in the avian nucleus laminaris leads to selective shortening of these dendrites due to their reabsorption into the cell body (Deitch and Rubel 1989a), a process that does not involve astrocytes (Deitch and Rubel 1989b). Such deafferentation leaves normally innervated dendrites on the same neuron unaffected (Sorensen and Rubel 2006). Similar results have been obtained for the deafferented cat lateral geniculate nucleus (Somogyi et al. 1987) and in gerbil superior olivary nucleus (Russell and Moore 1999). Interestingly, pharmacological block of transmission of boutons on specific dendrites leads to a decrease in their length (Sorensen and Rubel 2006). Following deafferentation, there is a significant shrinkage of the whole nucleus which has been attributed to a disappearance of boutons and dendrites (Griph and Westman 1977). This shrinkage is accompanied by very little change in the extent of extracellular space so that it may be attributed to the loss of the cellular constituents and/or their processes in the grey matter since there is little evidence for substantial changes in the size of the cell somas (Griph and Westman 1977).

It has been known for some time that the vacant synaptic sites on deafferented neurons in the hippocampus may be reinnervated by sprouting axon terminals of other axons (Frotscher et al. 1981; Hoff 1986; Matthews et al. 1976, 1979). However, this has not been shown to be the case in other brain regions such as the cerebellar cortex and the lateral geniculate nucleus, where deafferentation is followed by dendritic reduction (Hamori 1990). The conditions that lead to reinnervation of vacant sites by sprouting axons versus those in which the dendrites possessing these vacant sites are reabsorbed by the neurons are not known. Deafferentation of some nuclei leads to a substantial increase in glia, presumably due to an increase in the primary processes of astrocytes (Griph and Westman 1977), as reactive astrocytes are known to increase the diameter of their primary processes without changes in their length or in the extent of the large number of small processes (Wilhelmsson et al. 2006). To what extent this might require modification of the quantitative considerations in this work is not clear. It is also unclear as to whether there is substantial loss of axon collaterals as a consequence of the loss of synapses they normally form on dendrites. They have not been measured in the present work because of the great difficulty of following them in the grey matter, even though any loss of axon collaterals is expected to contribute significantly to a decrease of GMV (see Fig. 1.4). These caveats concerning the interpretation of the present observations require further research.

References

Alpar A, Ueberham U, Bruckner MK, Seeger G, Arendt T, Gartner U (2006) Different dendrite and dendritic spine alterations in basal and apical arbors in mutant human amyloid precursor protein transgenic mice. Brain Res 1099(1):189–198. https://doi.org/10.1016/j.brainres.2006.04.109

Balu DT, Basu AC, Corradi JP, Cacace AM, Coyle JT (2012) The NMDA receptor co-agonists, D-serine and glycine, regulate neuronal dendritic architecture in the somatosensory cortex. Neurobiol Dis 45(2):671–682. https://doi.org/10.1016/j.nbd.2011.10.006

Banasr M, Valentine GW, Li XY, Gourley SL, Taylor JR, Duman RS (2007) Chronic unpredictable stress decreases cell proliferation in the cerebral cortex of the adult rat. Biol Psychiatry 62 (5):496–504. https://doi.org/10.1016/j.biopsych.2007.02.006

Bannister NJ, Larkman AU (1995) Dendritic morphology of CA1 pyramidal neurones from the rat hippocampus: II. Spine distributions. J Comp Neurol 360(1):161–171. https://doi.org/10.1002/cne.903600112

Bell MA, Ball MJ (1985) Laminar variation in the microvascular architecture of normal human visual cortex (area 17). Brain Res 335(1):139–143

Benes FM, Parks TN, Rubel EW (1977) Rapid dendritic atrophy following deafferentation: an EM morphometric analysis. Brain Res 122(1):1–13

Bennett MR (2011) The prefrontal-limbic network in depression: a core pathology of synapse regression. Prog Neurobiol 93(4):457–467. https://doi.org/10.1016/j.pneurobio.2011.01.001

Bennur S, Shankaranarayana Rao BS, Pawlak R, Strickland S, McEwen BS, Chattarji S (2007) Stress-induced spine loss in the medial amygdala is mediated by tissue-plasminogen activator. Neuroscience 144(1):8–16. https://doi.org/10.1016/j.neuroscience.2006.08.075

Blinkov SM, Glezer II (1968) The human brain in figures and tables; a quantitative handbook. Basic Books, New York

Bolstad I, Leergaard TB, Bjaalie JG (2007) Branching of individual somatosensory cerebropontine axons in rat: evidence of divergence. Brain Struct Funct 212(1):85–93. https://doi.org/10.1007/s00429-007-0145-1

Braitenberg V, Schuz A (1998) Cortex: statistics and geometry of neuronal connectivity, 2nd edn. Springer, Berlin

Chen JR, Wang TJ, Huang HY, Chen LJ, Huang YS, Wang YJ, Tseng GF (2009) Fatigue reversibly reduced cortical and hippocampal dendritic spines concurrent with compromise of motor endurance and spatial memory. Neuroscience 161(4):1104–1113. https://doi.org/10.1016/j.neuroscience.2009.04.022

Chen Y, Rex CS, Rice CJ, Dube CM, Gall CM, Lynch G, Baram TZ (2010) Correlated memory defects and hippocampal dendritic spine loss after acute stress involve corticotropin-releasing hormone signaling. Proc Natl Acad Sci USA 107(29):13123–13128. https://doi.org/10.1073/pnas.1003825107

Conrad CD, LeDoux JE, Magarinos AM, McEwen BS (1999) Repeated restraint stress facilitates fear conditioning independently of causing hippocampal CA3 dendritic atrophy. Behav Neurosci 113(5):902–913

Cook SC, Wellman CL (2004) Chronic stress alters dendritic morphology in rat medial prefrontal cortex. J Neurobiol 60(2):236–248. https://doi.org/10.1002/neu.20025

Cragg BG (1967) The density of synapses and neurones in the motor and visual areas of the cerebral cortex. J Anat 101(Pt 4):639–654

D'Ambrosio R, Wenzel J, Schwartzkroin PA, McKhann GM 2nd, Janigro D (1998) Functional specialization and topographic segregation of hippocampal astrocytes. J Neurosci 18 (12):4425–4438

Dalla C, Whetstone AS, Hodes GE, Shors TJ (2009) Stressful experience has opposite effects on dendritic spines in the hippocampus of cycling versus masculinized females. Neurosci Lett 449 (1):52–56. https://doi.org/10.1016/j.neulet.2008.10.051

Deitch JS, Rubel EW (1989a) Changes in neuronal cell bodies in N. laminaris during deafferentation-induced dendritic atrophy. J Comp Neurol 281(2):259–268. https://doi.org/10.1002/cne.902810208

Deitch JS, Rubel EW (1989b) Rapid changes in ultrastructure during deafferentation-induced dendritic atrophy. J Comp Neurol 281(2):234–258. https://doi.org/10.1002/cne.902810207

Englisch HJ, Kunz G, Wenzel J (1974) Distribution of spines on the pyramidal neurons in the CA-1 region of the hippocampus in the rat. Z Mikrosk Anat Forsch 88(1):85–102

Foh E, Haug H, Konig M, Rast A (1973) Determination of quantitative parameters of the fine structure in the visual cortex of the cat, also a methodological contribution on measuring the neuropil (author's transl). Microsc Acta 75(2):148–168

Frotscher M, Nitsch C, Hassler R (1981) Synaptic reorganization in the rabbit hippocampus after lesion of commissural afferents. Anat Embryol (Berl) 163(1):15–30

Griph S, Westman J (1977) Volume composition of the lateral cervical nucleus in the cat. I. A stereological and electron microscopical study of normal and deafferentated animals. J Neurocytol 6(6):723–743

Hama K, Arii T, Katayama E, Marton M, Ellisman MH (2004) Tri-dimensional morphometric analysis of astrocytic processes with high voltage electron microscopy of thick Golgi preparations. J Neurocytol 33(3):277–285. https://doi.org/10.1023/B:NEUR.0000044189.08240.a2

Hamidi M, Drevets WC, Price JL (2004) Glial reduction in amygdala in major depressive disorder is due to oligodendrocytes. Biol Psychiatry 55(6):563–569. https://doi.org/10.1016/j.biopsych.2003.11.006

Hamori J (1990) Morphological plasticity of postsynaptic neurones in reactive synaptogenesis. J Exp Biol 153:251–260

Hill MN, Hillard CJ, McEwen BS (2011) Alterations in corticolimbic dendritic morphology and emotional behavior in cannabinoid CB1 receptor-deficient mice parallel the effects of chronic stress. Cereb Cortex 21(9):2056–2064. https://doi.org/10.1093/cercor/bhq280

Hof PR, Haroutunian V, Friedrich VL Jr, Byne W, Buitron C, Perl DP, Davis KL (2003) Loss and altered spatial distribution of oligodendrocytes in the superior frontal gyrus in schizophrenia. Biol Psychiatry 53(12):1075–1085

Hoff SF (1986) Lesion-induced transneuronal plasticity in the adult rat hippocampus. Neuroscience 19(4):1227–1233

Johnson SA, Wang JF, Sun X, McEwen BS, Chattarji S, Young LT (2009) Lithium treatment prevents stress-induced dendritic remodeling in the rodent amygdala. Neuroscience 163 (1):34–39. https://doi.org/10.1016/j.neuroscience.2009.06.005

Kassem MS, Lagopoulos J, Stait-Gardner T, Price WS, Chohan TW, Arnold JC et al (2013) Stress-induced grey matter loss determined by MRI Is primarily due to loss of dendrites and their synapses. Mol Neurobiol 47(2):645–661. https://doi.org/10.1007/s12035-012-8365-7

Knox CA, Oliveira A (1980) Brain aging in normotensive and hypertensive strains of rats. III. A quantitative study of cerebrovasculature. Acta Neuropathol 52(1):17–25

Konur S, Rabinowitz D, Fenstermaker VL, Yuste R (2003) Systematic regulation of spine sizes and densities in pyramidal neurons. J Neurobiol 56(2):95–112. https://doi.org/10.1002/neu.10229

Lehmenkuhler A, Sykova E, Svoboda J, Zilles K, Nicholson C (1993) Extracellular space parameters in the rat neocortex and subcortical white matter during postnatal development determined by diffusion analysis. Neuroscience 55(2):339–351

Liston C, Miller MM, Goldwater DS, Radley JJ, Rocher AB, Hof PR et al (2006) Stress-induced alterations in prefrontal cortical dendritic morphology predict selective impairments in perceptual attentional set-shifting. J Neurosci 26(30):7870–7874. https://doi.org/10.1523/JNEUROSCI.1184-06.2006

Lloyd SA, Wensley B, Faherty CJ, Smeyne RJ (2003) Regional differences in cortical dendrite morphology following in utero exposure to cocaine. Brain Res Dev Brain Res 147(1–2):59–66

Magarinos AM, Li CJ, Gal Toth J, Bath KG, Jing D, Lee FS, McEwen BS (2011) Effect of brain-derived neurotrophic factor haploinsufficiency on stress-induced remodeling of hippocampal neurons. Hippocampus 21(3):253–264. https://doi.org/10.1002/hipo.20744

Marcuzzo S, Dall'oglio A, Ribeiro MF, Achaval M, Rasia-Filho AA (2007) Dendritic spines in the posterodorsal medial amygdala after restraint stress and ageing in rats. Neurosci Lett 424 (1):16–21. https://doi.org/10.1016/j.neulet.2007.07.019

Matthews DA, Cotman C, Lynch G (1976) An electron microscopic study of lesion-induced synaptogenesis in the dentate gyrus of the adult rat. I. Magnitude and time course of degeneration. Brain Res 115(1):1–21

Matthews MA, St Onge MF, Faciane CL, Gelderd JB (1979) Spinal cord transection: a quantitative analysis of elements of the connective tissue matrix formed within the site of lesion following administration of piromen, cytoxan or trypsin. Neuropathol Appl Neurobiol 5(3):161–180

McEwen BS (2000a) Effects of adverse experiences for brain structure and function. Biol Psychiatry 48(8):721–731

McEwen BS (2000b) The neurobiology of stress: from serendipity to clinical relevance. Brain Res 886(1–2):172–189

McLaughlin KJ, Gomez JL, Baran SE, Conrad CD (2007) The effects of chronic stress on hippocampal morphology and function: an evaluation of chronic restraint paradigms. Brain Res 1161:56–64. https://doi.org/10.1016/j.brainres.2007.05.042

Minkwitz HG, Holz L (1975) The ontogenetic development of pyramidal neurons in the hippocampus (CA1) of the rat. J Hirnforsch 16(1):37–54

Nicholson C, Phillips JM (1981) Ion diffusion modified by tortuosity and volume fraction in the extracellular microenvironment of the rat cerebellum. J Physiol 321:225–257

Nimmerjahn A, Kirchhoff F, Helmchen F (2005) Resting microglial cells are highly dynamic surveillants of brain parenchyma in vivo. Science 308(5726):1314–1318. https://doi.org/10.1126/science.1110647

Oberheim NA, Takano T, Han X, He W, Lin JH, Wang F et al (2009) Uniquely hominid features of adult human astrocytes. J Neurosci 29(10):3276–3287. https://doi.org/10.1523/JNEUROSCI.4707-08.2009

Ogata K, Kosaka T (2002) Structural and quantitative analysis of astrocytes in the mouse hippocampus. Neuroscience 113(1):221–233

Pawlak R, Rao BS, Melchor JP, Chattarji S, McEwen B, Strickland S (2005) Tissue plasminogen activator and plasminogen mediate stress-induced decline of neuronal and cognitive functions in the mouse hippocampus. Proc Natl Acad Sci USA 102(50):18201–18206. https://doi.org/10.1073/pnas.0509232102

Pego JM, Morgado P, Pinto LG, Cerqueira JJ, Almeida OF, Sousa N (2008) Dissociation of the morphological correlates of stress-induced anxiety and fear. Eur J Neurosci 27(6):1503–1516. https://doi.org/10.1111/j.1460-9568.2008.06112.x

Pelvig DP, Pakkenberg H, Stark AK, Pakkenberg B (2008) Neocortical glial cell numbers in human brains. Neurobiol Aging 29(11):1754–1762. https://doi.org/10.1016/j.neurobiolaging.2007.04.013

Perez-Cruz C, Muller-Keuker JI, Heilbronner U, Fuchs E, Flugge G (2007) Morphology of pyramidal neurons in the rat prefrontal cortex: lateralized dendritic remodeling by chronic stress. Neural Plast 2007:46276. https://doi.org/10.1155/2007/46276

Pillai AG, de Jong D, Kanatsou S, Krugers H, Knapman A, Heinzmann JM et al (2012) Dendritic morphology of hippocampal and amygdalar neurons in adolescent mice is resilient to genetic

differences in stress reactivity. PLoS One 7(6):e38971. https://doi.org/10.1371/journal.pone.0038971
Pokorny J, Yamamoto T (1981) Postnatal ontogenesis of hippocampal CA1 area in rats. II. Development of ultrastructure in stratum lacunosum and moleculare. Brain Res Bull 7 (2):121–130
Qin M, Xia Z, Huang T, Smith CB (2011) Effects of chronic immobilization stress on anxiety-like behavior and basolateral amygdala morphology in Fmr1 knockout mice. Neuroscience 194:282–290. https://doi.org/10.1016/j.neuroscience.2011.06.047
Radley JJ, Sisti HM, Hao J, Rocher AB, McCall T, Hof PR et al (2004) Chronic behavioral stress induces apical dendritic reorganization in pyramidal neurons of the medial prefrontal cortex. Neuroscience 125(1):1–6. https://doi.org/10.1016/j.neuroscience.2004.01.006
Radley JJ, Rocher AB, Janssen WG, Hof PR, McEwen BS, Morrison JH (2005) Reversibility of apical dendritic retraction in the rat medial prefrontal cortex following repeated stress. Exp Neurol 196(1):199–203. https://doi.org/10.1016/j.expneurol.2005.07.008
Radley JJ, Rocher AB, Rodriguez A, Ehlenberger DB, Dammann M, McEwen BS, Morrison JH, Wearne SL, Hof PR (2008) Repeated stress alters dendritic spine morphology in the rat medial prefrontal cortex. J Comp Neurol 507(1):1141–1150
Russell FA, Moore DR (1999) Effects of unilateral cochlear removal on dendrites in the gerbil medial superior olivary nucleus. Eur J Neurosci 11(4):1379–1390
Schuz A, Palm G (1989) Density of neurons and synapses in the cerebral cortex of the mouse. J Comp Neurol 286(4):442–455. https://doi.org/10.1002/cne.902860404
Shansky RM, Hamo C, Hof PR, McEwen BS, Morrison JH (2009) Stress-induced dendritic remodeling in the prefrontal cortex is circuit specific. Cereb Cortex 19(10):2479–2484. https://doi.org/10.1093/cercor/bhp003
Somogyi J, Eysel U, Hamori J (1987) A quantitative study of morphological reorganization following chronic optic deafferentation in the adult cat dorsal lateral geniculate nucleus. J Comp Neurol 255(3):341–350. https://doi.org/10.1002/cne.902550303
Sorensen SA, Rubel EW (2006) The level and integrity of synaptic input regulates dendrite structure. J Neurosci 26(5):1539–1550. https://doi.org/10.1523/JNEUROSCI.3807-05.2006
Sousa N, Lukoyanov NV, Madeira MD, Almeida OF, Paula-Barbosa MM (2000) Reorganization of the morphology of hippocampal neurites and synapses after stress-induced damage correlates with behavioral improvement. Neuroscience 97(2):253–266
Spacek J, Hartmann M (1983) Three-dimensional analysis of dendritic spines. I. Quantitative observations related to dendritic spine and synaptic morphology in cerebral and cerebellar cortices. Anat Embryol (Berl) 167(2):289–310
Stepanyants A, Tamas G, Chklovskii DB (2004) Class-specific features of neuronal wiring. Neuron 43(2):251–259. https://doi.org/10.1016/j.neuron.2004.06.013
Stepanyants A, Martinez LM, Ferecsko AS, Kisvarday ZF (2009) The fractions of short- and long-range connections in the visual cortex. Proc Natl Acad Sci USA 106(9):3555–3560. https://doi.org/10.1073/pnas.0810390106
Sykova E, Vorisek I, Antonova T, Mazel T, Meyer-Luehmann M, Jucker M et al (2005a) Changes in extracellular space size and geometry in APP23 transgenic mice: a model of Alzheimer's disease. Proc Natl Acad Sci USA 102(2):479–484. https://doi.org/10.1073/pnas.0408235102
Sykova E, Vorisek I, Mazel T, Antonova T, Schachner M (2005b) Reduced extracellular space in the brain of tenascin-R- and HNK-1-sulphotransferase deficient mice. Eur J Neurosci 22 (8):1873–1880. https://doi.org/10.1111/j.1460-9568.2005.04375.x
Tata DA, Marciano VA, Anderson BJ (2006) Synapse loss from chronically elevated glucocorticoids: relationship to neuropil volume and cell number in hippocampal area CA3. J Comp Neurol 498(3):363–374. https://doi.org/10.1002/cne.21071
Trommald M, Jensen V, Andersen P (1995) Analysis of dendritic spines in rat CA1 pyramidal cells intracellularly filled with a fluorescent dye. J Comp Neurol 353(2):260–274. https://doi.org/10.1002/cne.903530208

Veena J, Srikumar BN, Mahati K, Bhagya V, Raju TR, Shankaranarayana Rao BS (2009) Enriched environment restores hippocampal cell proliferation and ameliorates cognitive deficits in chronically stressed rats. J Neurosci Res 87(4):831–843. https://doi.org/10.1002/jnr.21907

Vyas A, Mitra R, Shankaranarayana Rao BS, Chattarji S (2002) Chronic stress induces contrasting patterns of dendritic remodeling in hippocampal and amygdaloid neurons. J Neurosci 22 (15):6810–6818. https://doi.org/10.1523/JNEUROSCI.22-15-06810.2002

Vyas A, Bernal S, Chattarji S (2003) Effects of chronic stress on dendritic arborization in the central and extended amygdala. Brain Res 965(1–2):290–294

Vyas A, Pillai AG, Chattarji S (2004) Recovery after chronic stress fails to reverse amygdaloid neuronal hypertrophy and enhanced anxiety-like behavior. Neuroscience 128(4):667–673. https://doi.org/10.1016/j.neuroscience.2004.07.013

Watanabe Y, Gould E, McEwen BS (1992) Stress induces atrophy of apical dendrites of hippocampal CA3 pyramidal neurons. Brain Res 588(2):341–345

Wenzel J, Lammert G, Meyer U, Krug M (1991) The influence of long-term potentiation on the spatial relationship between astrocyte processes and potentiated synapses in the dentate gyrus neuropil of rat brain. Brain Res 560(1–2):122–131

Wilhelmsson U, Bushong EA, Price DL, Smarr BL, Phung V, Terada M et al (2006) Redefining the concept of reactive astrocytes as cells that remain within their unique domains upon reaction to injury. Proc Natl Acad Sci USA 103(46):17513–17518. https://doi.org/10.1073/pnas.0602841103

Yao X, Hrabetova S, Nicholson C, Manley GT (2008) Aquaporin-4-deficient mice have increased extracellular space without tortuosity change. J Neurosci 28(21):5460–5464. https://doi.org/10.1523/JNEUROSCI.0257-08.2008

Identification of the Core Neural Network Subserving PTSD in Animal Models and Their Modulation

3

3.1 Animal Models of PTSD

3.1.1 PTSD Diagnostic Criteria in the Diagnostic and Statistical Manual of Mental Disorders Version IV (DSM-IV) Applied to Animal Models

It is not possible to develop a full animal model of PTSD, given the human elements required to be satisfied according to DSM-IV, such as the traumatic event which should be accompanied by 'intense fear, helplessness or horror'. It is, however, possible to interpret the behaviour of animals in these terms without of course knowing that such interpretations are correct rather than being simply anthropomorphic. DSM-IV sets out six sets of criteria (listed A to F) that must be met before a diagnosis of PTSD can be specified for a patient. Within each of these sets, it is possible to discern behaviours that might be interpreted across mammalian species, so allowing the possibility of developing animal models of PTSD. Such diagnostic criteria for PTSD are the following, taken from DSM-IV (American Psychiatric Association 1994; Table 309.81; following their numbering):

A. The person/animal has been exposed to a traumatic event in which both of the following were present:
   1. The person/animal experienced, witnessed or was confronted with an event or events that involved actual or threatened death or serious injury or a threat to the physical integrity of self or others.
   2. The person's/animal's response involved intense fear, helplessness or horror. (Note that in the case of children, this may be expressed instead by disorganized or agitated behaviour.)

M. Bennett, J. Lagopoulos, *Stress, Trauma and Synaptic Plasticity*,
https://doi.org/10.1007/978-3-319-91116-8_3

B. The traumatic event is persistently re-experienced in one (or more) of the following ways:
    5. Physiological reactivity on exposure to internal or external cues that symbolize or resemble an aspect of the traumatic event.
C. Persistent avoidance of stimuli associated with the trauma and numbing of general responsiveness (not present before the trauma), as indicated by three (or more) of the following:
    2. Efforts to avoid activities, places or people that arouse recollections of the trauma
    3. Inability to recall an important aspect of the trauma (deterioration of memory)
D. Persistent symptoms of increased arousal (not present before the trauma), as indicated by two (or more) of the following:
    4. Hypervigilance
    5. Exaggerated startle response
E. Duration of the disturbance (symptoms in criteria B, C and D) is more than 1 month.
F. The disturbance causes clinically significant distress or impairment in social, occupational or other important areas of functioning.

These criteria have been chosen under DSM-IV for PTSD as they all specify a behaviour in humans that can also be observed in animals that have been traumatized, such as enhanced freezing in response to contextual fear conditioning and deterioration of memory (Yamamoto et al. 2009). Table 3.1 sets out the PTSD criteria described above (A to F), showing how various experiments on rodents have been interpreted by their authors in ways that might fulfil the different criteria after an animal has been traumatized. The 'trauma procedure' and 'time to test' the changes in behaviour of the animal following the procedure are given as well as the 'behaviour test outcome' describing the observations by each set of authors, together with the 'interpretation by authors' of their results. The listing of the authors under each of the criteria is on the basis of their behavioural observations, not their interpretation of these observations. It will be noted that in some cases the interpretation of the behaviour (interpretation by authors) conforms with the DSM-IV criteria but that there is a much better alignment between the observed behaviour (behaviour test outcome) and the criteria, presumably as the authors were not thinking of interpreting their observations in terms of these criteria. For example, in the case of criteria A2 (person/animal response to trauma involves disorganized/agitated behaviour), the observations are mostly interpreted using the description 'anxiety heightened' (fifth column), whereas if one looks simply at their behaviour on the + maze, it is clear that it could as equally be described as 'agitated behaviour' and so conform directly with criteria A2. In the case of criteria B5 (physiological reactivity on exposure to cues associated with an event), the actual behaviour of the animals on re-exposure to the context of the trauma is mostly interpreted as 'avoidance responses', which conforms to the criteria B5. The behaviour of animals under criteria C3 (an inability to recall, i.e. memory) is in general interpreted that way by the authors as is the animal behaviour under criteria D4 (increased arousal as in hypervigilance). The criteria F

Table 3.1 Application of PTSD diagnosis criteria (DSM-IV) to traumatized rodents

DSM-IV criteria	Trauma procedure	Time to test	Behaviour test outcome	Interpretation by authors	References
A2. Person/animal response to trauma involves disorganized/agitated behaviour	Exposed to cat once	3 weeks	Decreased time spent in open arm of + maze	Anxiety heightened	Adamec (1998)
	Exposed to cat on two occasions with 10 days interval	3 weeks	Decreased time spent in open arm of + maze	Anxiety heightened	Zoladz et al. (2012)
			Freezing in relation to a tome that was coupled with cat exposure	Fear memory elevated	
	Exposed to cat for 1 h on each of 31 days	3 weeks	Decreased time spent in open arm of + maze	Anxiety heightened	Zoladz et al. (2012)
	Unstable housing conditions		Increased startle response (piezoelectric)	Exaggerated startle like in PTSD	
	Exposed to predator urine when juvenile	9 weeks	Decreased time spent in open arm of + maze when adult	PTSD-like behaviour predisposition at 9 weeks	Cohen and Zohar (2004)
			Increased startle response (piezoelectric)		
	Exposed to predator and subjected to underwater trauma	1 week	Decreased time spent in open arms of + maze	Anxiety heightened	Cohen and Zohar (2004)
			Increased startle response (piezoelectric)	Fear memory elevated	
				Hypervigilance on startle response	
	An intense foot shock	4 weeks	Decreased time spent in open arms of + maze	Anxiety heightened	Louvart et al. (2005)
	Reminder of context of shock weekly for 3 weeks		Decrease of social behaviour (allogrooming)	Social behaviour altered	
	Alternating restraint and cold swims	1 week	Decreased time spent in open arms of + maze	Anxiety heightened	McGuire et al. (2010)
			Decrease of social behaviour (allogrooming, rearing)	Social interaction decreased	

(continued)

Table 3.1 (continued)

DSM-IV criteria	Trauma procedure	Time to test	Behaviour test outcome	Interpretation by authors	References
	Inescapable foot shock and predator odour in a transiently illuminated compartment with accompanying sound	1 week	Freezing behaviour when confined to the illuminated compartment in absence of any stressor	Anxiety heightened (decreased movement; longer latency to eating)	Hawley et al. (2013)
	Inescapable foot shock in illuminated compartment	2 weeks	Avoidance of illuminated compartment	Hyper-arousal and hypervigilance	Wakizono et al. (2007)
	Inescapable tail shock in dark compartment	1 week	Increased time to enter dark from light compartment	Passive avoidance	Paré (1996)
B5. Physiological reactivity on exposure to cues associated with an event	Exposure to predator odours on three occasions each day	4 weeks	Freezing behaviour on exposure to context only	A PTSD model for the impact on sexual behaviour	Hawley et al. (2013)
	Exposure to predator odour, then a foot shock	1 week	Longer time to achieve a first mount		
	Exposure to inescapable shock in a light chamber	2 weeks	Avoidance of light chamber	Avoidance responses	Koba et al. (2001)
	Exposure to cat's odour in fur skin	1 week	Avoidance of even clean cloth	Avoidance responses	Munoz-Abellan et al. (2010)
	Exposure to a single shock in the dark compartment of a light/dark shuttle box	3 weeks	Decreased time spent in open arms of + maze	Avoidance responses	Viviani et al. (2012)
			Less time spent in the dark side of the shuttle box		
C3. An inability to recall (memory)	Exposure to predator (cat) for 30 min	1 day	Location of a hidden escape platform in a radial-arm water maze impaired	Memory consolidation impaired, especially short-term memory	Park et al. (2008)

	Exposed to predator	1 week	Exploration of novel objects in an environment same as non-novel objects	Memory decline of ability to distinguish novel objects from non-novel objects	Goswami et al. (2012)
	Exposed to predator odour		Abolished ability to modulate a freezing response based on rewarding or aversive environments	Inability to identify safe environments	Cohen and Zohar (2004)
D4. Increased arousal as in hypervigilance	Exposed to 2 h restraint then 20 min forced swim, then exposed to either until unconscious in home cage	1 week	Startle response increases in amplitude	Hypervigilance and hyper-arousal	Yamamoto et al. (2009)
	Exposed to 2 h restraint plus 20 min forced swim	1 week	Enhanced freezing in response to fear conditioning	Anxiety heightened	Yamamoto et al. (2009)
	Exposed to inescapable shock	1 day	Avoidance of open fields and other rats	Fear memory	Chen et al. (2012a, b)
	Single restraint stress and forced swim test, then exposed to ether vapours Repeat 7 days later	7 days	Conditioned taste aversion (a novel taste associated with aversive experience) enhanced by the double stress event	Associative memory affected by stress	Brand et al. (2008)
	Exposed to inescapable electric foot shocks in one chamber (with a warning light), the other being in dark	2 weeks	Cross over into dark chamber when warning light on; cross over when no light on	Hyper-arousal, hypervigilance	Wakizono et al. (2007)
	Exposed to 2 h of restraint stress then 20 min forced swim	1 week	Enhanced freezing in response to fear conditioning	Anxiety heightened	Yamamoto et al. (2009)
	Exposed to electric foot shock and then predator odour		High level of freezing behaviour		Corley et al. (2012)

(continued)

Table 3.1 (continued)

DSM-IV criteria	Trauma procedure	Time to test	Behaviour test outcome	Interpretation by authors	References
D5. Increased arousal as in exaggerated startle response	Exposure to 30 min of intermittent tail shock for 2 days followed by exposure to tail shock apparatus context on third day	3 weeks	30% of rats developed non-habituation of startle response	PTSD model	Garrick et al. (2001)
	Exposed to odour of cat's hair once then again at 24 h		Increased freezing (in susceptible rats) during extinction tests (context of cats hair exposure only) and in acoustic startle response	PTSD-like phenotype	Nalloor et al. (2011)
	Exposure to 2 h sessions of 40 tail shocks for 3 days	2 weeks	Startle response to impulse auditory stimulus	Exaggerated startle response PTSD insights offered	Servatius et al. (1995)
E. Duration of symptoms in B, D and C above last for more than 1 month	Exposed for 3 s to an inescapable shock. At later times exposed to chamber without the shock	1 month	Memory tested using a Y-maze and a water T-maze task	Aversive memory reactivates. Hypo-activity/reminders of the stressors perpetuate the deleterious effects of stress on affective and cognitive processes	Winston et al. (2001)
	Exposed to acoustic startle (115 dB for 40 min at 20/day) for 3–5 days/weeks for 20 weeks, either preceded by light in a dark chamber or not	2 months	At 20 weeks, the startle response in light chamber greater by 30% than in the dark	No claim made. But clear that the startle stimulus itself is an unconditioned stimulus	Winston et al. (2001)
F. Impairment in social occupational and other functioning	Exposed to intermittent cold stress (4 °C, 6 h/day, 14 days) plus one day of acute social defeat stress	1–3 days	Increased in latency to feed Decrease in active buying behaviour	PTSD aspects modelled	Roth et al. (2012)

	Exposed to inescapable electric foot shocks. Two shock trains of 1 s/min for 5 min	4 weeks	Social avoidance (only visit the compartment next to one containing an opponent)	Social deficits in behaviour	Mikics et al. (2008)
			Failure to smell separated by perforated holes in a partition from another rat		
	Exposed to inescapable electric tail shock and social defeat by aggressive male	1 day	Submissive and defensive behaviour on encountering non-aggressive conspecifics	Trauma-related depression	Hammack et al. (2012)
	Exposed to inescapable shock, or escapable shock, or no shock	1 day	Decreased water and food consumption in comparison to escapable shock or no shock		Wagner et al. (1977)

Table 3.2 DSM-IV behavioural criteria for post-traumatic stress disorder (PTSD)

DSM-IV criteria	Behaviour of rodents	Author interpretation
A2. Personal response to trauma involves disorganized/agitated behaviour (8)	Decreased time spent in open arm of a plus maze (8/8)	Anxiety (7/8)
B5. Physiological reactivity of exposure to cues associated with the event (3)	Freezing behaviour when exposed to cues (2/3)	Anxiety (1/3)
		Hypervigilance (1/3)
		Fear heightened (1/3)
C2. Avoidance of places and people that arouse trauma recollections (8)	Avoidance of chamber where shock delivered (5/8)	Avoidance (4/8)
		Hypervigilance (1/8)
		Fear (1/8)
C3. An inability to recall (i.e. memory loss) (4)	Failure to remember site of hidden platform (1/4)	Memory decline (3/4)
	Failure to identify novel objects (1/4)	
	Failure to remember contextual clues (1/4)	
	Failure to remember taste aversion (1/4)	
D4. Increased arousal as in hypervigilance (5)	High level of freezing to a novel tone (4/5)	Anxiety (2/5)
		Fear (1/5)
		Hypervigilance (1/5)
D5. Increased arousal as in exaggerated startle response (7)	Startle response to auditory stimuli (7/7)	PTSD-like behaviour (4/7) hypervigilance (2/7)
E. Duration of symptoms in B, C and D above last for more than 1 month (2)	Memory of a T-maze (1/2)	Memory (1/2)
	Startle response in light chamber (1/2)	
F. Impairment in social occupation and other functioning (6)	Sniffing and grooming impairment (2/6)	Aspects of human PTSD (2/6)
	Burying behaviour impairment (1/6)	Social deficits (2/6)
	Social avoidance (1/6)	
	Submissive behaviour (1/6)	

(impairment in social, occupational and other functioning) is clearly fulfilled by the behaviour, but this is interpreted in different ways, including ‘social deficits in behaviour’. Criteria E (duration of symptoms in B, D and C criteria last for more than 1 month) is of particular importance, for which only one study has appropriate observations showing that this criteria D is fulfilled.

The list of observations in Table 3.1 is summarized in Table 3.2, which gives the frequency of observations and their interpretations in parentheses. It can be seen that there are ample observations of behaviour fulfilling the behavioural criteria of A2, C2, C3, D4, D5 and F but less for B5 and E. All in all these 43 papers on rodent behaviour following traumatic stress provide sufficient evidence for the hypothesis that such

Table 3.3 The type and total number of DSM-IV criteria met by the various animal studies

References	A1	A2	B5	C2	C3	D4	D5	E	F	DSM criteria met
Adamec (1998)	×	×								1
Zoladz et al. (2012)	×	×	×	×		×				4
Zoladz et al. (2008)	×	×		×		×	×			3
Cohen and Zohar (2004)	×		×	×		×				4
Louvart et al. (2005)	×	×	×	×		×				4
McGuire et al. (2010)	×					×			×	3
Hawley et al. (2013)	×		×						×	3
Wakizono et al. (2007)	×		×	×		x				4
Mackenzie et al. (2010)	×	×	×	×						3
Koba et al. (2001)	×		×	×						3
Munoz-Abellan et al. (2010)		×								1
Viviani et al. (2012)	×			×						2
Park et al. (2008)			×							1
Goswami et al. (2012)				×						1
Yamamoto et al. (2009)	×					×	×			2
Chen et al. (2012a, b)	×			×	×					2
Brand et al. (2008)				×	×					1
Corley et al. (2012)	×		×			×				3
Garrick et al. (2001)	×					×	×			2
Nalloor et al. (2011)	×		×				×			3
Servatius et al. (1995)	×						×			2
Winston et al. (2001)	×		×		×			×		4
Roth et al. (2012)	×								×	2
Mikics et al. (2008)	×								×	2
Hammack et al. (2012)	×								×	2
Wagner et al. (1977)	×								×	2

animals might provide a useful experimental model for investigating some aspects of PTSD. However, the proof for this in the end depends on whether rodent experiments provide insight into PTSD that leads to alleviation of some of its concomitants. This was already the case some 86 years ago when Pavlov (1927) discovered 'extinction' in animals which is also applicable to humans, as described below. When the results of the aforementioned studies are summarized on the basis of both behavioural observations as well as the author's interpretation of these observations, there are nine studies that fulfil two DSM-IV criteria for PTSD, seven studies that fulfil three and five studies that fulfil four of the six DSM-IV criteria for PTSD (see Table 3.3).

3.1.2 Acquisition and Extinction of Conditioned Fear in PTSD and Animal Models

After a number of conditioned stimulus (CS)–unconditioned stimulus (US) pairings, the CS elicits a conditioned response (CR) that increases in magnitude and frequency, referred to as 'acquisition'. When these CS–US pairings are followed by presentations of the CS alone or by unpaired CS and US presentations, the CR decreases, a phenomenon known as 'extinction'. Presentation of the CS at some time after the subject stopped responding might show renewed responding, indicating 'retention' of the CR, called 'spontaneous recovery'. Such recovery can be depressed by a 'renewal' CS presented in a small window of time following the period of acquisition. On the other hand, presentation of this CS in a novel context to that during the acquisition period followed by extinction in that novel context does not lead to any suppression of the CR in the original context, so that there is 'renewal' of the CR in this original context. Presentation of the US in the context of the original extinction will also have the effect of renewal, so there is 'reinstatement' of the CR.

These phenomena are spelt out in Fig. 3.1 which shows that spontaneous recovery of the extinguished fear response occurs gradually with time (abscissa) following extinction [repeated conditioned stimuli (CS) over 4 h in the absence of the unconditioned stimuli (US)] as well as in the absence of any further training; the extent of the retention grows with the duration of the extinction-to-test interval (either 100 or 1000 h shown). In this and all other cases shown, (b) to (e), acquisition of the fear response follows repeated CS/US pairs over 4 h succeeded by a 24 h interval, before a single CS reminder (b), (c) and (e) or no reminder (a) and (d). Figure 3.1 shows that spontaneous recovery, i.e. retention, is minimal even at 7000 h after the extinction period, if a reminder CS occurs 24 h after acquisition and extinction begins within 6 h after that. On the other hand, as shown in Fig. 3.1c, if extinction begins some 24 h after the reminder CS, then spontaneous recovery (A) of the response occurs so that retention is much the same as if no reminder CS were given [compare (c) with (b) and (a)]; on the abscissa the (A) refers to the context A in which the CS/US pairs are given during the acquisition, whereas (B) refers to context B in which both the reminder CS and the extinction CS are given. Figure 3.1d shows that retention of the response can be re-established at any time after the extinction is complete with a series of US, in the absence of the CS if the retention CS is given in the same context as the US, in this case the context A. Acquisition of a response may be carried out in two different contexts (CSA/US and CSB/US in Fig. 3.1e), with the reminder in context A (CSA) followed by extinction in both contexts (CSA and CSB), followed by renewal US 24 h later which in turn is followed 10 min later with a further extinction in both contexts (CSA and CSB). Under these circumstances, there is retention of the response in context B but not in A (Fig. 3.1e).

Table 3.4 shows a comparison of the acquisition, extinction, renewal and retention of fear response procedures used in animals and humans, as defined in Fig. 3.1. This table shows that the same procedures are now used to investigate fear responses in humans as were originally envisaged by Pavlov (1927) in his CS/US experiments

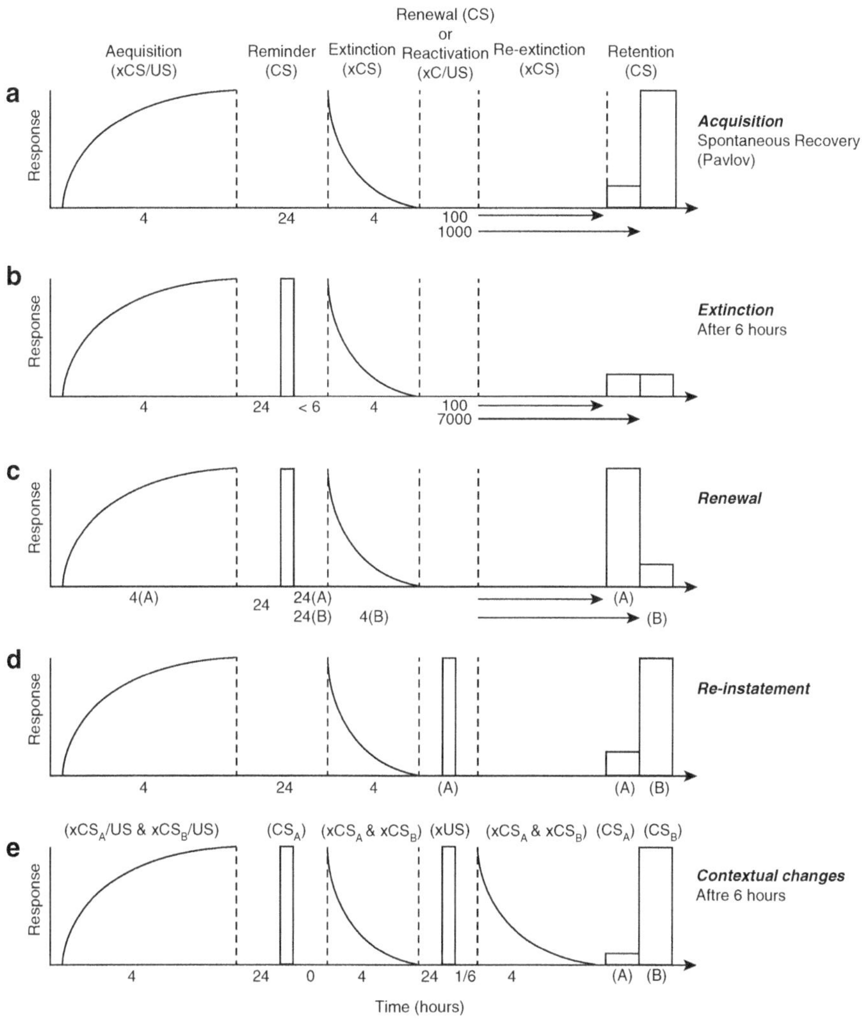

Fig. 3.1 Phenomena defining (**a**) acquisition, (**b**) extinction, (**c**) renewal, (**d**) reinstatement (**e**) and contextual changes (see Myers and Greenwood-Van Meerveld 2007)

on animals. Both in animals and humans, these procedures go under a variety of different names (see first column in Table 3.4); however, they are in fact the same procedures, as columns two and three in the table show.

Table 3.4 Comparison between animal and human behaviour described as involving 'extinguished fear responses' in a variety of circumstances

Definition of terms	Animal observations	Human observations
(A) Acquisition [presentation of CS/US pairs repetitively][a] (fear learning/conditioning (R))	Repeated CS (tone) + US (foot shock) (Monfils et al. 2009; Plendl and Wotjak 2010; Muravieva and Alberini 2010; Nader et al. 2000)	Repeated CS (colour slides) + US (electric shock or loud noise) (Schiller et al. 2010; Guthrie and Bryant 2006)
(B) Response	Freezing (Monfils et al. 2009; Plendl and Wotjak 2010; Muravieva and Alberini 2010; Nader et al. 2000)	Skin conductance change or eyelid startle to a loud noise (Schiller et al. 2010; Kindt et al. 2009; Guthrie and Bryant 2006)
(C) Reminder [presentation of a single CS] (retrieval (R)/ memory activation (H)[b]	Single CS [24–48 h after (A)] (Monfils et al. 2009; Muravieva and Alberini 2010; Nader et al. 2000)	Single CS [24 h after (A)] (Schiller et al. 2010; Kindt et al. 2009)
(D) Extinction [presentation of repeated CS] [consolidation; labile period (R; H)]	Repeated CS [10 min–24 h after (C)] (Monfils et al. 2009; Plendl and Wotjak 2010; Muravieva and Alberini 2010)	Repeated CS [10 min–24 h after (C)] (Schiller et al. 2010; Kindt et al. 2009)
(E) Renewal [presentation of CS] [reconsolidating memory (H)]	Single CS [24–48 h after (D)] (Monfils et al. 2009; Nader et al. 2000)	
(F) Reactivation [presentation of US] [memory reinstated (R), reconstituted (H), reconsolidated (R)]		Repeated US [immediately after (D)] (Kindt et al. 2009)
(G) Re-extinction [presentation as in (C)]	Repeated CS (Plendl and Wotjak 2010; Muravieva and Alberini 2010)	Repeated CS [immediately after (F)] (Schiller et al. 2010; Kindt et al. 2009)
(H) Retention [single CS in same or different context to that in (A)] [recovery (H)]	Single CS (Monfils et al. 2009)	Single CS (Schiller et al. 2010)

[a]Experiments carried out under the term
[b](R) and (H) refer to rat and human definitions, respectively

3.1.3 Summary of Parallel Changes in Grey Matter and White Matter in PTSD and Animal Models

Magnetic resonance imaging (MRI) studies of PTSD patients consistently reveal substantial loss of grey matter in the hippocampus ($-10 \pm 2\%$; mean $\pm$ SEM), as well as consistent decrease in the grey matter of the rostral anterior cingulate cortex (rACC; Table 3.5). However, the observations on the amygdala are more variable, for within seven sets of measurements, three showed an increase in grey matter, three a decrease and one no change, giving rise overall to no significant change ($-0.43 \pm 6\%$). These results for the grey matter changes are supported by those

Table 3.5 Structural and functional changes in grey matter (MRI/PET) in humans following PTSD

Structure	Grey matter volume (MRI)	Blood flow (rCBF)	Metabolism ($CMR_{glc(ox)}$)
Amygdala	Woon and hedges (2008)[a] (−)	Shin et al. (2004a, b) (+)	
	Rogers et al. (2009) (−16%)	Gold et al. (2011)[b] (0%)	
	Kuo et al. (2012) (+)	Osuch et al. (2008) (−)	
	Gilbertson et al. (2002) (+5%)	Phan et al. (2006) (−)	
	Lindauer et al. (2004) (0%)		
	Gurvits et al. (1996) (+15%)		
	Matsuoka et al. (2003) (−5.7%)		
rACC	Felmingham et al. (2009) (−)	Gold et al. (2011) (−4%)	Shin et al. (2009)[d] (+)
	Thomaes et al. (2010) (−)	King et al. (2009) (−)	
	Yamasue et al. (2003) (−)	Britton et al. (2005)[c] (−)	
	Rauch et al. (2003) (−)	Shin et al. (1999)[c] (−)	
	Araki et al. (2005) (−)	Bremner et al. (1999)[b] (−)	
vmPFC	Carrion and Wong (2012) (−)	Phan et al. (2006)[b] (−)	
	Woodward et al. (2009) (12%)	Bremner et al. (1999)[b] (−)	
		Kim et al. (2009) (−)	
		Shin et al. (2004a, b)	
Hippocampus	Admon et al. (2013) (−)	Bremner et al. (1999) (−)	Molina et al. (2010)[d] (−)
	Apfel et al. (2011) (−5.6%)	Kim et al. (2012) (−)	Kim et al. (2012)[d] (−)
	Felmingham et al. (2009) (−)		
	Thomaes et al. (2010) (−)		
	Woon et al. (2010) (−9.5%)		
	Bonne et al. (2008) (−9.3%)		
	Wang et al. (2010) (−10%)		
	Woon and hedges (2011) (−)		
	Rogers et al. (2009) (−8.5%)		
	Herringa et al. (2012) (−)		
	Stein et al. (1997) (−5%)		

(continued)

Table 3.5 (continued)

Structure	Grey matter volume (MRI)	Blood flow (rCBF)	Metabolism ($CMR_{glc(ox)}$)
	Gurvits et al. (1996) (−28%)		
	Gilbertson et al. (2002) (−10%)		
	Bremner et al. (1997) (−12%)		
	Bremner et al. (1995) (−8%)		
	Bonne et al. (2001) (0%)		
	Woon and hedges (2008) (−)		
	Zhang et al. (2011a) (−17.6%)		

[a]Meta-analysis
[b]Presented with stressful images
[c]Presented with trauma included scripts
[d]Resting state

measuring blood flow using positron emission tomography (PET) that consistently show decreases in the rACC and hippocampus of PTSD patients with variable results for the amygdala (Table 3.5). The ventral medial prefrontal cortex (vmPFC) also shows decreases in blood flow in these patients, but there are insufficient numbers of observations on the grey matter to relate them to these blood flow measurements. All in all the patterns of decreases in structure and function, indicated by these measurements, are consistent with considerable decreases in the rACC, hippocampus and probably the vmPFC, whereas it is likely that little in the way of consistent changes occurs in the amygdala. Too few measurements of the cortical metabolic rate of oxidized glucose ($CMR_{glc(ox)}$) are available to support or otherwise the observations on grey matter changes and blood flow (Table 3.5).

It is interesting to compare grey matter changes in the brains of rodents that have been subjected to chronic stress with that of PTSD patients. Table 3.6 shows that there are consistent decreases in the volume of grey matter determined with MRI in the hippocampus, although smaller than that reported for PTSD patients, and as for these patients, there are decreases in grey matter in the rACC (compare Tables 3.5 and 3.6). Also, as in patients, the amygdala does not give consistent results for grey matter changes one way or the other (Table 3.6). The few PET studies on chronically stressed rodents show decreases in $CMR_{glc(ox)}$ consistent with the grey matter changes (Table 3.6).

The observations on rodents suggest that restraint stress is likely to produce changes in the cortex which are analogous to those observed in stressed humans such as those with PTSD. This opens up the possibility of using this animal model both to probe the cellular changes in the cortex responsible for the grey matter volume changes and to identify the mechanisms responsible for the cellular changes

Table 3.6 Structural and functional changes in grey matter (MRI/PET) in rodents following chronic stress

Structure	Grey matter volume (MRI %)	Metabolism ($CMR_{glc(ox)}$)
Amygdala	Golub et al. (2011) (−7)	Jang et al. (2009) (−)
	Pego et al. (2008) (0)	
	Kassem et al. (2013) (0)	
rACC	Cerqueira et al. (2005a, b) (−18)	
	Kassem et al. (2013) (−16)	
Hippocampus	Golub et al. (2011) (−4)	Jang et al. (2009) (−)
	Czeh et al. (2001) (−7)	Sung et al. (2009) (−11%)
	Lee et al. (2009) (−3)	

that might be highly relevant to the amelioration of PTSD. In relation to the first question, Kassem et al. (2013) have recently shown that there are no changes in the number or volume of the somas of neurons, astrocytes or oligodendrocytes in this animal model. However, a loss of synaptic spine density of up to 60% occurs on different-order dendrites in the ACC and hippocampus (CA1) but not in the amygdala. The loss of spines is accompanied by decreases in cumulative dendritic length of neurons of over 40% in the ACC and hippocampus (CA1) giving rise to decreases in volume of dendrites of 2.6 mm^3 for the former and 0.6 mm^3 for the latter, with no change in the amygdala. These values are similar to the MRI-determined loss of grey matter volume following stress of 3.0 and 0.8 mm^3 in the ACC and hippocampus, respectively, with no changes in the amygdala or retrosplenal cortex (RSG). This quantitative study is the first to relate grey matter volume changes in the cortex measured with MRI to volume changes in cellular constituents of the grey matter and highlights the importance of determining the mechanisms responsible for the synaptic loss in these cortical areas in PTSD.

In regard to volumetric changes, there exists some debate, in particular for the hippocampus, as to whether the loss of grey matter is a result of the traumatic event or whether a smaller hippocampus preceded the trauma and predisposed the patient to PTSD. A study by Pitman and colleagues in monozygotic twins discordant for trauma exposure reported that smaller hippocampi did constitute a risk factor for developing PTSD, and this observation is further supported by several animal studies (in rodents and primates) that have reported how variations in hippocampal size/volume provide observable variations in behaviour resulting in PTSD symptoms (Lyons et al. 2001; Schwegler and Lipp 1983; Wimer et al. 1971).

Diffusion tensor imaging (DTI) has been used to determine fractional anisotropy (FA) changes in fibre pathways in the cortex of patients with PTSD and so changes in the integrity of axons constituting these pathways. Table 3.7 shows that in all cases but one there is a decrease in FA, with an average change for the posterior cingulate of 22 ± 4% (mean ± SEM) and a decrease of a similar amount of 21% in one case for the inferior longitudinal fasciculus. There are too few estimates of FA of the fibre pathways for stressed rodents (FA decreases of 22% have been reported for white matter associated with the hippocampus and increases of 12% in that

Table 3.7 Structural changes in major white matter fascicles (DTI) following PTSD in humans

Fascicles	Fractional anisotropy change (7%)
Prefrontal cortex/hippocampus	Admon et al. (2013) (−2.5)
SgACC/hippocampus	Fani et al. (2012) (−)
Cingulate (posterior)	Fani et al. (2012) (−)
Cingulum (posterior)	Wang et al. (2012b) (−25)
Cingulum (posterior)	Fani et al. (2012) (−15%)
Cingulum (anterior)	Kim et al. (2006a, b) (−27)
Cingulate (anterior)	Schuff et al. (2011) (−)
Precentral gyrus	Schuff et al. (2011) (−)
Superior temporal gyrus	Zhang et al. (2011b) (−)
Prefrontal gyrus	Zhang et al. (2011b) (−)
Superior frontal gyrus	Zhang et al. (2011b (−)
Medial frontal gyrus	Zhang et al. (2011b) (−)
Angular gyrus (posterior)	Schuff et al. (2011) (−)
Internal capsule (posterior)	Schuff et al. (2011) (−)
Cuneus	Zhang et al. (2011b) (−)
Corpus callosum	Abe et al. (2006) (−16)
Uncinate	Eluvathingal et al. (2006) (−8.6)
Inferior longitudinal fasciculus	Choi et al. (2012) (−21)

associated with the amygdala, (Ding et al. 2013) to make a comparison with the human observations. It is of considerable interest that the changes in the FA of the order of 21% in PTSD patients may be related to the grey matter changes determined in these patients with MRI. A quantitative cortical model has been developed, based on both computational and simulation approaches, which relates measured changes in cortical activity of grey matter with changes in the integrity of longitudinal fibre pathways (Bennett et al. 2013). The model consists of modules of up to 5000 neurons each, (80% excitatory and 20% inhibitory), with these having different degrees of synaptic connectedness both within a module as well as between modules. This cortical model predicts the quantitative extent of decrease in activity of cortical areas connected by fibre pathways that have deteriorated according to their FA measurements. In the case of the schizophrenia patients considered by Bennett et al. (2013), there was a 24% decrease in FA interpreted in the context of the model as a 24% decrease in the functioning of the inter-modular connections, through the complete loss of 24% of axons subserving the connections. Such a loss predicted a 7% decrease in the activity of the modules which is about the same as the decrease in activity observed, suggesting that deterioration of longitudinal fasciculi in schizophrenia explains the loss of activity in the grey matter. In the present case, the PTSD patients showed about a 22% decrease in FA of the posterior cingulum that might analogously be responsible for the loss of synapses in the grey matter and a decrease of about 10% of the grey matter (see Tables 3.6 and 3.7).

3.2 Neural Circuitry Subserving Animal Models of PTSD

3.2.1 The Core Synaptic Circuit: Amygdala, Medial Prefrontal Cortex and Hippocampus

Fear behaviour primarily involves connections between the amygdala, hippocampus and medial prefrontal cortex (mPFC) and to a lesser extent the orbitofrontal cortex and the perirhinal cortex. The amygdala, mPFC and hippocampus form the core synaptic circuitry involved in the acquisition, consolidation, retrieval of fear memory and fear extinction (Marek et al. 2013) (Fig. 3.2). This core synaptic circuitry is based on considerations concerning the hippocampal projection to the mPFC, in the rodent [both infralimbic (IL) and prelimbic (PL); (Parent et al. 2010)], as well as the reciprocal connections between these regions of the mPFC and the amygdala that allow for the gating of fear expression through projections to the inhibitory and excitatory circuits in the amygdala (Marek et al. 2013; Shin et al. 2005; Sotres-Bayon and Quirk 2010; Sotres-Bayon et al. 2012). The amygdala itself is divided into input and output regions. The input region consists of the basolateral amygdala (LA), where inputs from conditioned (CS) and unconditioned stimulus (US) converge. The output region consists of the central amygdala that is divided into three regions, the central lateral (CEl), the central medial (CEm) and the inhibitory intercalated region (ITC) (Fig. 3.2).

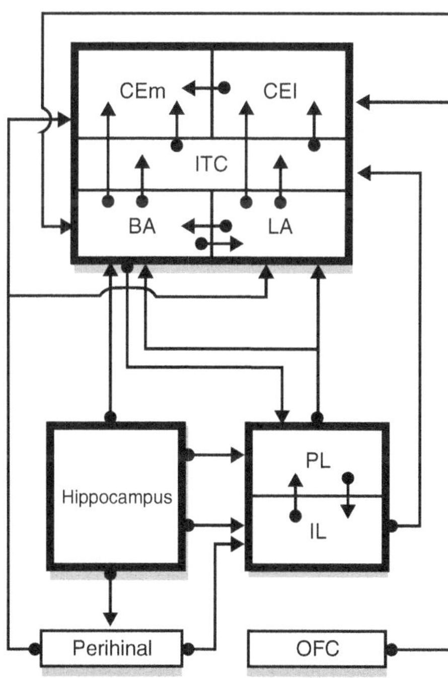

Fig. 3.2 Schematic drawing of the inter- and intra-connectivity of the major regions involved in fear conditioning and extinction (for description, see text) (after Fig. 1, Marek et al. 2013). The core synaptic circuitry is given in bold outline

The core synaptic circuit is involved in the formation of aversive stimuli associations, context-dependent modulation of these associations and extinction processes, respectively. Theta oscillations are responsible for the synchronization of activity in distributed sites in the mPFC, through phase locking, during acquisition of a CS (Paz et al. 2008). These oscillations probably provide the mechanism by which regions of the circuit are temporally bound into a single functioning unit (Lesting et al. 2011a, b). The oscillations occur with high specificity in the lateral nucleus of the amygdala (LA)–hippCA1–mPFC core network, with decreases of theta occurring throughout the core during extinction of CS fear learning but increasing during retrieval of CS fear. Interestingly, during extinction recall the theta coupling oscillations increase in LA–mPFC as well as in hippCA1–mPFC but not in hippCA1–LA, for which they remain at low levels. Gamma oscillations in the hippocampus appear to be responsible for relating perceptual features of an event to existing contextual information from different cortical regions (Lau and Zochowski 2011; Nyhus and Curran 2010; Yamaguchi et al. 2007).

3.2.1.1 The Local Inhibitory Intercalated Neurons, ITC, of the Amygdala

The acquisition, expression and extinction of conditioned fear is under the control of local inhibitory neurons in the amygdala, in particular the inhibitory neurons in the ITC (Ehrlich et al. 2009), as shown in Fig. 3.3. It is the increased inhibition of output neurons in the central amygdala (CEA) that leads to fear extinction, and this occurs as a consequence of potentiation of fear input synapses to GABAergic ITC neurons that in turn synapse on neurons in CEA (Amano et al. 2010). During extinction there is an increase in efficacy of synapses on ITCv (ventral ITC) GABAergic neurons due to infralimbic vmPFC; indeed the degree of bursting in infralimbic vmPFC neurons shortly after extinction predicts subsequent recall of extinction (Burgos-Robles et al. 2007). Thus, in the extinction context, fear is suppressed by restricting firing of CEm neurons through activity in the IL mPFC and the intercalated ITCv neurons (Fig. 3.3) (Knapska and Maren 2009). The hippocampus is involved in contextual regulation of fear expression with area CA1 required for contextual memory retrieval through its projection to extinction (E) neurons in basal amygdala (BA) (Ji and Maren 2008; Fig. 3.3). It is of considerable interest that, given the importance of GABAergic neurons in the acquisition, expression and extinction of conditioned fear, these neurons make synapses on a variety of GABAergic receptors. This opens up the possibility of modulating different components of the circuitry involved in fear with different antagonists. For example, the GABAergic synapses in the LA contain subunits a2bc2 whereas in CEm the GABAergic synapses contain the c1 subunit (Esmaeili et al. 2009) Fear (F) and extinction (E) neurons form different connections within BA as well as with ITCv and CEm neurons (Fig. 3.3). Within BA, F and E neurons are subjected to inhibitory inputs from different subsets of GABAergic interneurons, the latter being under the control of different excitatory inputs (as shown in the circuit diagram of Fig. 3.3. F cells project to CEm whereas E cells project to ITCv neurons as well as to inhibitory neurons in BA, which in turn synapse on F cells (Fig. 3.3). Hippocampal inputs from area CA1 form connections with E cells allowing for the context-dependent gating of extinction. Available

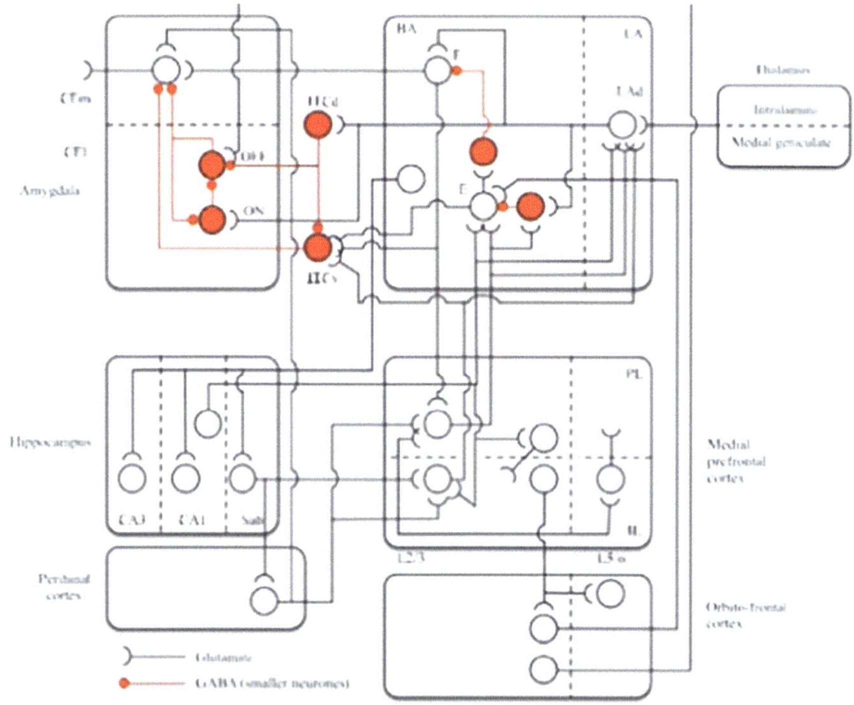

Fig. 3.3 Neurons and their synapses subserving the fear conditioning and extinction circuitry shown in Fig. 3.1. For description see text (after Pare and Duvarci 2012)

evidence suggests that the increased conditioned stimulus responsiveness of CEm output neurons after fear conditioning depends on two parallel mechanisms: disinhibition from CEl and ITCv inputs as well as excitation by glutamatergic BA F neurons.

In CEm disinhibition the CS presentations excite dorsal lateral nucleus (LAd) neurons, leading to the recruitment of ITCd (dorsal ITC) neurons and a subset of CEl cells, the (CEl on) cells (Fig. 3.3). ITCd cells then inhibit ITCv cells, disinhibiting CEm cells. In parallel, ITCd cells also inhibit subsets of CEl neurons, the (CEl off) neurons. The activation of (CEl on) cells by LAd inputs causes a further inhibition of CEl off neurons and disinhibition of CEm cells (Fig. 3.3). In CEm excitation the activation of LA neurons by the CS causes F neurons to fire. These cells project to CEm. The firing of E neurons is inhibited by the recruitment of local circuit interneurons, receiving inputs from LAd neurons.

In extinction, the reduced CS responsiveness of CEm neurons depends on increased feed-forward inhibition by ITCv neurons and disfacilitation from glutamatergic BA inputs. In CEm disfacilitation, the extinction of LAd responses to the CS results in a diminished recruitment of BA F neurons and a reduced activation of the interneurons that inhibits extinction E neurons in high-fear states.

The disinhibition of extinction E neurons causes increased excitation of a different set of interneurons, controlling fear cells. In CEm inhibition the reduced CS responsiveness of LAd neurons causes a disfacilitation of ITCd neurons, with consequent disinhibition of ITCv neurons. This effect coincides with an increased excitation of ITCv cells by inputs from BA extinction cells, thus resulting in an increased feed-forward inhibition of CEm cells. The disfacilitation of ITCd neurons also causes a disinhibition of subsets of CEl cells, possibly corresponding to CEl off neurons. This effect is reinforced by the reduced activation of CEl on neurons secondary to reduced LAd inputs.

In addition to these synaptic connections within the amygdala, there are projections from the amygdala to both the hippocampus and the PL-mPFC. In the amygdala BA neurons form synapses on CA3, CA1 and subiculum neurons and F neurons synapse on PL neurons as do perirhinal neurons. The orbitofrontal cortex sends projections to the amygdala where they make synaptic connections on E neurons and on off neurons in CEl (Pare and Duvarci 2012).

3.2.2 Input and Output of the Core Synaptic Circuit: Thalamus, Periaqueductal Grey and Hypothalamus

3.2.2.1 The Output Circuitry of the Amygdala

The only output from the amygdala is from the central amygdala, which projects into the periaqueductal grey (PAG) and the bed nucleus of the stria terminalis (BNST) of the hypothalamus (Fig. 3.4; Viviani et al. 2011). The major functions of the midbrain PAG include pain, analgesia, fear and anxiety (Behbehani 1995). The central nuclei receive convergent information from several other amygdaloid regions (Fig. 3.4; Pitkanen et al. 1997). CEm and BNST act in a complementary way, with CEm responsible for short- but not long-duration threat responses (i.e. phasic fear) and BNST mediating long- but not short-duration responses (i.e. sustained fear) (Walker et al. 2009).

The fear conditioning and extinction circuitry, together with its output circuitry (shown in Fig. 3.4), are involved in Pavlovian fear conditioning. Sensory information ascends to the amygdala through the midbrain and thalamus; auditory information also reaches the amygdala via the cortex. Convergence and association of the CS and US occur in the amygdala; contextual information processed by the hippocampus can also enter into association with the US in the amygdala. Conditioned and unconditioned fear responses (CRs and URs) are mediated by projections from the amygdala to an array of brain areas involved in autonomic and somatic defensive responses (Maren 2011).

3.2.2.2 The Input Circuitry of the Amygdala

Consider the case of an auditory CS and a nociceptive US. These are projected independently into lateral and CEl nuclei of the amygdala from the medial division of the medial geniculate nucleus and the posterior intralaminar nucleus, which act as sensory relays from the peribrachial nuclei (PB; subserving the US) and the inferior

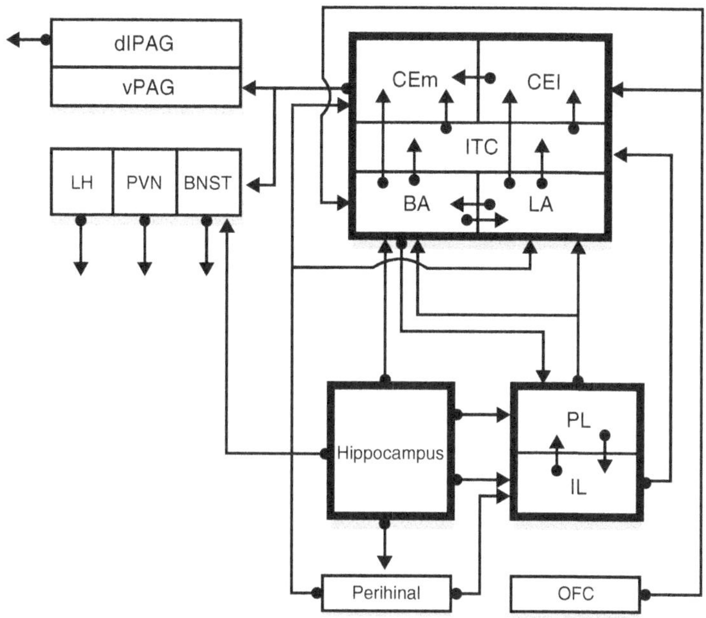

Fig. 3.4 The output of the 'fear conditioning and extinction circuitry' shown in Fig. 3.2. This is principally to the ventral periaqueductal grey (vPAG) and to the bed nucleus of the stria terminalis (BNST) of the hypothalamus. Other hypothalamic nuclei are the lateral hypothalamus (LH) and paraventricular nucleus (PVN). dlPAG is the dorsolateral periaqueductal grey. The core synaptic circuitry is given in bold outline (from Bennett et al. 2016)

nucleus (subserving the CS; Fig. 3.5; Weinberger 2011). Associative fear learning of the auditory CS is dependent on recruitment of disinhibitory microcircuits in the auditory cortex, receiving inputs from the medial geniculate and projecting to the LA (Fig. 3.5; Letzkus et al. 2011).

Retrieval of an extinction memory involves a major input from IL PFC to inhibitory neurons in the ITC whereas retrieval of fear memory involves PL PFC projections to the LA of the amygdala as well as hippocampal inputs to BA of the amygdala (Fig. 3.5; Knapska et al. 2012). If a tone–foot shock fear conditioning procedure is used and this is followed by extinction and testing in two different contexts (see Fig. 3.1c), then the extinguished tone outside of the extinction context recruits auditory and limbic areas that in turn modulate the interactions of the IL with the amygdala (Fig. 3.5; Bruchey et al. 2007).

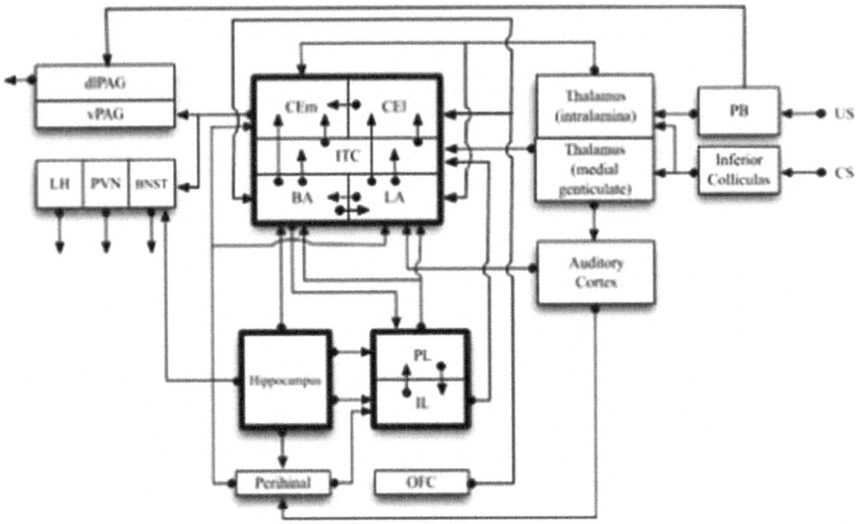

Fig. 3.5 The fear conditioning and extinction circuitry, together with its output circuitry (shown in Fig. 3.4) involved in Pavlovian fear conditioning. For description, see text (after Fig. 3.1 in Maren 2011). The core synaptic circuitry is given in bold outline

3.2.3 Modulation of the Core Synaptic Circuit by Basal Ganglia: The Ventral Tegmentum Area (VTA), Raphe Nucleus and Nucleus Accumbens (NAc)

Figure 3.6 shows the basal ganglia input to the core fear conditioning and extinction circuitry of Fig. 3.2. This principally consists of VTA and Raphe projections to both the PL vmPFC and the BA of the amygdala, with in addition the VTA projecting to the hippocampus and the NAc. The specifics of the dopaminergic and serotoninergic synapses subserved by the neurons of the VTA and Raphe, respectively, are shown in Fig. 3.7. The VTA-dopaminergic neurons (blue) form synapses in the mPFC on the somas of inhibitory (GABA) neurons (red circles; at D2 dopamine receptors) and in close relationship with glutamatergic synapses on dendrites of pyramidal neurons (triangles; at D1 dopamine receptors). The VTA-dopaminergic neurons also form synapses at D1 receptors on pyramidal neurons in the hippocampus and on neurons in the BLA, that is, BA and LA of the amygdala, and on NAc neurons (Wang et al. 2012a). The capacity for error prediction when learning to inhibit fear responses during extinction is dependent on dopaminergic release at synapses in the NAc (Holtzman-Assif et al. 2010). The D1 and D2 receptors are necessary for cue-dependent fear learning (Fadok et al. 2009).

The Raphe-serotonergic neurons (green) form synapses in the mPFC on inhibitory neurons (red circles; at 5-HT2A receptors) as well as on pyramidal neuron somas (at 5-HT1A receptors) and their dendrites (at 5-HT2A receptors). The serotonergic neurons in the Raphe also form synapses on glutamatergic neurons in the

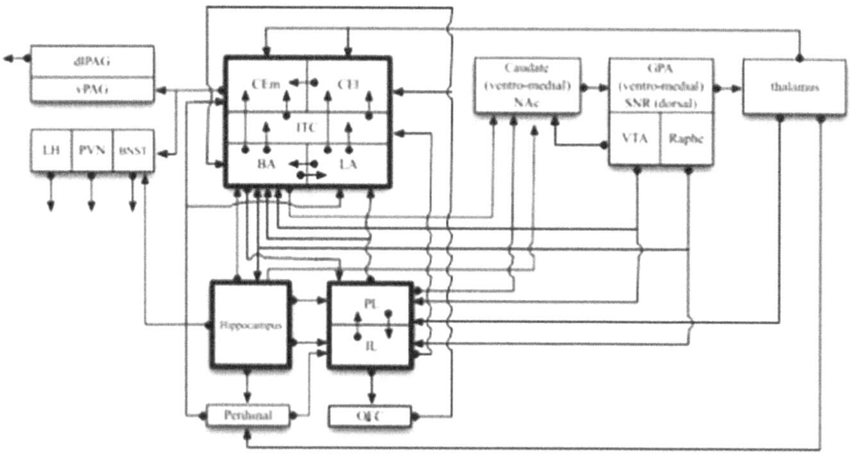

Fig. 3.6 The basal ganglia–thalamic feedback loop control of the fear conditioning and extinction circuitry, together with its output circuitry shown in Fig. 3.4. The loop consists of inputs to the ventromedial caudate and the nucleus accumbens (NAc) projecting to the ventromedial globus pallidus (GPA) and substantia nigra pars reticulate (SNr) and thence to the thalamus, which then projects back to the central nucleus of the amygdala (CEm and CEl) and to the prelimbic medial prefrontal cortex. The ventral tegmentum area (VTA) projects to both CEm and CEl as well as to the prelimbic medial prefrontal cortex and the Raphe nucleus does the same. The core synaptic circuitry is given in bold outline (from Bennett et al. 2016)

BLA (at 5-HT1A receptors). Both the dopaminergic and inhibitory (GABA) neurons in the VTA receive serotonergic synapses from the Raphe also. There is then a rich dopaminergic innervation of the whole fear conditioning and extinction circuitry and an extensive serotonergic innervation as well. Excitatory synapses from the BLA formed on NAc core neurons provide an important component of the circuit involved in cued instrumental tasks (Fig. 3.7) (Fadok et al. 2009; Jones et al. 2010; Vincent et al. 1993).

3.3 Transmitters and Growth Factors Mediating Modulation of the Core Synaptic Circuit in PTSD Animal Models

3.3.1 Dopamine and Serotonin Receptors/Transporters: Action at Synapses and their Nucleotide Polymorphisms in PTSD

Excitatory (glutamatergic) synapse from prefrontal cortical pyramidal neurons synapse on synaptic spines of inhibitory GABAergic neurons in the NAc (indicated by grey in the diagram of Fig. 3.8a). Dopaminergic terminals (indicated by vertical hatching in Fig. 3.8a) from VTA neurons synapse in close proximity to the spines, forming a triadic unit. Dopamine is released onto the dominant presynaptic D2 receptors to decrease the release of glutamate and onto postsynaptic D1 receptors located on the spine and adjacent parent dendrite to enhance excitatory glutamatergic

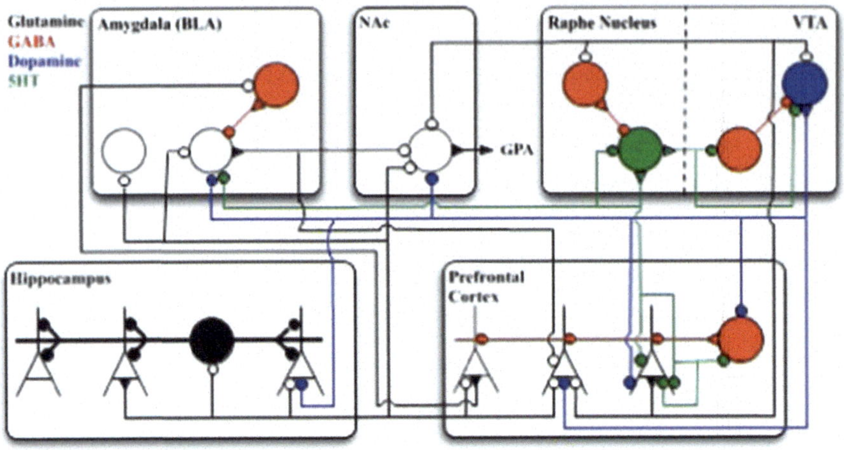

Fig. 3.7 The dopaminergic (originating in the VTA) and serotonergic (originating in the Raphe) synapses formed on neurons in the fear conditioning and extinction circuitry shown in Fig. 3.2 (from Bennett et al. 2016)

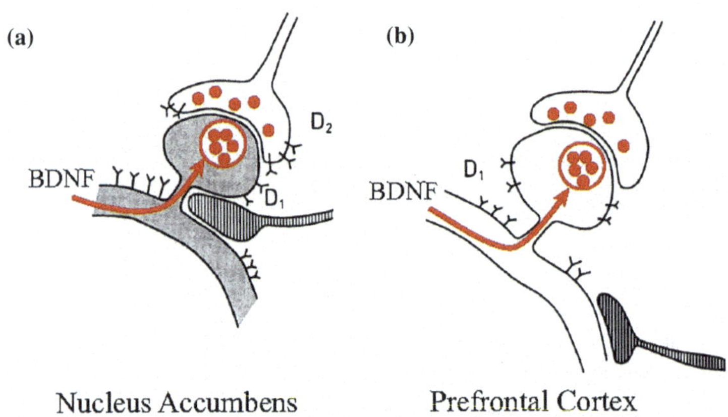

Fig. 3.8 The relationship between glutamatergic and dopaminergic synapses in nucleus accumbens (**a**) and prefrontal cortex (**b**) together with their D1 and D2 receptors. Arrows beginning in the dendrites and ending in the spines show the delivery of BDNF to the spines (in large vesicles) from the neuron soma (for spine growth) and from branch points of dendrites (for determining expression levels of dopamine receptors). Small synaptic vesicles are shown in the glutamatergic terminals (from Bennett et al. 2016)

transmission (Fig. 3.8a) (Fadok et al. 2009; Vincent et al. 1993). On the other hand, excitatory (glutamatergic) synapse from intracortical pyramidal neurons synapse on spines of other pyramidal neurons in mPFC together with dopaminergic terminals somewhat removed from the spine, on the parent dendrite (vertical hatching in

Table 3.8 Dopamine (D) and serotonin (S) receptor and transporter polymorphisms in PTSD

Receptor/transporter	Polymorphism	References
Dopamine D2 receptor	rs1800497	Comings et al. (1991)
Dopamine D2 receptor	rs1800497	Comings et al. (1991)
Dopamine D2 receptor	rs1079597	Gelernter et al. (1999)
Dopamine D2 receptor	rs1800498	Gelernter et al. (1999)
Dopamine D2 receptor	rs1800497	Young et al. (2002)
Dopamine D2 receptor	rs1799738	Voisey et al. (2009)
Dopamine D2 receptor	rs1800497	Voisey et al. (2009)
Dopamine D4 receptor	DRD4 exon III	Dragan and Oniszczenko (2009)
Dopamine transporter	DAT9	Drury et al. (2009)
Dopamine transporter	SLC6A3 (3′ tandem repeats)	Segman et al. (2002)
Serotonin 2A receptor	rs6311(5HT2A)	Lee et al. (2005)
Serotonin 2A receptor	1438A/G(5HT2A)	Mellman et al. (2009)
Serotonin transporter	SLC6A4 (5-HTTLPR)	Thakur et al. (2009)
Serotonin transporter	SLC6A4 (5-HTTLPR)	Koenen et al. (2009)
Serotonin transporter	SLC6A4 (5-HTTLPR)	Xie et al. (2009)
Serotonin transporter	rs4795541 (5-HTTLPR)	Grabe et al. (2009)
Serotonin transporter	(S′/S′) (5-HTTLPR)	Lee et al. (2005)
Serotonin transporter	SLC6A4 (5-HTTLPR)	Kilpatrick et al. (2007)
Serotonin transporter	SLC6A4 (5-HTTLPR)	Kolassa et al. (2010)
Serotonin transporter	S′/S′ (5-HTTLPR)	Wang et al. (2011)

Fig. 3.8b); dopamine D1 receptors (excitatory) occur on the spine and proximal dendrite rather than uniquely adjacent to the dopaminergic terminal (Fig. 3.8b).

There is then a rich dopaminergic and serotonergic innervation of the core synaptic circuitry involved in PTSD (Fig. 3.7), and the synapses that subserve this innervation (Fig. 3.8) possess a variety of transmitter receptors (such as D1, D2 and D4 dopamine receptors and 5HT2A serotonergic receptors) as well as transporters (such as the serotonin transporter 5-HTTLPR). Single-nucleotide polymorphisms (SNP) have been determined for all these receptors and transporters in patients suffering from PTSD (see Table 3.8), and some of these show epistatic interactions with the BDNF gene (Pezawas et al. 2008). Given the importance of the dopaminergic and serotonergic projections from the VTA and Raphe, respectively, to the functioning of the core synaptic circuitry involved in PTSD (see Figs. 3.6, 3.7 and 3.8), it seems likely that the genetic associations of PTSD with these SNPs arise as a consequence of the malfunctioning of this circuitry. This idea is supported by the changes in behaviour noted in mice with knockouts of dopamine and serotonin receptors and transporters. Anxiety is increased in mice, as measured using plus maze behaviour, following knockout of serotonin transporters, and extinction is compromised following knockout of dopamine D1 receptors and transporters (see Table 3.9).

Decreases in resting regional cerebral blood flow (rCBF), following decreases in grey matter in PTSD, have been noted above in the vmPFC and hippocampus with

Table 3.9 Dopamine (D) and serotonin (S) receptor and transporter knockout effects on anxiety behaviour in mice

Receptor/ transporter	Knockout	Behavioural effect	References
Dopamine D1 receptor	D1–/–	Extinction of CS delayed Retention prolonged	El-Ghundi et al. (2001)
Dopamine D1 receptor	D1–/–	Spatial learning delayed	El-Ghundi et al. (1999) and Falzone et al. (2002)
Dopamine D4 receptor	D4–/–	Avoidance of fear US increased	Falzone et al. (2002)
Dopamine transporter	DAT–/–	Extinction of CS delayed	Hironaka et al. (2004)
Serotonin transporter	5-HTT–/–	Anxiety increased (plus maze measured)	Adamec et al. (2006)
Serotonin transporter	5-HTT–/–	Anxiety increased (plus maze measured)	Holmes and Hariri (2003)

equivocal observation for the amygdala (Table 3.5), these areas constituting the core synaptic circuit in PTSD (Fig. 3.2). If PTSD subjects are presented with trauma-related scripts, imagery or audios, they show considerable deficits in rCBF and activity determined with fMRI during the presentations across a range of cortical areas, especially vmPFC and rACC. This highlights that there are deficits outside the core synaptic circuitry that come into play under conditions in which trauma-related events are presented to PTSD patients.

The hippocampal–VTA loop (Fig. 3.6) hypothesis postulates that if information arrives in the hippocampus that is not already stored in long-term memory, then a 'novelty' signal is conveyed to the NAc and thence to the palladium (GPA) and onto the VTA. This contributes to the firing of dopaminergic neurons there which project both into the mPFC, the BA and back to the hippocampus where they produce enhancement of LTP and learning, so establishing a long-term memory of novel information (see Fig. 5 in Lisman and Grace 2005). Dopamine released in dorsal CA1 onto D1 receptors initiates the expression of BDNF necessary for long-term memory (O'Carroll et al. 2006; Rossato et al. 2009); for a recent review, see Bekinschtein et al. (2014).

Dopamine D1 receptors gate long-term changes in synaptic strength at glutamatergic synapses on CA1 hippocampal pyramidal neurons (Lemon and Manahan-Vaughan 2006; Navakkode et al. 2007), and this is necessary for changes in the hippocampus necessary to support long-term memories (O'Carroll et al. 2006). D1 dopamine receptor antagonists delivered to dorsal hippocampus some hours after a fearful experience block long-term memory of it (Rossato et al. 2009).

3.3.2 Brain-Derived Neurotrophic Factor (BDNF): Action on Dopamine Receptors and Their Nucleotide Polymorphisms in PTSD

In Fig. 3.8, secretion of BDNF from the postsynaptic spine is shown due to activity-dependent transcription and translation in the postsynaptic neuron. BDNF is processed in Golgi outposts localized at the branch point of a few dendrites from where it is transported to the spine, where it determines expression levels of dopamine receptors (Lu et al. 2008). In addition, this dendritically synthesized BDNF is a key regulator of spine head growth and spine pruning, whereas somatically synthesized BDNF actually promotes the formation of new spines (Orefice et al. 2013). The regulation of dopamine receptors by BDNF opens up the possibility of controlling receptor expression through modulation of BDNF. This has not been possible because of BDNF possessing poor plasma stability, restricted nervous system penetration as well as pleiotropic actions that derive from their concomitant binding to multiple receptors (Longo and Massa 2013). This might be overcome in the future through the use of small molecule ligands for the BDNF receptor (Knowles et al. 2013) or even virally mediated changes in BDNF gene expression (Berton et al. 2006).

The BDNF Val66Met SNP affects the release of BDNF in the IL mPFC, which is probably due to decreases in both NMDA, and GABA receptor-mediated synaptic transmission (Pattwell et al. 2012). The Val66Met SNP leads to decreased neuronal dendritic complexity in vmPFC and hence to a smaller volume of vmPFC (Yu et al. 2009) as well as a decrease in hippocampal activity (Table 3.5). This is associated with a decrease in extinction learning, but not in acquisition or retention of aversive memories in rodents (Yu et al. 2009) and with anxiety-related behaviours (Soliman et al. 2010). The Val66Met SNP then gives rise to similar changes in vmPFC as well as behavioural changes in humans and mice (Table 3.10). A recent study has demonstrated a significant interaction between the BDNF allele and childhood adversity in the CA2/3 and CA4/DG hippocampal subfield volumes. In this study, the carriers of the MET allele that did not experience childhood adversity had larger CA4/DG and CA2/3 volumes, and conversely those with childhood adversity had smaller CA4/DG and CA2/3 volumes, both compared to Val-allele homozygotes, which nicely highlight how stress by gene interactions impacts hippocampal volume (Frodl et al. 2014).

It has been noted that the Val66Met SNP in the BDNF gene is associated with a decrease in extinction learning (Yu et al. 2009). It is then of interest that direct enhancement of the action of BDNF in the vmPFC reduces conditioned fear for some 48 h after auditory fear conditioning, even in the absence of any extinction training (Peters et al. 2010). In general, changes in BDNF mRNA, protein or receptor expression in the mPFC or hippocampus are associated with changes in extinction and to some extent acquisition (see Table 3.11).

Table 3.11 shows that learning during fear acquisition is dependent on BDNF mRNA increases in the hippocampus (Takei et al. 2011) and that extinction of this conditioned fear is reduced if there is reduction in BDNF mRNA, reflecting that

Table 3.10 Effects of BDNF polymorphisms on behaviour and medial prefrontal cortex during extinction in humans and mice

BDNF polymorphism	Human	Mouse	References
Val 66 met	vmPFC activity (fMRI) decreased	IL PFC activity (electrical) decreased	Pattwell et al. (2012)
Val 66 met	vmPFC activity (fMRI) decreased	Freezing decreased	Soliman et al. (2010)
	Amygdala activity (fMRI) increased		
	Skin conductance decreased		
Val 66 met		Aversive memory decreased	Yu et al. (2009)
		IL PFC activity (c-fos) decreased	
Val 66 met	Skin conductance decreased		Lonsdorf et al. (2010)

Table 3.11 Effects of changes in BDNF in different brain regions on acquisition and extinction

Brain region	Acquisition	Extinction	References
Amygdala			
BLA neuron TrkB reduction		Retention impaired	Chhatwal et al. (2006)
BLA neuron TrkB reduction	Fear conditioning accompanied by		Rattiner et al. (2004a, b)
Medial prefrontal cortex (mPFC)			
mPFC (female) increased methylation exon IV		Resistant to extinction	Baker-Andresen et al. (2013)
mPFC (prelimbic) BDNF knockout	Failure to consolidate	Extinction intact	Choi et al. (2010)
Cortical BDNF exons Ia and IV mRNA increased		Extinction accompanied by	Bredy et al. (2007)
mPFC (infralimbic) BDNF increased	Reduction of		Peters et al. (2010)
Hippocampus			
Hippocampal BDNF mRNA decreased		Reduced extinction	Heldt et al. (2007)
Hippocampal CA1 BDNF mRNA decreased	Increase in stress (predator)		Kozlovsky et al. (2007)
Hippocampal BDNF mRNA increased	Fear conditioning accompanied by		Takei et al. (2011)
Hippocampal exon-specific BDNF mRNA increased	Fear learning accompanied by		Lubin et al. (2008)
Hippocampal BDNF protein synthesis increased	Associative learning accompanied by		Bekinschtein et al. (2007)

extinction involves a new memory (Heldt et al. 2007). Long-term memory dependence on mPFC also depends on BDNF, with increases in BDNF mRNA here accompanying on the one hand the new memory associated with extinction of conditioned fear (Bredy et al. 2007) so that female mice that have low BDNF due to methylation of BDNF exon IV mRNA in mPFC are resistant to extinction (Baker-Andresen et al. 2013). On the other hand, the BDNF knockout in mPFC leads to failure to consolidate learned fear (Choi et al. 2010). In the BA, decreases in the functioning of BDNF through decreases in TrkB receptors lead to impairment of retention of extinction (Chhatwal et al. 2006). As all of these sites, hippocampus, mPFC and BA receive a rich dopaminergic innervation from the VTA (see Fig. 3.7), it follows that any decreases in D1 receptors on the function of these structures will lead to changes in the expression of BDNF and so long-term memories associated with each of them.

During the consolidation of fear learning, following fear conditioning, there is a selective increase in BDNF transcripts containing exons I and III in the amygdala (Rattiner et al. 2004a, b). Preventing BDNF signalling, using the BDNF scavenger TrkB-Fc, engenders social avoidance (Duclot and Kabbaj 2013). Indeed BDNF gene expression and activation of its high-affinity receptor (TrkB) in the amygdala, hippocampus and prefrontal cortex are necessary for the formation of emotional memories, including fear memories and their extinction (for a review, see Andero and Ressler 2012; Kirtley and Thomas 2010; Psotta et al. 2013; also see Fig. 3.4). There is evidence for epigenetic mechanisms regulating the expression of the BDNF gene associated with extinction of conditioned fear, which is accompanied by an increase in histone H4 acetylation around the BDNF P4 gene promoter with increases in BDNF exon I and IV mRNA expression in prefrontal cortex (Bredy et al. 2007). Furthermore, contextual fear learning induces in the hippocampus changes in BDNF DNA methylation and alters local chromatin structure leading to differential regulation of exon-specific BDNF mRNAs (I, IV, VI, IX; Lubin et al. 2008; Takei et al. 2011). In the amygdala (LA), DNA methylation (and histone H3 acetylation) accompanies reconsolidation of fear memories (Maddox and Schafe 2011).

Small-molecule ligands targeting the BDNF receptor TrkB are now being realized (for a review, see Longo and Massa 2013). These may be used to modify various aspects of the BDNF-dependent synaptic mechanisms (Lu et al. 2013), including those involved in fear learning and its extinction, discussed above. However, much needs to be done to realize strategies, utilizing these molecules that effectively target appropriate synapses in the core amygdala, hippocampal and prefrontal network to ameliorate PTSD.

3.4 Conclusion

The answer to fine important questions concerning PTSD is illuminated by the research reviewed above. First, there is good evidence that the behavioural criteria for a diagnosis of PTSD are applicable to that of traumatized animals and in general these are met. Furthermore, the phenomena of acquisition, extinction and reactivation of fear behaviour, originally discovered in animal experiments, are also successfully applicable to humans. Second, the deficits in grey matter found in PTSD are also observed in traumatized animals, lending neuropsychological support to the use of animals to probe what has gone awry in PTSD. Third, the core synaptic circuitry identified in animals as mediating behaviour following trauma is amygdala, vmPFC and hippocampus, modulated by the basal ganglia; the jury is at present out on whether this is the case in PTSD as the observations using fMRI and blood flow measurements are equivocal and open to technical objections. It is then of the first importance to refine the imaging studies so as to answer this question concerning the applicability of the core synaptic circuitry to humans. Imaging genetics would be of considerable value in this context (see, e.g. Meyer et al. 2014).

Nevertheless, the effects of basal ganglia in controlling glutamatergic transmission through dopaminergic and serotonergic mechanisms in the core circuitry provide a ready explanation for why modifying these mechanisms delays extinction in animal models and predisposes towards PTSD. Finally, changes of BDNF in the core synaptic circuitry have significant effects on acquisition and extinction in animal experiments, and SNPs in the BDNF gene predispose to PTSD. All these observations have their explanation in the way in which BDNF controls the disposition of dopaminergic/serotonergic receptors and transporters and thereby synaptic transmission in the core synaptic circuit. For example failure of extinction could arise in humans from failure of D1 receptor incorporation consequent on BDNF SNPs (Fig. 3.8b), leading to decline in glutamate transmission in IL vmPFC (Fig. 3.7) and hence to decline in ITCv excitation and disinhibition of CEm neurons (Fig. 3.3) with failure of extinction. But other causal pathways are possible at the present stage in our knowledge of the core synaptic circuitry (Fig. 3.2).

References

Abe O, Yamasue H, Kasai K, Yamada H, Aoki S, Iwanami A et al (2006) Voxel-based diffusion tensor analysis reveals aberrant anterior cingulum integrity in posttraumatic stress disorder due to terrorism. Psychiatry Res 146(3):231–242. https://doi.org/10.1016/j.pscychresns.2006.01.004

Adamec RE (1998) Evidence that NMDA-dependent limbic neural plasticity in the right hemisphere mediates pharmacological stressor (FG-7142)-induced lasting increases in anxiety-like behavior. Study 1—role of NMDA receptors in efferent transmission from the cat amygdala. J Psychopharmacol 12(2):122–128. https://doi.org/10.1177/026988119801200202

Adamec R, Burton P, Blundell J, Murphy DL, Holmes A (2006) Vulnerability to mild predator stress in serotonin transporter knockout mice. Behav Brain Res 170(1):126–140. https://doi.org/10.1016/j.bbr.2006.02.012

Admon R, Leykin D, Lubin G, Engert V, Andrews J, Pruessner J, Hendler T (2013) Stress-induced reduction in hippocampal volume and connectivity with the ventromedial prefrontal cortex are related to maladaptive responses to stressful military service. Hum Brain Mapp 34(11):2808–2816. https://doi.org/10.1002/hbm.22100

Amano T, Unal CT, Pare D (2010) Synaptic correlates of fear extinction in the amygdala. Nat Neurosci 13(4):489–494. https://doi.org/10.1038/nn.2499

American Psychiatric Association (1994) Diagnostic and statistical manual of mental disorders, 4th edn. American Psychiatric Association, Arlington, VA

Andero R, Ressler KJ (2012) Fear extinction and BDNF: translating animal models of PTSD to the clinic. Genes Brain Behav 11(5):503–512. https://doi.org/10.1111/j.1601-183X.2012.00801.x

Apfel BA, Ross J, Hlavin J, Meyerhoff DJ, Metzler TJ, Marmar CR et al (2011) Hippocampal volume differences in Gulf War veterans with current versus lifetime posttraumatic stress disorder symptoms. Biol Psychiatry 69(6):541–548. https://doi.org/10.1016/j.biopsych.2010.09.044

Araki T, Kasai K, Yamasue H, Kato N, Kudo N, Ohtani T et al (2005) Association between lower P300 amplitude and smaller anterior cingulate cortex volume in patients with posttraumatic stress disorder: a study of victims of Tokyo subway sarin attack. NeuroImage 25(1):43–50. https://doi.org/10.1016/j.neuroimage.2004.11.039

Baker-Andresen D, Flavell CR, Li X, Bredy TW (2013) Activation of BDNF signaling prevents the return of fear in female mice. Learn Mem 20(5):237–240. https://doi.org/10.1101/lm.029520.112

Behbehani MM (1995) Functional characteristics of the midbrain periaqueductal gray. Prog Neurobiol 46(6):575–605

Bekinschtein P, Cammarota M, Igaz LM, Bevilaqua LR, Izquierdo I, Medina JH (2007) Persistence of long-term memory storage requires a late protein synthesis- and BDNF- dependent phase in the hippocampus. Neuron 53(2):261–277. https://doi.org/10.1016/j.neuron.2006.11.025

Bekinschtein P, Cammarota M, Medina JH (2014) BDNF and memory processing. Neuropharmacology 76 Pt C:677–683. https://doi.org/10.1016/j.neuropharm.2013.04.024

Bennett MR, Farnell L, Gibson WG (2013) Fiber pathway pathology, synapse loss and decline of cortical function in schizophrenia. PLoS One 8(4):e60518. https://doi.org/10.1371/journal.pone.0060518

Bennett MR, Hatton SN, Lagopoulos J (2016) Stress, trauma and PTSD: translational insights into the core synaptic circuitry and its modulation. Brain Struct Funct 221(5):2401–2426. https://doi.org/10.1007/s00429-015-1056-1

Berton O, McClung CA, Dileone RJ, Krishnan V, Renthal W, Russo SJ et al (2006) Essential role of BDNF in the mesolimbic dopamine pathway in social defeat stress. Science 311 (5762):864–868. https://doi.org/10.1126/science.1120972

Bonne O, Brandes D, Gilboa A, Gomori JM, Shenton ME, Pitman RK, Shalev AY (2001) Longitudinal MRI study of hippocampal volume in trauma survivors with PTSD. Am J Psychiatry 158(8):1248–1251. https://doi.org/10.1176/appi.ajp.158.8.1248

Bonne O, Vythilingam M, Inagaki M, Wood S, Neumeister A, Nugent AC et al (2008) Reduced posterior hippocampal volume in posttraumatic stress disorder. J Clin Psychiatry 69 (7):1087–1091

Brand L, Groenewald I, Stein DJ, Wegener G, Harvey BH (2008) Stress and re-stress increases conditioned taste aversion learning in rats: possible frontal cortical and hippocampal muscarinic receptor involvement. Eur J Pharmacol 586(1–3):205–211. https://doi.org/10.1016/j.ejphar.2008.03.004

Bredy TW, Wu H, Crego C, Zellhoefer J, Sun YE, Barad M (2007) Histone modifications around individual BDNF gene promoters in prefrontal cortex are associated with extinction of conditioned fear. Learn Mem 14(4):268–276. https://doi.org/10.1101/lm.500907

Bremner JD, Randall P, Scott TM, Bronen RA, Seibyl JP, Southwick SM et al (1995) MRI-based measurement of hippocampal volume in patients with combat-related posttraumatic stress disorder. Am J Psychiatry 152(7):973–981. https://doi.org/10.1176/ajp.152.7.973

Bremner JD, Randall P, Vermetten E, Staib L, Bronen RA, Mazure C et al (1997) Magnetic resonance imaging-based measurement of hippocampal volume in posttraumatic stress disorder related to childhood physical and sexual abuse—a preliminary report. Biol Psychiatry 41 (1):23–32

Bremner JD, Narayan M, Staib LH, Southwick SM, McGlashan T, Charney DS (1999) Neural correlates of memories of childhood sexual abuse in women with and without posttraumatic stress disorder. Am J Psychiatry 156(11):1787–1795. https://doi.org/10.1176/ajp.156.11.1787

Britton JC, Phan KL, Taylor SF, Fig LM, Liberzon I (2005) Corticolimbic blood flow in posttraumatic stress disorder during script-driven imagery. Biol Psychiatry 57(8):832–840. https://doi.org/10.1016/j.biopsych.2004.12.025

Bruchey AK, Shumake J, Gonzalez-Lima F (2007) Network model of fear extinction and renewal functional pathways. Neuroscience 145(2):423–437. https://doi.org/10.1016/j.neuroscience.2006.12.014

Burgos-Robles A, Vidal-Gonzalez I, Santini E, Quirk GJ (2007) Consolidation of fear extinction requires NMDA receptor-dependent bursting in the ventromedial prefrontal cortex. Neuron 53(6):871–880. https://doi.org/10.1016/j.neuron.2007.02.021

Carrion VG, Wong SS (2012) Can traumatic stress alter the brain? Understanding the implications of early trauma on brain development and learning. J Adolesc Health 51(2 Suppl):S23–S28. https://doi.org/10.1016/j.jadohealth.2012.04.010

Cerqueira JJ, Catania C, Sotiropoulos I, Schubert M, Kalisch R, Almeida OF et al (2005a) Corticosteroid status influences the volume of the rat cingulate cortex—a magnetic resonance imaging study. J Psychiatr Res 39(5):451–460. https://doi.org/10.1016/j.jpsychires.2005.01.003

Cerqueira JJ, Pego JM, Taipa R, Bessa JM, Almeida OF, Sousa N (2005b) Morphological correlates of corticosteroid-induced changes in prefrontal cortex-dependent behaviors. J Neurosci 25(34):7792–7800. https://doi.org/10.1523/JNEUROSCI.1598-05.2005

Chen DY, Bambah-Mukku D, Pollonini G, Alberini CM (2012a) Glucocorticoid receptors recruit the CaMKIIalpha-BDNF-CREB pathways to mediate memory consolidation. Nat Neurosci 15(12):1707–1714. https://doi.org/10.1038/nn.3266

Chen X, Li Y, Li S, Kirouac GJ (2012b) Early fear as a predictor of avoidance in a rat model of post-traumatic stress disorder. Behav Brain Res 226(1):112–117. https://doi.org/10.1016/j.bbr.2011.09.004

Chhatwal JP, Stanek-Rattiner L, Davis M, Ressler KJ (2006) Amygdala BDNF signaling is required for consolidation but not encoding of extinction. Nat Neurosci 9(7):870–872. https://doi.org/10.1038/nn1718

Choi DC, Maguschak KA, Ye K, Jang SW, Myers KM, Ressler KJ (2010) Prelimbic cortical BDNF is required for memory of learned fear but not extinction or innate fear. Proc Natl Acad Sci USA 107(6):2675–2680. https://doi.org/10.1073/pnas.0909359107

Choi P, Ren M, Phan TG, Callisaya M, Ly JV, Beare R et al (2012) Silent infarcts and cerebral microbleeds modify the associations of white matter lesions with gait and postural stability: population-based study. Stroke 43(6):1505–1510. https://doi.org/10.1161/STROKEAHA.111.647271

Cohen H, Zohar J (2004) An animal model of posttraumatic stress disorder: the use of cut-off behavioral criteria. Ann N Y Acad Sci 1032:167–178. https://doi.org/10.1196/annals.1314.014

Comings DE, Comings BG, Muhleman D, Dietz G, Shahbahrami B, Tast D et al (1991) The dopamine D2 receptor locus as a modifying gene in neuropsychiatric disorders. JAMA 266(13):1793–1800

Corley MJ, Caruso MJ, Takahashi LK (2012) Stress-induced enhancement of fear conditioning and sensitization facilitates extinction-resistant and habituation-resistant fear behaviors in a novel animal model of posttraumatic stress disorder. Physiol Behav 105(2):408–416. https://doi.org/10.1016/j.physbeh.2011.08.037

Czeh B, Michaelis T, Watanabe T, Frahm J, de Biurrun G, van Kampen M et al (2001) Stress-induced changes in cerebral metabolites, hippocampal volume, and cell proliferation are prevented by antidepressant treatment with tianeptine. Proc Natl Acad Sci USA 98(22):12796–12801. https://doi.org/10.1073/pnas.211427898

Ding AY, Li Q, Zhou IY, Ma SJ, Tong G, McAlonan GM, Wu EX (2013) MR diffusion tensor imaging detects rapid microstructural changes in amygdala and hippocampus following fear conditioning in mice. PLoS One 8(1):e51704. https://doi.org/10.1371/journal.pone.0051704

Dragan WL, Oniszczenko W (2009) The association between dopamine D4 receptor exon III polymorphism and intensity of PTSD symptoms among flood survivors. Anxiety Stress Coping 22(5):483–495. https://doi.org/10.1080/10615800802419407

Drury SS, Theall KP, Keats BJ, Scheeringa M (2009) The role of the dopamine transporter (DAT) in the development of PTSD in preschool children. J Trauma Stress 22(6):534–539. https://doi.org/10.1002/jts.20475

Duclot F, Kabbaj M (2013) Individual differences in novelty seeking predict subsequent vulnerability to social defeat through a differential epigenetic regulation of brain-derived neurotrophic factor expression. J Neurosci 33(27):11048–11060. https://doi.org/10.1523/jneurosci.0199-13.2013

Ehrlich I, Humeau Y, Grenier F, Ciocchi S, Herry C, Luthi A (2009) Amygdala inhibitory circuits and the control of fear memory. Neuron 62(6):757–771. https://doi.org/10.1016/j.neuron.2009.05.026

El-Ghundi M, Fletcher PJ, Drago J, Sibley DR, O'Dowd BF, George SR (1999) Spatial learning deficit in dopamine D(1) receptor knockout mice. Eur J Pharmacol 383(2):95–106

El-Ghundi M, O'Dowd BF, George SR (2001) Prolonged fear responses in mice lacking dopamine D1 receptor. Brain Res 892(1):86–93

Eluvathingal TJ, Chugani HT, Behen ME, Juhasz C, Muzik O, Maqbool M et al (2006) Abnormal brain connectivity in children after early severe socioemotional deprivation: a diffusion tensor imaging study. Pediatrics 117(6):2093–2100. https://doi.org/10.1542/peds.2005-1727

Esmaeili A, Lynch JW, Sah P (2009) GABAA receptors containing gamma1 subunits contribute to inhibitory transmission in the central amygdala. J Neurophysiol 101(1):341–349. https://doi.org/10.1152/jn.90991.2008

Fadok JP, Dickerson TM, Palmiter RD (2009) Dopamine is necessary for cue-dependent fear conditioning. J Neurosci 29(36):11089–11097. https://doi.org/10.1523/JNEUROSCI.1616-09.2009

Falzone TL, Gelman DM, Young JI, Grandy DK, Low MJ, Rubinstein M (2002) Absence of dopamine D4 receptors results in enhanced reactivity to unconditioned, but not conditioned, fear. Eur J Neurosci 15(1):158–164

Fani N, King TZ, Jovanovic T, Glover EM, Bradley B, Choi K et al (2012) White matter integrity in highly traumatized adults with and without post-traumatic stress disorder. Neuropsychopharmacology 37(12):2740–2746. https://doi.org/10.1038/npp.2012.146

Felmingham K, Williams LM, Whitford TJ, Falconer E, Kemp AH, Peduto A, Bryant RA (2009) Duration of posttraumatic stress disorder predicts hippocampal grey matter loss. Neuroreport 20(16):1402–1406. https://doi.org/10.1097/WNR.0b013e3283300fbc

Frodl T, Skokauskas N, Frey EM, Morris D, Gill M, Carballedo A (2014) BDNF Val66Met genotype interacts with childhood adversity and influences the formation of hippocampal subfields. Hum Brain Mapp 35(12):5776–5783. https://doi.org/10.1002/hbm.22584

Garrick T, Morrow N, Shalev AY, Eth S (2001) Stress-induced enhancement of auditory startle: an animal model of posttraumatic stress disorder. Psychiatry 64(4):346–354

Gelernter J, Southwick S, Goodson S, Morgan A, Nagy L, Charney DS (1999) No association between D2 dopamine receptor (DRD2) "A" system alleles, or DRD2 haplotypes, and posttraumatic stress disorder. Biol Psychiatry 45(5):620–625

Gilbertson MW, Shenton ME, Ciszewski A, Kasai K, Lasko NB, Orr SP, Pitman RK (2002) Smaller hippocampal volume predicts pathologic vulnerability to psychological trauma. Nat Neurosci 5(11):1242–1247. https://doi.org/10.1038/nn958

Gold AL, Shin LM, Orr SP, Carson MA, Rauch SL, Macklin ML et al (2011) Decreased regional cerebral blood flow in medial prefrontal cortex during trauma-unrelated stressful imagery in Vietnam veterans with post-traumatic stress disorder. Psychol Med 41(12):2563–2572. https://doi.org/10.1017/S0033291711000730

Golub Y, Kaltwasser SF, Mauch CP, Herrmann L, Schmidt U, Holsboer F (2011) Reduced hippocampus volume in the mouse model of Posttraumatic Stress Disorder. J Psychiatr Res 45(5):650–659. https://doi.org/10.1016/j.jpsychires.2010.10.014

Goswami S, Samuel S, Sierra OR, Cascardi M, Pare D (2012) A rat model of post-traumatic stress disorder reproduces the hippocampal deficits seen in the human syndrome. Front Behav Neurosci 6:26. https://doi.org/10.3389/fnbeh.2012.00026

Grabe HJ, Spitzer C, Schwahn C, Marcinek A, Frahnow A, Barnow S (2009) Serotonin transporter gene (SLC6A4) promoter polymorphisms and the susceptibility to posttraumatic stress disorder in the general population. Am J Psychiatry 166(8):926–933. https://doi.org/10.1176/appi.ajp.2009.08101542

Gurvits TV, Shenton ME, Hokama H, Ohta H, Lasko NB, Gilbertson MW et al (1996) Magnetic resonance imaging study of hippocampal volume in chronic, combat-related posttraumatic stress disorder. Biol Psychiatry 40(11):1091–1099. https://doi.org/10.1016/S0006-3223(96)00229-6

Guthrie RM, Bryant RA (2006) Extinction learning before trauma and subsequent posttraumatic stress. Psychosom Med 68(2):307–311

Hammack SE, Cooper MA, Lezak KR (2012) Overlapping neurobiology of learned helplessness and conditioned defeat: implications for PTSD and mood disorders. Neuropharmacology 62 (2):565–575. https://doi.org/10.1016/j.neuropharm.2011.02.024

Hawley WR, Grissom EM, Belkin MN, James TF, Dohanich GP (2013) Decreased sexual motivation and heightened anxiety in male Long-Evans rats are correlated with the memory for a traumatic event. Arch Sex Behav 42(4):659–668. https://doi.org/10.1007/s10508-012-0017-5

Heldt SA, Stanek L, Chhatwal JP, Ressler KJ (2007) Hippocampus-specific deletion of BDNF in adult mice impairs spatial memory and extinction of aversive memories. Mol Psychiatry 12 (7):656–670. https://doi.org/10.1038/sj.mp.4001957

Herringa R, Phillips M, Almeida J, Insana S, Germain A (2012) Post-traumatic stress symptoms correlate with smaller subgenual cingulate, caudate, and insula volumes in unmedicated combat veterans. Psychiatry Res 203(2–3):139–145. https://doi.org/10.1016/j.pscychresns.2012.02.005

Hironaka N, Ikeda K, Sora I, Uhl GR, Niki H (2004) Food-reinforced operant behavior in dopamine transporter knockout mice: enhanced resistance to extinction. Ann N Y Acad Sci 1025:140–145. https://doi.org/10.1196/annals.1316.018

Holmes A, Hariri AR (2003) The serotonin transporter gene-linked polymorphism and negative emotionality: placing single gene effects in the context of genetic background and environment. Genes Brain Behav 2(6):332–335

Holtzman-Assif O, Laurent V, Westbrook RF (2010) Blockade of dopamine activity in the nucleus accumbens impairs learning extinction of conditioned fear. Learn Mem 17(2):71–75

Jang DP, Lee SH, Park CW, Lee SY, Kim YB, Cho ZH (2009) Effects of fluoxetine on the rat brain in the forced swimming test: a [F-18]FDG micro-PET imaging study. Neurosci Lett 451 (1):60–64. https://doi.org/10.1016/j.neulet.2008.12.024

Ji J, Maren S (2008) Differential roles for hippocampal areas CA1 and CA3 in the contextual encoding and retrieval of extinguished fear. Learn Mem 15(4):244–251. https://doi.org/10.1101/lm.794808

Jones JL, Day JJ, Wheeler RA, Carelli RM (2010) The basolateral amygdala differentially regulates conditioned neural responses within the nucleus accumbens core and shell. Neuroscience 169 (3):1186–1198. https://doi.org/10.1016/j.neuroscience.2010.05.073

Kassem MS, Lagopoulos J, Stait-Gardner T, Price WS, Chohan TW, Arnold JC et al (2013) Stress-induced grey matter loss determined by MRI Is primarily due to loss of dendrites and their synapses. Mol Neurobiol 47(2):645–661. https://doi.org/10.1007/s12035-012-8365-7

Kilpatrick DG, Koenen KC, Ruggiero KJ, Acierno R, Galea S, Resnick HS et al (2007) The serotonin transporter genotype and social support and moderation of posttraumatic stress disorder and depression in hurricane-exposed adults. Am J Psychiatry 164(11):1693–1699. https://doi.org/10.1176/appi.ajp.2007.06122007

Kim JH, Budde MD, Liang HF, Klein RS, Russell JH, Cross AH, Song SK (2006a) Detecting axon damage in spinal cord from a mouse model of multiple sclerosis. Neurobiol Dis 21(3):626–632. https://doi.org/10.1016/j.nbd.2005.09.009

Kim SJ, Jeong DU, Sim ME, Bae SC, Chung A, Kim MJ et al (2006b) Asymmetrically altered integrity of cingulum bundle in posttraumatic stress disorder. Neuropsychobiology 54 (2):120–125. https://doi.org/10.1159/000098262

Kim Y, Teylan MA, Baron M, Sands A, Nairn AC, Greengard P (2009) Methylphenidate-induced dendritic spine formation and DeltaFosB expression in nucleus accumbens. Proc Natl Acad Sci USA 106(8):2915–2920. https://doi.org/10.1073/pnas.0813179106

Kim SY, Chung YK, Kim BS, Lee SJ, Yoon JK, An YS (2012) Resting cerebral glucose metabolism and perfusion patterns in women with posttraumatic stress disorder related to sexual assault. Psychiatry Res 201(3):214–217. https://doi.org/10.1016/j.pscychresns.2011.08.007

Kindt M, Soeter M, Vervliet B (2009) Beyond extinction: erasing human fear responses and preventing the return of fear. Nat Neurosci 12(3):256–258

King AP, Abelson JL, Britton JC, Phan KL, Taylor SF, Liberzon I (2009) Medial prefrontal cortex and right insula activity predict plasma ACTH response to trauma recall. NeuroImage 47 (3):872–880. https://doi.org/10.1016/j.neuroimage.2009.05.088

Kirtley A, Thomas KL (2010) The exclusive induction of extinction is gated by BDNF. Learn Mem 17(12):612–619. https://doi.org/10.1101/lm.1877010

Knapska E, Maren S (2009) Reciprocal patterns of c-Fos expression in the medial prefrontal cortex and amygdala after extinction and renewal of conditioned fear. Learn Mem 16(8):486–493. https://doi.org/10.1101/lm.1463909

Knapska E, Macias M, Mikosz M, Nowak A, Owczarek D, Wawrzyniak M et al (2012) Functional anatomy of neural circuits regulating fear and extinction. Proc Natl Acad Sci USA 109 (42):17093–17098. https://doi.org/10.1073/pnas.1202087109

Knowles JK, Simmons DA, Nguyen TV, Vander Griend L, Xie Y, Zhang H et al (2013) Small molecule p75NTR ligand prevents cognitive deficits and neurite degeneration in an Alzheimer's mouse model. Neurobiol Aging 34(8):2052–2063. https://doi.org/10.1016/j.neurobiolaging.2013.02.015

Koba T, Kodama Y, Shimizu K, Nomura S, Sugawara M, Kobayashi Y, Ogasawara T (2001) Persistent behavioural changes in rats following inescapable shock stress: a potential model of posttraumatic stress disorder. World J Biol Psychiatry 2(1):34–37

Koenen KC, Aiello AE, Bakshis E, Amstadter AB, Ruggiero KJ, Acierno R et al (2009) Modification of the association between serotonin transporter genotype and risk of posttraumatic stress disorder in adults by county-level social environment. Am J Epidemiol 169(6):704–711. https://doi.org/10.1093/aje/kwn397

Kolassa IT, Ertl V, Eckart C, Glockner F, Kolassa S, Papassotiropoulos A et al (2010) Association study of trauma load and SLC6A4 promoter polymorphism in posttraumatic stress disorder: evidence from survivors of the Rwandan genocide. J Clin Psychiatry 71(5):543–547. https://doi.org/10.4088/JCP.08m04787blu

Kozlovsky N, Matar MA, Kaplan Z, Kotler M, Zohar J, Cohen H (2007) Long-term down-regulation of BDNF mRNA in rat hippocampal CA1 subregion correlates with PTSD-like behavioural stress response. Int J Neuropsychopharmacol 10(6):741–758. https://doi.org/10.1017/S1461145707007560

Kuo JR, Kaloupek DG, Woodward SH (2012) Amygdala volume in combat-exposed veterans with and without posttraumatic stress disorder a cross-sectional study. Arch Gen Psychiatry 69 (10):1080–1086

Lau T, Zochowski M (2011) Interaction between connectivity and oscillatory currents in a heterogeneous neuronal network. Phys Rev E Stat Nonlinear Soft Matter Phys 83(5 Pt 1):051908. https://doi.org/10.1103/PhysRevE.83.051908

Lee HJ, Lee MS, Kang RH, Kim H, Kim SD, Kee BS et al (2005) Influence of the serotonin transporter promoter gene polymorphism on susceptibility to posttraumatic stress disorder. Depress Anxiety 21(3):135–139. https://doi.org/10.1002/da.20064

Lee T, Jarome T, Li SJ, Kim JJ, Helmstetter FJ (2009) Chronic stress selectively reduces hippocampal volume in rats: a longitudinal magnetic resonance imaging study. Neuroreport 20 (17):1554–1558. https://doi.org/10.1097/WNR.0b013e328332bb09

Lemon N, Manahan-Vaughan D (2006) Dopamine D1/D5 receptors gate the acquisition of novel information through hippocampal long-term potentiation and long-term depression. J Neurosci 26(29):7723–7729. https://doi.org/10.1523/JNEUROSCI.1454-06.2006

Lesting J, Geiger M, Narayanan RT, Pape HC, Seidenbecher T (2011a) Impaired extinction of fear and maintained amygdala-hippocampal theta synchrony in a mouse model of temporal lobe epilepsy. Epilepsia 52(2):337–346. https://doi.org/10.1111/j.1528-1167.2010.02758.x

Lesting J, Narayanan RT, Kluge C, Sangha S, Seidenbecher T, Pape HC (2011b) Patterns of coupled theta activity in amygdala-hippocampal-prefrontal cortical circuits during fear extinction. PLoS One 6(6):e21714. https://doi.org/10.1371/journal.pone.0021714

Letzkus JJ, Wolff SB, Meyer EM, Tovote P, Courtin J, Herry C, Luthi A (2011) A disinhibitory microcircuit for associative fear learning in the auditory cortex. Nature 480(7377):331–335. https://doi.org/10.1038/nature10674

Lindauer RJ, Vlieger EJ, Jalink M, Olff M, Carlier IV, Majoie CB et al (2004) Smaller hippocampal volume in Dutch police officers with posttraumatic stress disorder. Biol Psychiatry 56 (5):356–363. https://doi.org/10.1016/j.biopsych.2004.05.021

Lisman JE, Grace AA (2005) The hippocampal-VTA loop: controlling the entry of information into long-term memory. Neuron 46(5):703–713. https://doi.org/10.1016/j.neuron.2005.05.002

Longo FM, Massa SM (2013) Small-molecule modulation of neurotrophin receptors: a strategy for the treatment of neurological disease. Nat Rev Drug Discov 12(7):507–525

Lonsdorf TB, Ruck C, Bergstrom J, Andersson G, Ohman A, Lindefors N, Schalling M (2010) The COMTval158met polymorphism is associated with symptom relief during exposure-based cognitive-behavioral treatment in panic disorder. BMC Psychiatry 10:99. https://doi.org/10.1186/1471-244X-10-99

Louvart H, Maccari S, Ducrocq F, Thomas P, Darnaudery M (2005) Long-term behavioural alterations in female rats after a single intense footshock followed by situational reminders. Psychoneuroendocrinology 30(4):316–324. https://doi.org/10.1016/j.psyneuen.2004.09.003

Lu Y, Christian K, Lu B (2008) BDNF: a key regulator for protein synthesis-dependent LTP and long-term memory? Neurobiol Learn Mem 89(3):312–323. https://doi.org/10.1016/j.nlm.2007.08.018

Lu B, Nagappan G, Guan X, Nathan PJ, Wren P (2013) BDNF-based synaptic repair as a disease-modifying strategy for neurodegenerative diseases. Nat Rev Neurosci 14(6):401–416. https://doi.org/10.1038/nrn3505

Lubin FD, Roth TL, Sweatt JD (2008) Epigenetic regulation of BDNF gene transcription in the consolidation of fear memory. J Neurosci 28(42):10576–10586. https://doi.org/10.1523/JNEUROSCI.1786-08.2008

Lyons DM, Yang C, Sawyer-Glover AM, Moseley ME, Schatzberg AF (2001) Early life stress and inherited variation in monkey hippocampal volumes. Arch Gen Psychiatry 58(12):1145–1151

Mackenzie L, Nalivaiko E, Beig MI, Day TA, Walker FR (2010) Ability of predator odour exposure to elicit conditioned versus sensitised post traumatic stress disorder-like behaviours, and forebrain deltaFosB expression, in rats. Neuroscience 169(2):733–742. https://doi.org/10.1016/j.neuroscience.2010.05.005

Maddox SA, Schafe GE (2011) Epigenetic alterations in the lateral amygdala are required for reconsolidation of a Pavlovian fear memory. Learn Mem 18(9):579–593. https://doi.org/10.1101/lm.2243411

Marek R, Strobel C, Bredy TW, Sah P (2013) The amygdala and medial prefrontal cortex: partners in the fear circuit. J Physiol 591(10):2381–2391. https://doi.org/10.1113/jphysiol.2012.248575

Maren S (2011) Seeking a spotless mind: extinction, deconsolidation, and erasure of fear memory. Neuron 70(5):830–845. https://doi.org/10.1016/j.neuron.2011.04.023

Matsuoka Y, Yamawaki S, Inagaki M, Akechi T, Uchitomi Y (2003) A volumetric study of amygdala in cancer survivors with intrusive recollections. Biol Psychiatry 54(7):736–743

McGuire J, Herman JP, Horn PS, Sallee FR, Sah R (2010) Enhanced fear recall and emotional arousal in rats recovering from chronic variable stress. Physiol Behav 101(4):474–482. https://doi.org/10.1016/j.physbeh.2010.07.013

Mellman TA, Alim T, Brown DD, Gorodetsky E, Buzas B, Lawson WB et al (2009) Serotonin polymorphisms and posttraumatic stress disorder in a trauma exposed African American population. Depress Anxiety 26(11):993–997. https://doi.org/10.1002/da.20627

Meyer D, Bonhoeffer T, Scheuss V (2014) Balance and stability of synaptic structures during synaptic plasticity. Neuron 82(2):430–443

Mikics E, Toth M, Varju P, Gereben B, Liposits Z, Ashaber M et al (2008) Lasting changes in social behavior and amygdala function following traumatic experience induced by a single series of foot-shocks. Psychoneuroendocrinology 33(9):1198–1210. https://doi.org/10.1016/j.psyneuen.2008.06.006

Molina ME, Isoardi R, Prado MN, Bentolila S (2010) Basal cerebral glucose distribution in long-term post-traumatic stress disorder. World J Biol Psychiatry 11(2 Pt 2):493–501. https://doi.org/10.3109/15622970701472094

Monfils MH, Cowansage KK, Klann E, LeDoux JE (2009) Extinction-reconsolidation boundaries: key to persistent attenuation of fear memories. Science 324(5929):951–955

Munoz-Abellan C, Armario A, Nadal R (2010) Do odors from different cats induce equivalent unconditioned and conditioned responses in rats? Physiol Behav 99(3):388–394. https://doi.org/10.1016/j.physbeh.2009.12.008

Muravieva EV, Alberini CM (2010) Limited efficacy of propranolol on the reconsolidation of fear memories. Learn Mem 17(6):306–313

Myers B, Greenwood-Van Meerveld B (2007) Corticosteroid receptor-mediated mechanisms in the amygdala regulate anxiety and colonic sensitivity. Am J Physiol Gastrointest Liver Physiol 292(6):G1622–G1629

Nader K, Schafe GE, Le Doux JE (2000) Fear memories require protein synthesis in the amygdala for reconsolidation after retrieval. Nature 406(6797):722–726

Nalloor R, Bunting K, Vazdarjanova A (2011) Predicting impaired extinction of traumatic memory and elevated startle. PLoS One 6(5):e19760. https://doi.org/10.1371/journal.pone.0019760

Navakkode S, Sajikumar S, Frey JU (2007) Synergistic requirements for the induction of dopaminergic D1/D5-receptor-mediated LTP in hippocampal slices of rat CA1 in vitro. Neuropharmacology 52(7):1547–1554. https://doi.org/10.1016/j.neuropharm.2007.02.010

Nyhus E, Curran T (2010) Functional role of gamma and theta oscillations in episodic memory. Neurosci Biobehav Rev 34(7):1023–1035. https://doi.org/10.1016/j.neubiorev.2009.12.014

O'Carroll CM, Martin SJ, Sandin J, Frenguelli B, Morris RG (2006) Dopaminergic modulation of the persistence of one-trial hippocampus-dependent memory. Learn Mem 13(6):760–769. https://doi.org/10.1101/lm.321006

Orefice LL, Waterhouse EG, Partridge JG, Lalchandani RR, Vicini S, Xu B (2013) Distinct roles for somatically and dendritically synthesized brain-derived neurotrophic factor in morphogenesis of dendritic spines. J Neurosci 33(28):11618–11632. https://doi.org/10.1523/JNEUROSCI.0012-13.2013

Osuch EA, Willis MW, Bluhm R, CSTS Neuroimaging Study Group, Ursano RJ, Drevets WC (2008) Neurophysiological responses to traumatic reminders in the acute aftermath of serious motor vehicle collisions using [15O]-H2O positron emission tomography. Biol Psychiatry 64(4):327–335. https://doi.org/10.1016/j.biopsych.2008.03.010

Paré WP (1996) Enhanced retrieval of unpleasant memories influenced by shock controllability, shock sequence, and rat strain. Biol Psychiatry 39(9):808–813

Pare D, Duvarci S (2012) Amygdala microcircuits mediating fear expression and extinction. Curr Opin Neurobiol 22(4):717–723. https://doi.org/10.1016/j.conb.2012.02.014

Parent MA, Wang L, Su J, Netoff T, Yuan LL (2010) Identification of the hippocampal input to medial prefrontal cortex in vitro. Cereb Cortex 20(2):393–403. https://doi.org/10.1093/cercor/bhp108

Park CR, Zoladz PR, Conrad CD, Fleshner M, Diamond DM (2008) Acute predator stress impairs the consolidation and retrieval of hippocampus-dependent memory in male and female rats. Learn Mem 15(4):271–280. https://doi.org/10.1101/lm.721108

Pattwell SS, Bath KG, Perez-Castro R, Lee FS, Chao MV, Ninan I (2012) The BDNF Val66Met polymorphism impairs synaptic transmission and plasticity in the infralimbic medial prefrontal cortex. J Neurosci 32(7):2410–2421. https://doi.org/10.1523/JNEUROSCI.5205-11.2012

Pavlov IP (1927) Conditioned reflexes: an investigation of the physiological activity of the Cerebral Cortex. Oxford University Press, London

Paz R, Bauer EP, Pare D (2008) Theta synchronizes the activity of medial prefrontal neurons during learning. Learn Mem 15(7):524–531. https://doi.org/10.1101/lm.932408

Pego JM, Morgado P, Pinto LG, Cerqueira JJ, Almeida OF, Sousa N (2008) Dissociation of the morphological correlates of stress-induced anxiety and fear. Eur J Neurosci 27(6):1503–1516. https://doi.org/10.1111/j.1460-9568.2008.06112.x

Peters J, Dieppa-Perea LM, Melendez LM, Quirk GJ (2010) Induction of fear extinction with hippocampal-infralimbic BDNF. Science 328(5983):1288–1290. https://doi.org/10.1126/science.1186909

Pezawas L, Meyer-Lindenberg A, Goldman AL, Verchinski BA, Chen G, Kolachana BS et al (2008) Evidence of biologic epistasis between BDNF and SLC6A4 and implications for depression. Mol Psychiatry 13(7):709–716. https://doi.org/10.1038/mp.2008.32

Phan KL, Britton JC, Taylor SF, Fig LM, Liberzon I (2006) Corticolimbic blood flow during nontraumatic emotional processing in posttraumatic stress disorder. Arch Gen Psychiatry 63 (2):184–192. https://doi.org/10.1001/archpsyc.63.2.184

Pitkanen A, Savander V, LeDoux JE (1997) Organization of intra-amygdaloid circuitries in the rat: an emerging framework for understanding functions of the amygdala. Trends Neurosci 20 (11):517–523

Plendl W, Wotjak CT (2010) Dissociation of within- and between-session extinction of conditioned fear. J Neurosci 30(14):4990–4998. https://doi.org/10.1523/JNEUROSCI.6038-09.2010

Psotta L, Lessmann V, Endres T (2013) Impaired fear extinction learning in adult heterozygous BDNF knock-out mice. Neurobiol Learn Mem 103:34–38. https://doi.org/10.1016/j.nlm.2013.03.003

Rattiner LM, Davis M, French CT, Ressler KJ (2004a) Brain-derived neurotrophic factor and tyrosine kinase receptor B involvement in amygdala-dependent fear conditioning. J Neurosci 24 (20):4796–4806. https://doi.org/10.1523/JNEUROSCI.5654-03.2004

Rattiner LM, Davis M, Ressler KJ (2004b) Differential regulation of brain-derived neurotrophic factor transcripts during the consolidation of fear learning. Learn Mem 11(6):727–731. https://doi.org/10.1101/lm.83304

Rauch SL, Shin LM, Segal E, Pitman RK, Carson MA, McMullin K et al (2003) Selectively reduced regional cortical volumes in post-traumatic stress disorder. Neuroreport 14(7):913–916. https://doi.org/10.1097/01.wnr.0000071767.24455.10

Rogers MA, Yamasue H, Abe O, Yamada H, Ohtani T, Iwanami A et al (2009) Smaller amygdala volume and reduced anterior cingulate gray matter density associated with history of post-traumatic stress disorder. Psychiatry Res 174(3):210–216. https://doi.org/10.1016/j.pscychresns.2009.06.001

Rossato JI, Bevilaqua LR, Izquierdo I, Medina JH, Cammarota M (2009) Dopamine controls persistence of long-term memory storage. Science 325(5943):1017–1020. https://doi.org/10.1126/science.1172545

Roth MK, Bingham B, Shah A, Joshi A, Frazer A, Strong R, Morilak DA (2012) Effects of chronic plus acute prolonged stress on measures of coping style, anxiety, and evoked HPA-axis reactivity. Neuropharmacology 63(6):1118–1126. https://doi.org/10.1016/j.neuropharm.2012.07.034

Schuff N, Zhang Y, Zhan W, Lenoci M, Ching C, Boreta L et al (2011) Patterns of altered cortical perfusion and diminished subcortical integrity in posttraumatic stress disorder: an MRI study. NeuroImage 54(Suppl 1):S62–S68. https://doi.org/10.1016/j.neuroimage.2010.05.024

Schiller D, Monfils MH, Raio CM, Johnson DC, Ledoux JE, Phelps EA (2010) Preventing the return of fear in humans using reconsolidation update mechanisms. Nature 463(7277):49–53. https://doi.org/10.1038/nature08637

Schwegler H, Lipp HP (1983) Hereditary covariations of neuronal circuitry and behavior: correlations between the proportions of hippocampal synaptic fields in the regio inferior and two-way avoidance in mice and rats. Behav Brain Res 7(1):1–38

Segman RH, Cooper-Kazaz R, Macciardi F, Goltser T, Halfon Y, Dobroborski T, Shalev AY (2002) Association between the dopamine transporter gene and posttraumatic stress disorder. Mol Psychiatry 7(8):903–907. https://doi.org/10.1038/sj.mp.4001085

Servatius RJ, Ottenweller JE, Natelson BH (1995) Delayed startle sensitization distinguishes rats exposed to one or three stress sessions: further evidence toward an animal model of PTSD. Biol Psychiatry 38(8):539–546. https://doi.org/10.1016/0006-3223(94)00369-E

Shin LM, McNally RJ, Kosslyn SM, Thompson WL, Rauch SL, Alpert NM et al (1999) Regional cerebral blood flow during script-driven imagery in childhood sexual abuse-related PTSD: a PET investigation. Am J Psychiatry 156(4):575–584. https://doi.org/10.1176/ajp.156.4.575

Shin LM, Orr SP, Carson MA, Rauch SL, Macklin ML, Lasko NB et al (2004a) Regional cerebral blood flow in the amygdala and medial prefrontal cortex during traumatic imagery in male and female Vietnam veterans with PTSD. Arch Gen Psychiatry 61(2):168–176. https://doi.org/10.1001/archpsyc.61.2.168

Shin LM, Shin PS, Heckers S, Krangel TS, Macklin ML, Orr SP et al (2004b) Hippocampal function in posttraumatic stress disorder. Hippocampus 14(3):292–300. https://doi.org/10.1002/hipo.10183

Shin LM, Wright CI, Cannistraro PA, Wedig MM, McMullin K, Martis B et al (2005) A functional magnetic resonance imaging study of amygdala and medial prefrontal cortex responses to overtly presented fearful faces in posttraumatic stress disorder. Arch Gen Psychiatry 62 (3):273–281. https://doi.org/10.1001/archpsyc.62.3.273

Shin LM, Lasko NB, Macklin ML, Karpf RD, Milad MR, Orr SP et al (2009) Resting metabolic activity in the cingulate cortex and vulnerability to posttraumatic stress disorder. Arch Gen Psychiatry 66(10):1099–1107. https://doi.org/10.1001/archgenpsychiatry.2009.138

Soliman F, Glatt CE, Bath KG, Levita L, Jones RM, Pattwell SS et al (2010) A genetic variant BDNF polymorphism alters extinction learning in both mouse and human. Science 327 (5967):863–866. https://doi.org/10.1126/science.1181886

Sotres-Bayon F, Quirk GJ (2010) Prefrontal control of fear: more than just extinction. Curr Opin Neurobiol 20(2):231–235. https://doi.org/10.1016/j.conb.2010.02.005

Sotres-Bayon F, Sierra-Mercado D, Pardilla-Delgado E, Quirk GJ (2012) Gating of fear in prelimbic cortex by hippocampal and amygdala inputs. Neuron 76(4):804–812. https://doi.org/10.1016/j.neuron.2012.09.028

Stein MB, Koverola C, Hanna C, Torchia MG, McClarty B (1997) Hippocampal volume in women victimized by childhood sexual abuse. Psychol Med 27(4):951–959

Sung KK, Jang DP, Lee S, Kim M, Lee SY, Kim YB et al (2009) Neural responses in rat brain during acute immobilization stress: a [F-18]FDG micro PET imaging study. NeuroImage 44 (3):1074–1080. https://doi.org/10.1016/j.neuroimage.2008.09.032

Takei S, Morinobu S, Yamamoto S, Fuchikami M, Matsumoto T, Yamawaki S (2011) Enhanced hippocampal BDNF/TrkB signaling in response to fear conditioning in an animal model of posttraumatic stress disorder. J Psychiatr Res 45(4):460–468. https://doi.org/10.1016/j.jpsychires.2010.08.009

Thakur GA, Joober R, Brunet A (2009) Development and persistence of posttraumatic stress disorder and the 5-HTTLPR polymorphism. J Trauma Stress 22(3):240–243. https://doi.org/10.1002/jts.20405

Thomaes K, Dorrepaal E, Draijer N, de Ruiter MB, van Balkom AJ, Smit JH, Veltman DJ (2010) Reduced anterior cingulate and orbitofrontal volumes in child abuse-related complex PTSD. J Clin Psychiatry 71(12):1636–1644. https://doi.org/10.4088/JCP.08m04754blu

Vincent SL, Khan Y, Benes FM (1993) Cellular distribution of dopamine D1 and D2 receptors in rat medial prefrontal cortex. J Neurosci 13(6):2551–2564

Viviani D, Charlet A, van den Burg E, Robinet C, Hurni N, Abatis M et al (2011) Oxytocin selectively gates fear responses through distinct outputs from the central amygdala. Science 333 (6038):104–107. https://doi.org/10.1126/science.1201043

Viviani D, Haegler P, Strasser DS, Steiner MA (2012) Sex comparison on long-lasting behavioral and physiological disturbances induced by single shock experience in rats. Physiol Behav 107 (2):243–251. https://doi.org/10.1016/j.physbeh.2012.06.018

Voisey J, Swagell CD, Hughes IP, Morris CP, van Daal A, Noble EP et al (2009) The DRD2 gene 957C>T polymorphism is associated with posttraumatic stress disorder in war veterans. Depress Anxiety 26(1):28–33. https://doi.org/10.1002/da.20517

Wagner HR 2nd, Hall TL, Cote IL (1977) The applicability of inescapable shock as a source of animal depression. J Gen Psychol 96(2d Half):313–318. https://doi.org/10.1080/00221309.1977.9920828

Wakizono T, Sawamura T, Shimizu K, Nibuya M, Suzuki G, Toda H et al (2007) Stress vulnerabilities in an animal model of post-traumatic stress disorder. Physiol Behav 90 (4):687–695. https://doi.org/10.1016/j.physbeh.2006.12.008

Walker DL, Miles LA, Davis M (2009) Selective participation of the bed nucleus of the stria terminalis and CRF in sustained anxiety-like versus phasic fear-like responses. Prog Neuro-Psychopharmacol Biol Psychiatry 33(8):1291–1308. https://doi.org/10.1016/j.pnpbp.2009.06.022

Wang Z, Neylan TC, Mueller SG, Lenoci M, Truran D, Marmar CR et al (2010) Magnetic resonance imaging of hippocampal subfields in posttraumatic stress disorder. Arch Gen Psychiatry 67(3):296–303. https://doi.org/10.1001/archgenpsychiatry.2009.205

Wang Z, Baker DG, Harrer J, Hamner M, Price M, Amstadter A (2011) The relationship between combat-related posttraumatic stress disorder and the 5-HTTLPR/rs25531 polymorphism. Depress Anxiety 28(12):1067–1073. https://doi.org/10.1002/da.20872

Wang WS, Kang S, Liu WT, Li M, Liu Y, Yu C et al (2012a) Extinction of aversive memories associated with morphine withdrawal requires ERK-mediated epigenetic regulation of brain-derived neurotrophic factor transcription in the rat ventromedial prefrontal cortex. J Neurosci 32 (40):13763–13775. https://doi.org/10.1523/JNEUROSCI.1991-12.2012

Wang YC, Ho UC, Ko MC, Liao CC, Lee LJ (2012b) Differential neuronal changes in medial prefrontal cortex, basolateral amygdala and nucleus accumbens after postweaning social isolation. Brain Struct Funct 217(2):337–351. https://doi.org/10.1007/s00429-011-0355-4

Weinberger NM (2011) The medial geniculate, not the amygdala, as the root of auditory fear conditioning. Hear Res 274(1–2):61–74. https://doi.org/10.1016/j.heares.2010.03.093

Wimer CC, Wimer RE, Roderick TH (1971) Some behavioral differences associated with relative size of hippocampus in the mouse. J Comp Physiol Psychol 76(1):57–65

Winston CR, Leavell BJ, Ardayfio PA, Beard C, Commissaris RL (2001) A nonextinction procedure for long-term studies of classically conditioned enhancement of acoustic startle in the rat. Physiol Behav 73(1–2):9–17

Woodward SH, Schaer M, Kaloupek DG, Cediel L, Eliez S (2009) Smaller global and regional cortical volume in combat-related posttraumatic stress disorder. Arch Gen Psychiatry 66 (12):1373–1382. https://doi.org/10.1001/archgenpsychiatry.2009.160

Woon FL, Hedges DW (2008) Hippocampal and amygdala volumes in children and adults with childhood maltreatment-related posttraumatic stress disorder: a meta-analysis. Hippocampus 18 (8):729–736. https://doi.org/10.1002/hipo.20437

Woon F, Hedges DW (2011) Gender does not moderate hippocampal volume deficits in adults with posttraumatic stress disorder: a meta-analysis. Hippocampus 21(3):243–252. https://doi.org/10.1002/hipo.20746

Woon FL, Sood S, Hedges DW (2010) Hippocampal volume deficits associated with exposure to psychological trauma and posttraumatic stress disorder in adults: a meta-analysis. Prog Neuro-Psychopharmacol Biol Psychiatry 34(7):1181–1188. https://doi.org/10.1016/j.pnpbp.2010.06.016

Xie P, Kranzler HR, Poling J, Stein MB, Anton RF, Brady K et al (2009) Interactive effect of stressful life events and the serotonin transporter 5-HTTLPR genotype on posttraumatic stress disorder diagnosis in 2 independent populations. Arch Gen Psychiatry 66(11):1201–1209. https://doi.org/10.1001/archgenpsychiatry.2009.153

Yamaguchi Y, Sato N, Wagatsuma H, Wu Z, Molter C, Aota Y (2007) A unified view of theta-phase coding in the entorhinal-hippocampal system. Curr Opin Neurobiol 17(2):197–204. https://doi.org/10.1016/j.conb.2007.03.007

Yamamoto S, Morinobu S, Takei S, Fuchikami M, Matsuki A, Yamawaki S, Liberzon I (2009) Single prolonged stress: toward an animal model of posttraumatic stress disorder. Depress Anxiety 26(12):1110–1117. https://doi.org/10.1002/da.20629

Yamasue H, Kasai K, Iwanami A, Ohtani T, Yamada H, Abe O et al (2003) Voxel-based analysis of MRI reveals anterior cingulate gray-matter volume reduction in posttraumatic stress disorder due to terrorism. Proc Natl Acad Sci USA 100(15):9039–9043. https://doi.org/10.1073/pnas.1530467100

Young RM, Lawford BR, Noble EP, Kann B, Wilkie A, Ritchie T et al (2002) Harmful drinking in military veterans with post-traumatic stress disorder: association with the D2 dopamine receptor A1 allele. Alcohol Alcohol 37(5):451–456

Yu H, Wang Y, Pattwell S, Jing D, Liu T, Zhang Y et al (2009) Variant BDNF Val66Met polymorphism affects extinction of conditioned aversive memory. J Neurosci 29 (13):4056–4064. https://doi.org/10.1523/JNEUROSCI.5539-08.2009

Zhang J, Tan Q, Yin H, Zhang X, Huan Y, Tang L et al (2011a) Decreased gray matter volume in the left hippocampus and bilateral calcarine cortex in coal mine flood disaster survivors with recent onset PTSD. Psychiatry Res 192(2):84–90. https://doi.org/10.1016/j.pscychresns.2010.09.001

Zhang L, Zhang Y, Li L, Li Z, Li W, Ma N et al (2011b) Different white matter abnormalities between the first-episode, treatment-naive patients with posttraumatic stress disorder and generalized anxiety disorder without comorbid conditions. J Affect Disord 133(1–2):294–299. https://doi.org/10.1016/j.jad.2011.03.040

Zoladz PR, Conrad CD, Fleshner M, Diamond DM (2008) Acute episodes of predator exposure in conjunction with chronic social instability as an animal model of post-traumatic stress disorder. Stress 11(4):259–281. https://doi.org/10.1080/10253890701768613

Zoladz PR, Fleshner M, Diamond DM (2012) Psychosocial animal model of PTSD produces a long-lasting traumatic memory, an increase in general anxiety and PTSD-like glucocorticoid abnormalities. Psychoneuroendocrinology 37(9):1531–1545. https://doi.org/10.1016/j.psyneuen.2012.02.007

Modulation of the Core Neural Network in Stress: The Role of Brain-Derived Neurotrophic Factor and LTP

4

4.1 BDNF Gene Transcription Controlled by Glucocorticoid and Mineralocorticoid Receptors

4.1.1 The BDNF Gene

The rat BDNF gene consists of 4 short 50 exons and a 30 exon encoding the mature BDNF protein (Timmusk et al. 1993). Quantitative PCR analysis of BDNF mRNA containing these five upstream exons indicates that each of the alternative transcripts is most abundant in the hippocampus, intermediate in the substantia nigra and cerebellum and least abundant in the striatum, although the magnitude of these differences in expression varies indicating that BDNF gene transcription in the mature brain is regulated by alternate promoters that are differentially active across these regions (Bishop et al. 1994). Thus the BDNF gene possesses differential exon regulation and usage with different subcellular distributions as well as different distributions in different parts of the brain. The gene consists of 8 untranslated 50 exons each possessing their own promoters with the possibility of being connected to a 30 coding exon, thus forming a tripartite transcript that allows for different splice variants of BDNF mRNA. So the rodent BDNF transcripts consist of a protein-coding exon spliced to one of eight noncoding exons as well as a transcript containing only the protein-coding exon. This structure of the rodent BDNF gene is shown in Fig. 4.1 [see Fig. 3 in Boulle et al. (2012)]. Exons are represented in this figure as boxes and the introns as lines with the exon numbers indicated in Roman numerals. The 30 coding exon (exon IX) contains two polyadenylation sites (poly A). The red box shows the region of exon IX coding for the pro-BDNF protein. It is the variety of transcripts made possible by this arrangement of the BDNF gene that allows for specificity to different regions of the brain (Aid et al. 2007). Furthermore,

---

M. Bennett, J. Lagopoulos, *Stress, Trauma and Synaptic Plasticity*,
https://doi.org/10.1007/978-3-319-91116-8_4

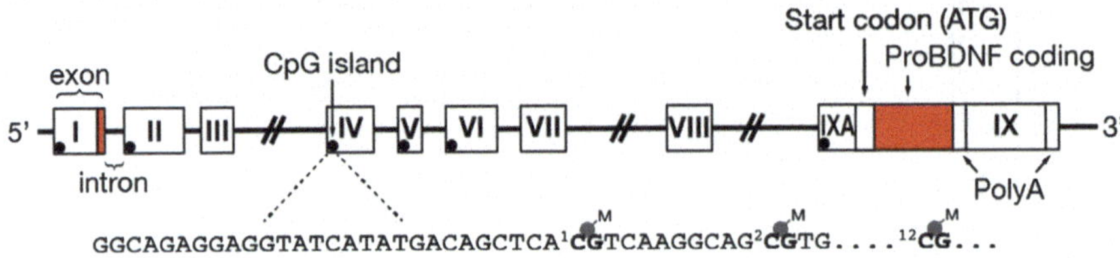

Fig. 4.1 Structure of the rodent BDNF gene. Exons are represented as boxes and the introns as lines. Numbers of the exons are indicated in Roman numerals. The coding exon (exon IX) contains two polyadenylation sites (poly A). The start codon (ATG) that marks the initiation of transcription is indicated. The red box shows the region of exon IX coding for the pro-BDNF protein. Some exons, like exon II and IX, contain different transcript variants with alternative splice donor sites. Also shown is part of the BDNF exon IV sequence in adults with adverse infant experiences showing cytosine methylation (M) at 3 of the 12 CG dinucleotide sites (numbered with superscripts). See Boulle et al. (2012)

these transcripts of BDNF genes in rodents are not only specific to different regions of the cortex and archicortex, but their expression in these regions is under control of the epigenetic processes of histone deacetylation and DNA methylation. The flexibility provided by different splice variants of BDNF mRNA, with differences in their 50 or 30 extremities, together with epigenetic modulation of transcription of the BDNF gene, provides for a wide variety of regulatory processes in determining the formation and regression of dendritic spines in different brain regions in response to stress.

Such epigenetic changes in the BDNF gene are indicated by the differential expression of BDNF exons in the brains of rats under different stress-behaviour paradigms. For instance, expression of exons I and III are increased in the amygdala after stress and fear conditioning. Following the extinction of such fear, there is an increase in exons I and IV in the prefrontal cortex (Bredy et al. 2007; Ou and Gean 2007). Context exposure leads to an increase in exons I and VI in the hippocampus, whereas consolidation of fear learning leads to increases in exon IV (Lubin et al. 2008). On maternal separation of rodents, there is an upregulation of exon II by postnatal day 14 and of exons IV and V at postnatal day 21 (Zocchi and Sassone-Corsi 2012). Following chronic stress through restraint, there is an increase of exons I and II but a decrease of exons III and IV in hippocampal areas CA1, CA3, CA4 and the dentate gyrus (Hansson et al. 2006; Nair et al. 2007).

4.1.2 Glucocorticoid Receptors, Mineralocorticoid Receptors, Their Chaperones and Co-activators

A multiplicity of glucocorticoid receptor (GR) isoforms allows for quite specific tissue actions for the participation of GR in different functions. The variety of GR isoforms arises from the fact that the GR gene possesses nine different exons together with three promoters resulting in a variety of isoforms with different 5′-untranslated regions. The number of these is further enhanced by the exons possessing several alternative splice sites. For example, the alpha and beta isoforms of exon IX arise as a consequence of alternative splicing at the 5′-untranslated regions (Yudt and Cidlowski 2002).

Pulses of cortisol in humans and corticosterone in rodents (the glucocorticoid hormones) enter the brain and follow an ultradian pattern (i.e. occur more frequently than once in a 24-h cycle). The frequency of these pulses of released glucocorticoids increases under stress, during which they may also summate to give large amplitude responses, that is, a prolonged non-pulsatile increase can occur. As the mineralocorticoid receptors (MR) bind glucocorticoids with high affinity (as opposed to the GR which do so only at low affinity), the binding of these receptors to DNA is contingent on the levels of glucocorticoid hormones determined in part by these ultradian patterns of release (Lightman 2008). At normal basal levels of cortisol, the MR are almost fully occupied because of their high affinity, whereas the opposite is the case for the GR. There is then little in the way of an extra response available through the MR when a surge in cortisol occurs (Mitra et al. 2009).

After binding to ligands that are agonists, the cytoplasmic GR and MR each exchange certain chaperones (including FKBP4 for FKBP5; see below Sect. 4.1) and cross the nuclear membrane where these chaperones together with heat-shock proteins (such as HSP90) dissociate from the GR/MR with which they have been complexed allowing the receptors to ultimately modulate, when subsequently complexed with co-activators (containing histone acetyltransferase P300/CBP), the transcription activity of their respective genes (Fig. 4.2). This is achieved through binding their cognate DNA-binding sequences, the GR and MR response elements. It should be noted that the binding of MR to the DNA promoter region leads to an extreme reorganization of the chromatin that allows additional DNA-binding transcription factors and other co-regulators to scan the site. The modulation of TrkB mRNA in the hippocampus by ligand-bound GR and MR presumably uses these mechanisms (Schaaf et al. 2000).

Thus promoter-bound GR and MR initiate transcription of downstream coding sequences by associating with numerous co-activator complexes as well as chromatin remodelling factors, which are critical determinants of the action of the ligand-bound receptors.

Ligand-bound GR and MR then act to modulate the expression of various target genes with different transcriptional potencies (van der Laan and Meijer 2008). The target genes of GR and MR are cell-type specific, although it is not clear at this time whether this holds for cells of the same type (e.g. pyramidal neurons) but in different parts of the brain, nor whether it holds for the BDNF gene at all (Wang et al. 2004).

It is possible that the different expression patterns of BDNF exons throughout the brain following stress are partly due to differences in the extent to which the GR and MR engage these co-activators (Meijer 2002). Steroid receptors can bind to their glucocorticoid response elements (GRE) on the DNA or interfere with signaling of other transcription factors via protein–protein interaction (Meijer 2002). Indeed two different isoforms of steroid receptor co-activator induce opposing effects on expression of CRF genes. But none of the above is specific for GR on BDNF regulation (Stranahan et al. 2011). Suppression of BDNF expression in leptin-deficient mice is

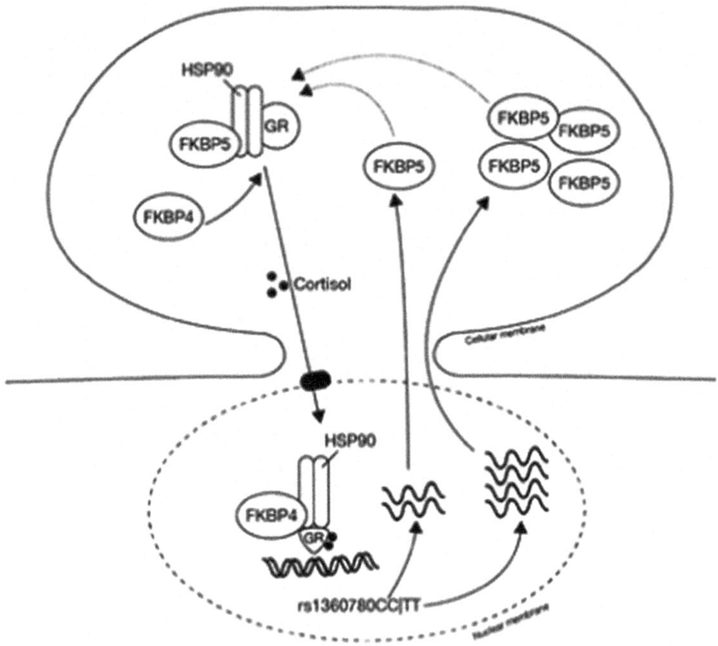

Fig. 4.2 Schematic representation of the possible molecular differences leading to FKBP5 polymorphism-dependent ultrashort feedback loop activation. Shown are both situations in which individuals homozygous for the alleles associated with less FKBP5 induction and protective against stress-related psychiatric disorders (the allele rs1360780CC) as well as the situation in individuals homozygous for the high-induction or high-risk alleles (such as rs1360780TT). Presumably through differential binding to the glucocorticoid receptor response element or differential recruitment of co-factor, the GR leads to less (two rv) or more (five rv) induction of FKBP5 mRNA and thus protein. If more FKBP5 is present after GR activation, more of the GR complexes bind this co-chaperone and not FKBP4, thus holding more of the receptors in a state with less affinity for cortisol and decreasing the amount of GR translocating to the nucleus through the nuclear membrane. See Binder (2009)

brain region specific as it is reduced by glucocorticoids in the hippocampus but not in the hypothalamus.

The glucocorticoid ligand-GR-co-chaperone complex modulates the transcription of different genes involved in metabolism and inflammation by binding to the genomic GRE, with very little in common between the different modulated gene sets in different cells [So et al. (2007); note that their Table 1 lists these glucocorticoid binding regions and does not list BDNF as a target gene].

The action of glucocorticoids in a particular part of the brain depends on several factors, including ligand concentration and exposed receptor levels in the part of the brain under consideration as well as concentration of different chaperone types and on recruitment of co-activators and co-repressors (van der Laan and Meijer 2008).

4.1.3 Co-chaperones FKBP5 and FKBP4: Polymorphisms and Stress

The co-chaperone FKBP5 generally attenuates the activity of GR and MR but increases the activity of the androgen receptor (AD). According to the model shown in Fig. 4.2, cytoplasmic hormone-free GR associates with a chaperone complex that consists of HSP90 dimer and FKBP5 or FKBP4. The dimer itself complexes with the hormone-binding domain of the GR. Binding of glucocorticoid to the GR transfers the equilibrium of the chaperone complex towards the FKBP4-containing ones. This facilitates dimerization and transport of the ligand-bound GR into the nucleus. GR-enhanced expression of the FKBP5 in turn moves the equilibrium back towards the FKBP5-containing complexes in the cytoplasm, as a consequence of significant excess of FKBP5 over FKBP4, resulting in attenuation of the action of the glucocorticoid as a consequence of the lower capacity of the FKBP5-containing complexes to bind glucocorticoids and cross into the nucleus (Jaaskelainen et al. 2011). It is the binding of co-chaperone FKBP5 HSP90 that determines within the GR complex its sensitivity for cortisol; high levels of FKBP5 binding lower the sensitivity of the GR to cortisol and also restrict the complex from moving into the nucleus from the cytoplasm. Intron hormone response elements in the FKBP5 gene, acted on by the GR complex, increase FKBP5 mRNA and protein expression so providing the feedback pathway within a neuron for GR complex sensitivity (Binder 2009; Jaaskelainen et al. 2011).

There are certain SNPs in the gene for FKBP5 that result in increased FKBP5 protein production so altering the sensitivity of the GR complex for cortisol and transfer of the activated GR complex to the nucleus. This occurs as the co-chaperone FKBP4, which increases the affinity of the GR complex for cortisol over that when FKBP5 binds the complex, as well as enhancing transport of the GR complex to the nucleus, is displaced by excess FKBP5 arising as a consequence of the SNPs in its gene. The result is a decrease in the normal negative feedback control in the hypothalamic pituitary axis (HPA) of cortisol release primarily through desensitization of the GR receptor complex leading to high levels of cortisol. There is then a linear increase in the plasma cortisol level with an increase in FKBP5 expression (Yehuda et al. 2010).

There is an association between SNPs rs3800373 and rs1360780TT in the FKBP5 gene and peri-traumatic stress disorder in traumatized children, characterized by, for example, their relative insensitivity to pain (Boscarino et al. 2011; Koenen et al. 2005), as well as in abused children at risk of developing post-traumatic stress disorder (PTSD) symptoms when they are adults (Binder et al. 2008). It seems likely that the polymorphism rs1360780TT has deleterious consequences as a result of excess induction of FKBP5 by the GR complex, so leading to decreases in FKBP4 binding and so less effective feedback in the stress hormone HPA (Binder 2009). On the other hand, subjects that possess homozygous alleles of the FKBP5 gene, such as rs1360780CC or rs9296158, possess lower levels of FKBP5 induction, and so this increases the effectiveness of feedback in the HPA axis consequent on a decrease in FKBP5 protein (Mehta et al. 2011). PTSD is experienced more readily in subjects that have been subjected to childhood abuse if they carry the alleles giving rise to high levels of FKBP5 (Binder et al. 2008; Pitman et al. 2012; Xie et al. 2010), but not if they carry the alleles responsible for low levels of FKBP5, for in this case their risk for PTSD is comparable to those subjects that have not experienced childhood abuse (Binder 2009). Even in animals that do not carry the above SNPs, there are increased levels of FKBP5 in the cytoplasm of neurons in the prefrontal cortex, hippocampus and central nucleus of the amygdala if they have experienced chronic stress (Guidotti et al. 2013; Scharf et al. 2011).

The expression of the co-chaperone FKBP5 gene is under control of EphB2, which is normally bound to the NR1 subunit of the NMDA receptor. EphB2 is cleaved from NR1 by the serine protease neuropsin in the amygdala, an action that is accelerated under conditions of stress. Thus if neuropsin is at low levels in an animal, there is little cleavage of EphB2 from NR1, with consequent lower levels of FKBP5 and relatively lower levels of anxiety in a stressful situation because of the sustained negative feedback control in the HPA system (Attwood et al. 2011). This system is also compromised as a consequence of polymorphisms that change the interaction of chromatin with the FKBP5 gene transcription start site and long-range enhancers (Klengel et al. 2013). Following childhood trauma, FKBP5 DNA is subject to demethylation, giving rise to restrictions in the normal regulation of the HPA.

4.2 Evidence that BDNF Gene Transcription Controlled by Epigenetic Changes

BDNF expression is under the control of GR (Suri and Vaidya 2013). BDNF is subject to epigenetic control, that is, to changes in BDNF gene expression that remain stable during cell divisions, but do not involve changes in the BDNF DNA sequence (Spijker and van Rossum 2012). This has been most thoroughly investigated for the nine noncoding exons and common coding exon of the rodent BDNF gene (Aid et al. 2007; Roth and Sweatt 2011; Timmusk et al. 1993).

4.2.1 Epigenetic Changes in the BDNF Gene

'CpG dinucleotide sites' refer to adjacent cytosine and guanine residues next to each other and linked by phosphodiester bonds in the backbone of DNA. The 'p' is to distinguish cytosine followed by guanine from a cytosine base paired to a guanine. The rodent BDNF exon IV possesses 12 CpG sites that are potentially methylated and if so bind MeCP2 (methyl-CpG binding protein 2) that then represses transcription of the gene. Such methylation involves transfer of a methyl group from S adenosylmethionine to the 50 position of the cytosine pyramidal ring by the DNA methyltransferase. MeCP2 contains a methyl docking domain that allows it to bind to CpG sites and is released by histone deacetylation with SIRT1, which is a nicotinamine adenine dinucleotide (NAD)-dependent histone deacetylase (Boulle et al. 2012; Zocchi and Sassone-Corsi 2012). CpG sites at promoters are in general

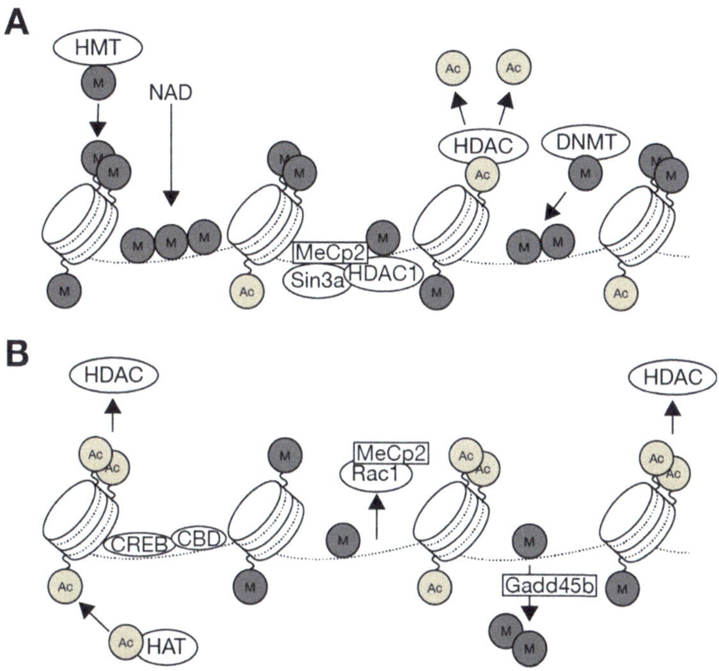

Fig. 4.3 Epigenetic mechanism associated with repression and activation of BDNF exon IV transcription. The BDNF exon IV displays 12 distinct CpG sites, which can be methylated and interact selectively with MeCp2 to form complexes that repress gene transcription (see also Fig. 4.1). Histone methyltransferases (HMT) are responsible for adding methyl groups at histone tails [panel (**a**)], whereas histone deacetylases (HDAC) remove acetylation at histone tails [panel (**b**)], both processes that repress gene transcription. Moreover, low levels of nicotinamine adenine dinucleotide (NAD) promote DNA methylation at the BDNF locus. BDNF gene activation is associated with increased histone H3 and H4 acetylation, which is mediated by histone acetyl transferase (HAT) activity. DNA demethylation might be facilitated by growth arrest and DNA damage proteins such as Gadd45b. An increased binding of CREB to its specific binding protein, CREB binding protein (CBP), is also associated with an increase in BDNF gene transcription. See Boulle et al. (2012)

unmethylated if the gene is expressed; however, if methylated, the gene is not expressed. Even if the CpG site at a gene promoter is unmethylated, it is still dependent on the chromatin microenvironment state which in turn is dependent on histone modifications (Lubin 2011). Thus BDNF gene transcription can also be suppressed through the mechanisms of adding methyl groups to histone tails (H3K27), by means of histone methyltransferases (HMTs), or by removing acetylation of the histone tails by means of histone deacetylases (HDACs) (Fig. 4.3) (Boulle et al. 2012). Conversely, BDNF gene activation is associated with increased histone H3 and H4 acetylation that is mediated by histone acetyltransferase (HAT) activity.

There are well-established pathways for producing epigenetic changes in the BDNF gene. One of these is repressor element 1 silencing transcription factor (REST), which can silence both coding and noncoding neuronal genes via epigenetic mechanisms, including those for BDNF (Rodenas-Ruano et al. 2012). The activity of REST is enhanced when MeCP2 binds to the REST promoter, even though this is unmethylated, causing modification of the histone profile with consequent decrease in BDNF expression. This is the reason why patients with Rett syndrome have lower levels of BDNF, for they are deficient in MeCP2 (Abuhatzira et al. 2007).

Transcription of the BDNF gene is facilitated by action potentials (depolarization) through epigenetic mechanisms. The normal inhibition of BDNF gene expression through MeCP2 binding to the BDNF exon IV promoter is relieved following depolarization as a consequence of release of MeCP2 due to calcium-dependent phosphorylation (Lubin et al. 2008). BDNF exons I and IV mRNAs increase in neurons concomitantly with decreases in DNA methyltransferase mRNAs following action potential/depolarization (Sharma et al. 2008). It is interesting that MeCP2 levels are under homeostatic control by BDNF as this induces the micro RNA132 (see Sect. 4.3.1) which in turn controls translation of MeCP2 (Klein et al. 2007).

Finally, a well-established procedure for producing epigenetic changes in BDNF involves environmental changes. If mice have had adequate care when young, very few of the 12 CpG sites of the BDNF exon IV in the prefrontal cortex are cytosine methylated. On the other hand, if mice have been abused and neglected when young, then they possess many sites that are cytosine methylated in the prefrontal cortex (Roth et al. 2009; Roth and Sweatt 2011). Chromatin inactivation and/or inhibition of transcription factor binding has the effect of silencing genes, such as the GR gene, and this is associated with methylation of the 50 cytosine of CpG dinucleotide. Such methylation occurs at the transcription factor NGF1-A binding site of the I7 promoter of GR of newborn mice that have not had caring mothers (Turner et al. 2010).

4.2.2 Epigenetic Changes in the BDNF Gene Following Different Behavioural Paradigms

PTSD involves a failure to extinguish fearful memories. Extinction is modulated by levels of BDNF that in turn are regulated by (a) the HPA pathway through the release of glucocorticoids, (b) the endocannabinoid receptor-mediated pathway, (c) the PAC1 receptor-mediated pathway and (d) adenylate-cyclase-activating polypeptide-mediated pathways. Epigenetic mechanisms also control BDNF levels with extinction enhanced by histone deacetylase inhibitors and partial NMDA receptor agonists such as D-cycloserine. In the case of the glucocorticoids, chronic restraint stress together with social isolation leads to increases in corticosterone to a level at which there is binding to GR complex followed by movement of the complex into the nucleus in pyramidal neurons of both the prefrontal cortex and hippocampus. In contrast, following restraint stress alone, the corticosterone levels are insufficient, and so binding to the GR complex does not occur, with failure of the complex to transfer to the nucleus (Djordjevic et al. 2009). Following elevated corticosterone levels, there is a very significant decrease in BDNF exons II and IV mRNA, whereas exons I and III mRNA remain unaffected, at least in the hippocampus and prefrontal cortex (Dwivedi et al. 2006). The pathway by which

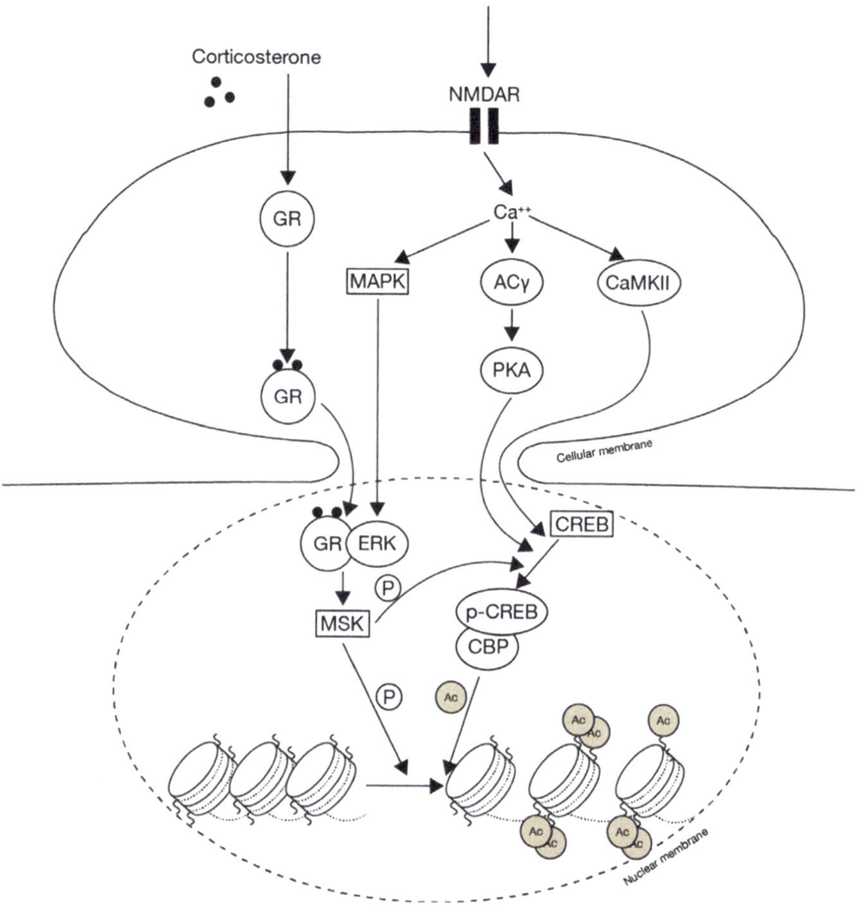

Fig. 4.4 Schematic representation of the signaling and epigenetic pathways in granule neurons of the dentate gyrus thought to be involved in the consolidation process of memory formation after a psychologically stressful challenge. Activation of NMDAR results in stimulation of the MAPK/ERK signaling cascade, the AC•/PKA cascade and the CaMKII cascade. In conjunction with activated GR, these signaling cascades result in the activation of MSK and ERK leading to the formation of dual histone acetylation marks along the c-Fos promoter and subsequently induction of gene transcription. Signaling via CREB also leads to the same outcome. The induction of gene transcription is thought to be instrumental in the consolidation of memory formation in various stressful learning events. See Trollope et al. (2012)

corticosterone exerts its effects on BDNF expression has been partly determined (Fig. 4.4). The activation of GR following the stressful event as well as of NMDA receptors leads, for granule cells in the dentate gyrus, to activation of MSK responsible for dual H3S10p/K14 as histone marks along the promoters of various genes, such as the BDNF gene (Reul and Chandramohan 2007; Trollope et al. 2012).

4.2.2.1 Childhood Stress

Sustained alterations in the expression of the BDNF gene are under the control of environmental factors through epigenetic factors, particularly when young. Deleterious environmental factors during development can lead to sustained decreases in the expression of BDNF throughout adulthood, as well as for FKBP5 and the serotonin transporter, with consequent difficulties in the response of patients to behavioural or pharmaceutical interventions (Boulle et al. 2012; van Winkel et al. 2013). In the case of animal models, it is well established that enhanced caring in early development, such as occurs when mothers share care of their offspring, leads to increases in the level of BDNF gene acetylation in adulthood. The implications of this are that soon after a stressful challenge in adulthood, these animals have increased expression of BDNF protein (Branchi et al. 2011).

4.2.2.2 Restraint Stress

An occurrence of restraint stress in rodents leads in the hippocampus to decreases in acetylation of histone H3 for promoters I and IV with VI giving decreases in exons I and IV BDNF mRNA (Fuchikami et al. 2011). BDNF mRNA is decreased by restraint-induced stress for periods as short as 8 h in the hippocampus (areas CA1, CA3 and the dentate gyrus) (Schaaf et al. 2000; Smith et al. 1995). On the other hand, chronic restraint stress does not change the level of BDNF mRNA in the prefrontal cortex (Chiba et al. 2012).

4.2.2.3 Predator and Social Defeat Stress

A different behavioural paradigm, that of exposure to a predator, gives DNA methylation of BDNF promoter IV in the hippocampus (dorsal area CA1) and a decrease in BDNF exon IV mRNA (Roth et al. 2011). Predator exposure also methylates DNA of promoters in the hippocampus (ventral, CA3) leading to a decrease in BDNF exon IV but no changes in the medial prefrontal cortex or in the BLA (Roth et al. 2011). Some animals show extreme behaviour on exposure to a predator, and this is accompanied by a decrease in expression of BDNF mRNA in hippocampal region CA1, a decrease that does not occur in animals that do not show such extremes (Kozlovsky et al. 2007). Social defeat stress gives rise to histone H3K27 methylation at the BDNF promoters of exons IV and VI, resulting in a decrease in total BDNF mRNA in the hippocampus (Tsankova et al. 2006). Early-life adversity gives rise to DNA methylation of BDNF promoter IV and a decrease in total BDNF mRNA in the prefrontal cortex (Roth et al. 2009). In contrast following voluntary exercise and an enriched environment, there is an increase in total BDNF mRNA in the hippocampus as a consequence of increases in histone acetylation at promoter IV or decreases in its methylation (Boulle et al. 2012).

4.2.2.4 Fear Conditioning (Memory)

Of particular interest in relation to regulation of BDNF exon mRNAs is that following fear conditioning, the levels of these mRNAs change significantly. For example, in the BLA, following light-shock associative fear conditioning, the levels of exon I and III mRNAs are greatly increased, while that for exons II and IV stay the

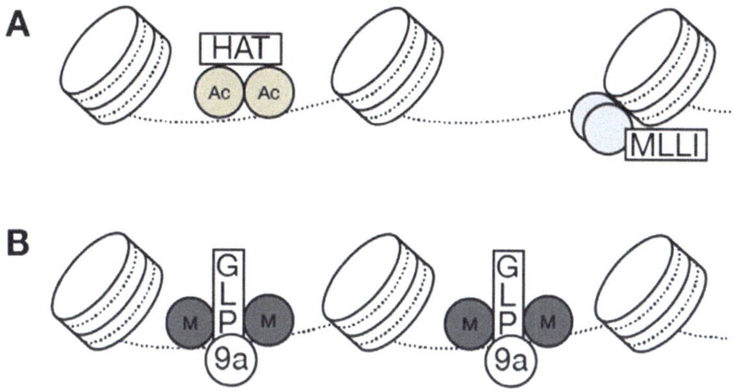

Fig. 4.5 Model for G9a/GLP complex transcriptional activity in the hippocampus during fear memory consolidation. Shown [panels (**a**) and (**b**)] is the role of G9a/GLP in the regulation of chromatin remodelling during long-term memory consolidation. Regulation of histone lysine methylation mediates active and repressive transcriptional regulation of genes in the hippocampus. The changes in chromatin structure result in transcriptional gene silencing in the hippocampus. H3K9me2 dimethylation is associated with transcriptional silencing (not shown). The G9a/GLP complex methyltransferase is specific for producing this modification. Abbreviations: *Ac* acetylation, *M* methylation, *MLLI* histone H3 lysine 4 methyltransferase (which regulates memory formation), *H3K9me2* histone H3 lysine 9 dimethylation, *HAT* histone acetyltransferase, *G9a/GLP* G9a/G9a-like protein (GLP) complex methyltransferase (from Bennett and Lagopoulos 2014)

same (Rattiner and Davis 2004). In the hippocampus following fear conditioning, there is demethylation of cytosine residues in BDNF exon IV in the hippocampus that are involved in memory of the fearful event. Subsequent inhibition of NMDA receptors changes the pattern and extent of cytosine residues that are demethylated which results in a decrease in retention of the memory (Lubin et al. 2008). Thus memory of a fearful event can be in part extinguished by the NMDA receptor agonist D-cycloserine (Kaplan and Moore 2011).

The memory of the context of a fearful situation is accompanied by increases in histone H3 lysine 9 demethylation (H3K9me2) in the hippocampus, altering transcription of BDNF. The BDNF exon IV gene promoter in the hippocampal area CA1 is modulated by H3K9me2 that in turn is catalysed by G9a/G9a-like protein (GLP) complex lysine dimethyltransferase (Fig. 4.5). Inhibition of G9a/GLP, for example, using bPIX, then changes consolidation of the context of fearful memories (Gupta-Agarwal et al. 2012). A single incidence of sustained stress (e.g. placing a rat in restraint for 2 h followed by 20 min of forced swimming) with contextual fear conditioning consisting of a single foot shock leads to enhanced expression of BDNF mRNA and protein in the hippocampus as well as an increase in the acetylation of histones H3 and H4 at the exon I and IV promoters, all of which are implicated in retaining a memory of the contextual fear conditioning (Fig. 4.3) (Takei et al. 2011). Fearful auditory memories are accompanied in the BLA by changes in histone acetylation of H3. Inhibiting deacetylase activity in the lateral nucleus reinstates histone acetylation and consolidates the fearful auditory memory (Maddox and Schafe 2011).

4.2.2.5 Suicide

Suicide subjects have similar levels of BDNF and show similar epigenetic changes in the gene as those with PTSD. Thus suicide subjects have low levels in the hippocampus and prefrontal cortex of BDNF mRNA (Dwivedi 2009; Dwivedi et al. 2003), differential methylation of the BDNF promoter of exon IV in some parts of the cortex (Keller et al. 2010) and in general possess BDNF Val/66Met (rs6265; 196G > A) (Zai et al. 2012).

In suicide subjects, there is a decrease in binding of the GR gene transcriptional factor NGF1-A and so of GR gene transcription together with increased methylation of the NR3C1 promoter, with consequent decrease in GR mRNA (McGowan et al. 2009; Tsai et al. 2011). This leads to a failure of the normal negative feedback regulation on the release of cortisol.

4.3 BDNF Control of Dendritic Spines

BDNF changes spine morphology in a synaptic activity-dependent manner, so that with normal transmission, BDNF increases the properties of stubby, type I spines, with miniature synaptic potentials capable in the presence of BDNF to establish defined morphological spine types (Tyler and Pozzo-Miller 2003). Spines in the absence of spontaneous electrical activity are significantly smaller than normal, at least on the proximal domain of dendrites (Harvey et al. 2005). On the other hand, simultaneous electrical activity and glutamate release onto the spines give an immediate increase in size, which is dependent on BDNF (Tanaka et al. 2008).

BDNF synthesized in dendrites regulates the morphology of spines (An et al. 2008). This dendritically synthesized BDNF is secreted as proBDNF (Nagappan et al. 2009), and subsequent activation of TrkB receptors induces the synthesis of a number of proteins that determines the morphology of the spines, including Arc, Homer2 and LIMK1 (Kang and Schuman 1996; Schratt et al. 2004; Yin et al. 2002) which promote actin polymerization (Bramham and Wells 2007) and so enlargement of spine heads (Sala et al. 2001). It follows that inhibiting BDNF synthesis reduces the size of spine heads and impairs long-term potentiation of synaptic transmission (An et al. 2008; Waterhouse and Xu 2009). In Sect. 4.3, detailed consideration is given to the mechanisms of BDNF/TrkB control of dendritic spines.

Membrane receptors for mature BDNF, the TrkB receptors, exist amongst many other membrane receptors, such as voltage-gated ionic channel, to control the actin cytoskeleton in dendritic spines and so their formation and regression (Bennett 2009). The following receptor and ion channel pathways occur on the spines' plasmalemma (Bennett 2009): (1) Sema 3A acting on plexin A receptors to decrease RAS, ERK and so WASP nucleation of G-actin around Arp2/3. (2) NRG-1 acting on ErbB4/ErbB2 receptors, anchored to the PSD-95 scaffolding protein and Rho GTPase to increase WASP as well as release of SFK to act on the NR2B subunit of the NMDA receptor to phosphorylate it, so enhancing calcium influx through NMDA receptors. (3) Glutamate acting on NMDA receptors to enhance calcium entry into the cytosol and via CaMKK, CaMK2, bPIX, GIT1, RAC, PAK and LIMK1 to inhibit cofilin depolymerization of ADP-actin in the F-actin filaments. (4) Voltage-dependent calcium channel, CAV1.3a, which is anchored by PDZ and SHANK in the cytosol and through HOMER releases calcium from calcium stores. (5) Dopamine acting on D2 dopamine receptors, which, through PP1, activates calcium–calmodulin kinase 2 to modulate CAV1.3a and phosphorylates the NR1 subunit of NMDA. (6) Dopamine acting on D2 dopamine receptors, which, through Gp and PLCb, releases IP3 to activate calcium release from internal stores; this calcium activates pCREB. (7) Glutamate acting on NMDA receptors activates the RhoA, ROCK and profiling pathway to provide G-actin bound to profilin. (8) Ephrin binding to the EphB receptor activates Kalirin, which then acts on RAC1/PAK to excite LIMK1 and so inhibits cofilin depolymerization of F-actin. (9) BDNF acting on TrkB receptors to activate Rho GTPases RhoA, Cde42 and Rac1 and hence their downstream target proteins profilin, Arp2/3 and cofilin (Fig. 4.6). It is the action of BDNF in controlling spine growth and regression that is examined in detail below.

[The following are the definitions of the acronyms used above: BDNF, brain-derived neurotrophic factor; bPIX, guanine nucleotide exchange factor for RAC (GEF); CaMK2, calcium–calmodulin-dependent kinase 2; CaMKK, calcium–calmodulin-dependent protein kinase kinase; cofilin, severs and depolymerizes ADP-actin; D2, dopamine D2 receptor; EphB, ephrin receptor B; ErbB2, receptors for neuregulin; ErbB4, receptors for neuregulin; ERK, extracellular signal-regulated kinases; GKAP, guanylate kinase-associated protein; Gp, G-protein; HOMER, scaffolding protein; IP3, inositol triphosphate; kalirin, Rho GEF; LIMK, LIM kinase, phosphorylates ADF/cofilin; NMDA, *N*-methyl-D-aspartate; NR1, NR2A and NR2B, subunits of the NMDA receptor; NRG-1, neuregulin 1; PAK, downstream effector of RAC (sometimes called P21-activated kinase); pCREB, phosphorylated cyclic AMP response element-binding protein; PDZ, protein domain; PLCb, protein lipase Cb; plexin A, receptor for Sema 3A; PP1, protein phosphatase 1; profilin, actin regulatory molecule; PSD-95, postsynaptic density 95, a scaffolding protein; RAC, Rho GTPase; RAS, Rho GTPase; RhoA, Rho GTPase; Rho GTPase, Rho family GTPases, a subgroup of the superfamily of GTPases; ROCK, Rho-associated kinase; sema 3A, semaphorin 3A; SFK, src family kinase; SHANK, scaffolding molecule; TrkB, BDNF receptor; WASP, Wiskott–Aldrich syndrome protein that triggers actin polymerization via Arp2/3 complex].

The principal receptors at the plasmalemma that activate signaling cascades, which then lead to changes in F-actin treadmilling and spine shape, are neurotransmitter receptors as well as voltage-dependent calcium ion channels on the one hand and surface membrane receptors that activate intracellular signaling cascades that control activity of Rho GTPases on the other. The main transmitter-activated calcium permeable channels are the NMDA receptors, which are modulated by ErbB4 receptors, and the main receptors controlling the Rho GTPases are ephrin, BDNF and plexin. NMDA receptor activation by glutamate leads to the largest calcium influx that any receptor can generate. This calcium can in turn activate

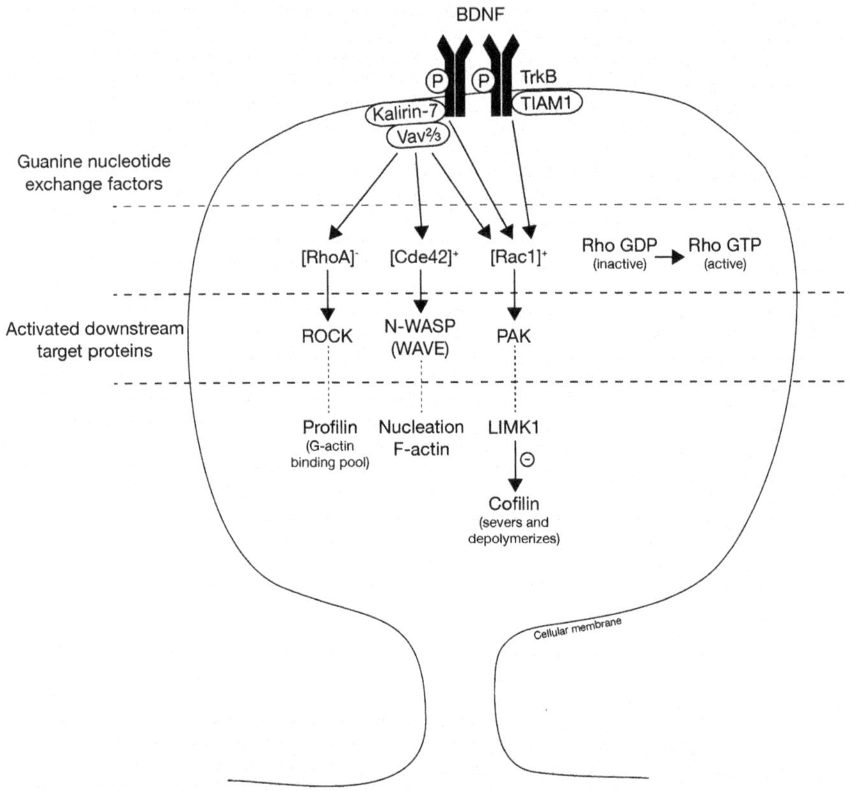

Fig. 4.6 Model for the role of Vav2/3 guanine nucleotide exchange factors in BDNF/TrkB-induced dendritic spine growth and functional synapse plasticity through altering actin dynamics in the spine head. The Rho GTPases Rac1 and Cde42 enhance spine growth, whereas RhoA decreases it (indicated by the + and − signs). Rac1, Cde42 and RhoA act on their downstream target proteins PAK (p21-activated kinase), WASP (Wiskott–Aldrich syndrome protein) and ROCK (Rho kinase). These then regulate proteins that affect actin filament extension, branching or myosin contractility such as profilin, cofilin and Arp2/3 (not shown). There is considerable crosstalk between the activated downstream target proteins which is also not shown (from Bennett and Lagopoulos 2014)

CaMKII, which in turn phosphorylates and activates calmodulin-dependent kinase kinase (CaMKK, CaMK1). These kinases form a multimolecular complex with the guanine nucleotide exchange factor bPIX. Phosphorylation of this results in activation of Rac1 and Pak1. Pak1-mediated phosphorylation then activates LIM kinase (LIMK), which inactivates actin-depolymerizing protein ADP/cofilin, so promoting stabilization of F-actin and giving spine enlargement. In hippocampal neurons, an actin-regulated pathway is directed by Rho small GTPases, such as Rho A. Rho A recruits and activates its specific kinase ROCK, which in turn complexes with profilin Ia, so modifying actin stability through a RhoA/ROCK/profilin IIa pathway. It is known that RhoA associates with ionotropic glutamate receptors (iGluRs) and metabotropic glutamate receptors (mGluRs) at the plasma membrane of dendritic spines, with activation of the iGluRs leading to detachment of RhoA from these receptors and their recruitment to the mGluRs. This triggers local reduction of RhoA activity at the iGluRs, resulting in inactivation of RhoA-specific kinase ROCK there and so disruption of the ROCK-profilin IIa, producing rapid changes in the underlying actin cytoskeleton and remodelling of spine shape. Activation of the NMDA receptors can also lead to increases in activity of LIMK, and hence inhibition of ADP/cofilin and spine enlargement, by the GTPase RhoA through RhoA-specific kinase (ROCK). This pathway can also increase profilin and hence availability of G ATP-actin for F-actin polymerization. ROCK can also activate myosin regulatory light chain (MLC) and simultaneously inhibit myosin light chain phosphatase (MLCP), so leading to myosin motor activity and increased retrograde flow of F-actin with concomitant spine shortening (for a model of this process, see Bennett et al. 2011). Collectively, glutamate stimulation of NMDA receptors can, through both CaKMII and RhoA, lead to inhibition of ADF/cofilin (with perhaps a small enhancement from the phosphatase calcineurin) and increase profilin, which probably more than offsets in most instances the increased retrograde flow of F-actin due to enhanced myosin motor activity contingent on the increase in ROCK.

BDNF modulates GABAergic synaptic transmission (Bariohay et al. 2008) by either inhibiting spontaneous inhibitory postsynaptic currents (Ohbuchi et al. 2009) at some synapse or increasing it at others through a BDNF/TrkB/P13K/PKC signaling pathway (Porcher et al. 2011), these differences perhaps reflecting whether a potassium chloride cotransporter 2 is active or not (Huang et al. 2012).

4.3.1 BDNF/TrkB Changes in ERK-Mediated Control of Dendritic Spines

In the BLA, chronic restraint stress increases the arborization and length of dendrites of pyramidal neurons, with the opposite occurring for pyramidal neurons in the CA3 region of the hippocampus (Hill et al. 2011; Johnson et al. 2009; Vyas et al. 2002). In the central nucleus of the amygdala, there is no decrease, and perhaps an increase, in the dendrites of pyramidal neurons following restraint stress (Fig. 4.7) (Vyas et al. 2003).

4.3.1.1 Control of Dendritic Spines Through the ERK Pathway

Signaling pathways evoked by neurotrophins and their receptors involve a number of different second messengers to effect synaptic plasticity (Numakawa et al. 2013). Mammalian neurotrophins, BDNF, NGF, NT-3 and NT-4/5, stimulate intracellular signaling through binding to Trk receptors and a common receptor p75NTR. The mature neurotrophins bind to their specific Trks with high affinity, whereas the pro-neurotrophins interact with p75NTR with high affinity. Activation of Trks stimulates three main signaling cascades, (a) PLCgamma-MAPK/ERK, (b) phosphatidylinositol 13-kinase (P13K)/Akt and (c) MAP kinase Ras/ERK pathways, to exert effects on neurons (Fig. 4.8) (Santos et al. 2010).

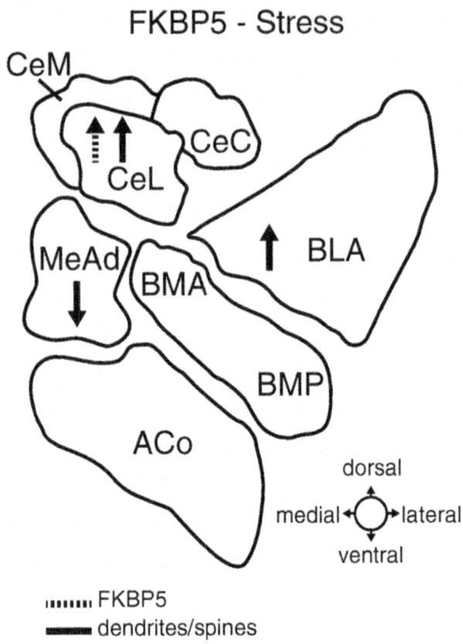

Fig. 4.7 Schematic of the effects of restraint and social stress on levels of the glucocorticoid receptor co-chaperone FKBP5 (see also Fig. 4.2) in different nuclei of the amygdala following stress together with the enhancement or decrease in dendrites and their spines. Abbreviations: *ACo* anterior cortical nucleus of the amygdala; *BLA* basolateral nucleus of the amygdala, anterior part; *BMA* basomedial nucleus of the amygdala, anterior part; *BMP* basomedial nucleus of the amygdala, posterior part; *CeC* central nucleus of the amygdala, capsular division; *CeL* central nucleus of the amygdala, lateral division; *CeM* central nucleus of the amygdala, medial division. Changes in dendrites and spines in BLA from Hill et al. (2011), Johnson et al. (2009) and Vyas et al. (2002). For CeL from Vyas et al. (2003). For the MeAd from Bennur et al. (2007). FKBP5 changes from Scharf et al. (2011). Diagram adopted from Fung et al. (2011)

Glucocorticoids reduce expression levels of proteins necessary for dendritic spines by impairing Shp2 association with TrkB which normally maintains ERK activation (Fig. 4.8) (Numakawa et al. 2013). Glucocorticoid receptor expression also determines TrkB-induced PLCgamma activation via interactions with TrkB (Fig. 4.8).

Synaptic spine proteins that are upregulated by BDNF are inhibited by GR agonists as a consequence of their suppressing ERK signaling through Shp2 and TrkB interaction (Numakawa et al. 2013). Normally BDNF increases ERK activation and synaptic protein expression, with this diminished by GR agonists. In addition, the association of GR with TrkB is important in controlling BDNF-stimulated PLCgamma signaling (Numakawa et al. 2013) (Fig. 4.8). Thus BDNF acting on TrkB receptors increases the density of apical dendrites of pyramidal neurons in the CA1 region of the hippocampus by activating the ERK signaling pathway (Alonso et al. 2004).

Stimulation of NMDA receptors at high frequency has two effects on dendritic spines. On the one hand, it causes GTPase activation leading to actin polymerization that is responsible for shortening of the spine neck and enlarging of the spine head. In

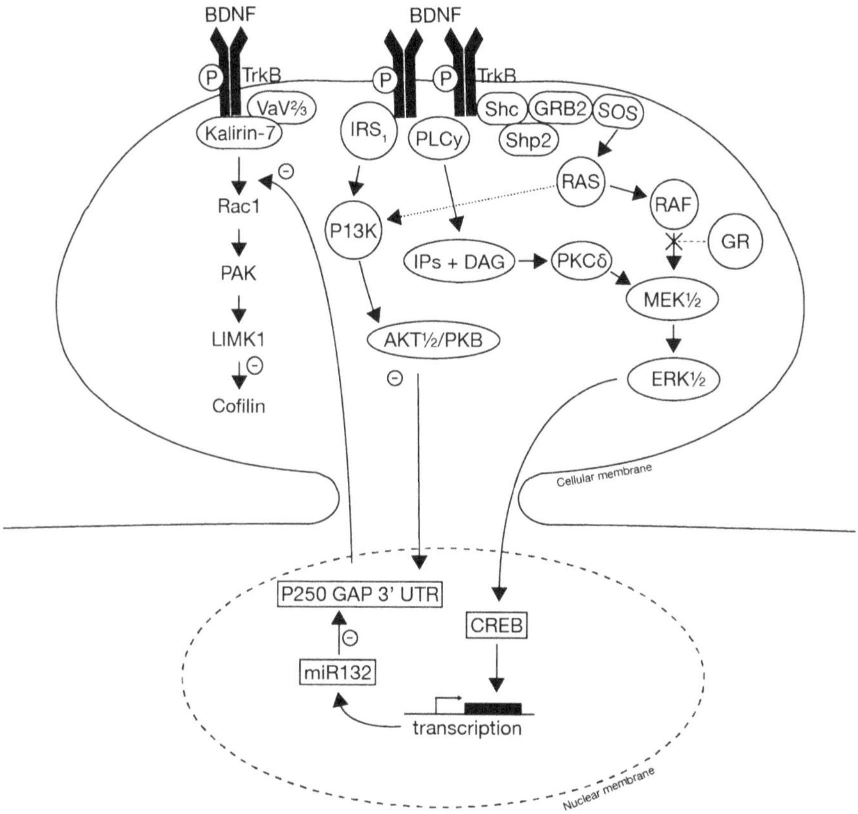

Fig. 4.8 Glucocorticoids reduce expression levels of synaptic protein. The association of TrkB–Shp2 (Shp2 interaction with TrkB shown to be indirect via binding to other adapter molecules) maintains ERK1/2 activation, while GRs negatively influence activated ERK1/2. ERK1/2-mediated expression of synaptic proteins such as NR2A and synapsin I following activation of CREB may therefore be downregulated by activation of GR (see Kumamaru et al. 2011; Numakawa et al. 2013)

parallel with this, the transcription of Arc/Arg 3.1 mRNA is accelerated, and it is transported into dendrites, where on activation of ERK it is dissociated from the dendritic microtubule-based transport system to associate with the polymerized actin in the dendritic spine (Huang et al. 2007).

4.3.2 BDNF/TrkB Changes in Small GTPase-Mediated Control of Dendritic Spines

4.3.2.1 RhoA, Rac1, Cde42 and Vav2/3

The Vav2/3 family of Rac/RhoA guanine nucleotide exchange factors play an important role in BDNF/TrkB signaling, inducing dendritic spine growth and functional synapse plasticity (Figs. 4.6 and 4.9) (Hale et al. 2011). Thus there is severe

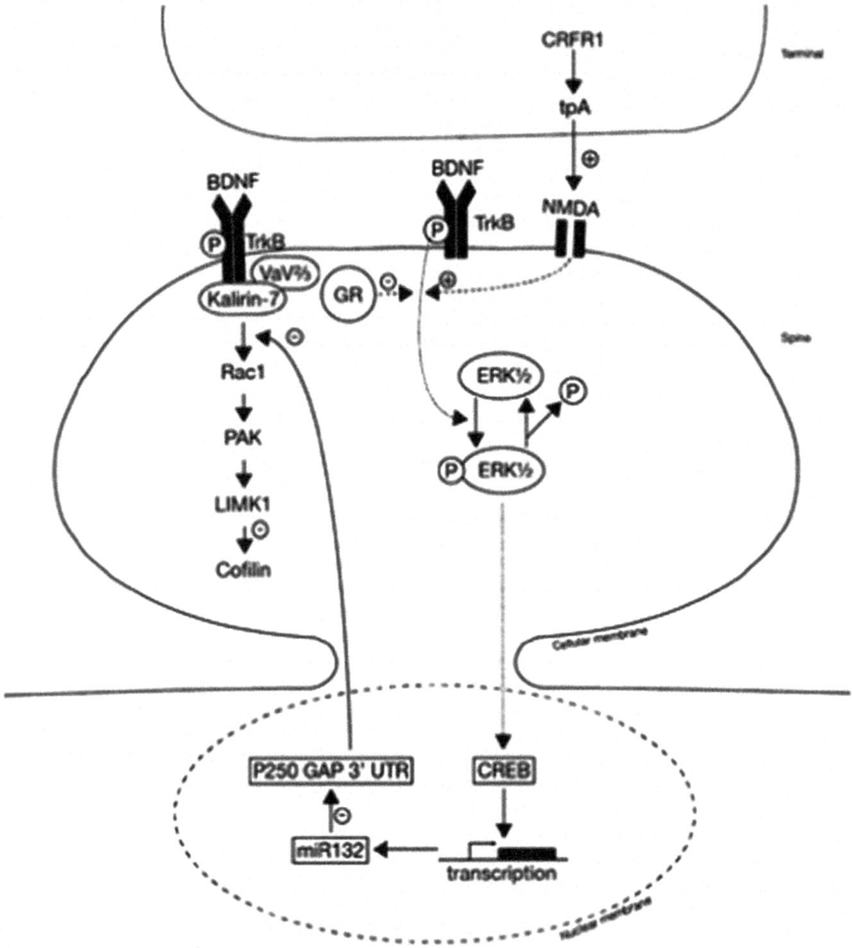

Fig. 4.9 The ERK1/2 pathway is important for the upregulation of miR132 during BDNF stimulation (see also Fig. 4.8). Pretreatment with GR agonists decreases BDNF-increased ERK1/2 activation and miR132 expression. Possible mechanisms the tpA may use to promote stress-induced plasticity are also shown here. Stress causes an upregulation of CRF, which through its CRFR1 receptor facilitates tpA release during depolarization. In order to facilitate neuronal plasticity, extracellular tpA may use plasminogen-dependent or plasminogen-independent mechanisms, such as NMDA receptor activation and activation of BDNF. The above events increase intracellular concentrations of calcium and cause phosphorylation of EKR1/2, activation of CREB and transcription of numerous genes involved in facilitation of stress-induced functional or structural plasticity. These events may mediate some of the behavioural signatures of stress, such as an increase in fear and anxiety (see Kawashima et al. 2010; Skrzypiec et al. 2008)

reduction of rapid dendritic spine growth on pyramidal neurons in the hippocampus of Vav-deficient mice (Hale et al. 2011) (Figs. 4.6 and 4.9).

RhoA/Rho kinase, Rac1 and Cde42 are involved in different processes of spine morphogenesis, motility and stability through their final action on actin and microtubule dynamics with subsequent cytoskeletal reorganization that determine spine growth. Rac1 and Cde42 are essential for spine head morphing, promoting spine head growth and stabilization. The initial growth of dendrites and of their spines is under control of Cde42, which accelerates the polymerization of actin at synapses (Shen et al. 2006), through activation of ADF/cofilin (Chen et al. 2006), whereas their late growth is controlled by Rac1 (Fig. 4.6) (Vadodaria et al. 2013). On the other hand RhoA/Rho kinases inhibit protrusive motility to convert dynamic filopodia into stable spines (Fig. 4.6) (Elia et al. 2006; Tashiro et al. 2000; Tashiro and Yuste 2004, 2008). Kalirin-7, a Rho GEF, is essential for activation of Rac1, RhoA and RhoG which it does via its two Rho GEF domains (Rabiner et al. 2005) (Fig. 4.6). Activation of *N*-methyl-D-aspartate receptor (NMDAR) leads to inactivation of Rho/Rho-kinase signaling, so contributing to dendritic arbour growth by allowing the extension of dendrites (Govek et al. 2005).

The activated downstream target proteins for RhoA, Cde42 and Rac1 are, respectively, ROCK, N-WASP (neuronal Wiskott–Aldrich syndrome protein) and PAK p21-activated kinase (Govek et al. 2005) (Fig. 4.6). However, these downstream target proteins are almost certainly controlled by other Rho GTPases than those mentioned so far, for there are at present 19 identified Rho GTPase proteins that have not been mentioned above in the context of actin dynamics in dendritic spines, namely, RhoA, Rac and Cde42, whereas at least Rif, RhoD, Wrch1 and TC10 should also be considered (Fig. 4.6).

4.3.3 BDNF/TrkB Changes in mRNA Modulation of ERK-Mediated Control of Dendritic Spines

BDNF acting on presynaptic TrkB receptors induces rapid effects on glutamate release (Carvalho et al. 2008; D'Amore et al. 2013; Merighi et al. 2008) through the TrkB-Src-PLC signaling pathway (Zhang et al. 2013). However, the principal role of BDNF/TrkB changes appears to be on the dendritic spines, and this is now considered.

4.3.3.1 microRNA

mRNA also regulate spine growth for Rac1 itself and are under the negative control of p250GAP, the translation of which is also under the negative control of the mRNA miR132 (Fig. 4.9). BDNF upregulates miR132 but not mRNA 9, 16, 124, 128a, 128b, 134 or 138. Transfection of exogenous miR132 upregulates glutamate receptors NR2A, NR2B and GluR1 via the MAPK/ERK pathway (Fig. 4.9) (Kawashima et al. 2010). The use of antisense RNA to inhibit miR132 function decreases the BDNF-dependent increase in these glutamate receptors. miR132, under the control of cAMP response element-binding (CREB) protein, inhibits p250GAP and thereby enhances the growth of dendritic spines (Wayman et al. 2008).

4.3.4 BDNF/TrkB Changes to TRCP3 Channel-Mediated Control of Dendritic Spines

The activity-dependent release of BDNF from mossy fibres leads to activation of the transient receptor potential canonical 3 (TRCP3) channels in hippocampal neurons, as a consequence of activation of TrkB receptors, with subsequent changes in the intracellular calcium concentrations (Li et al. 2010). The pathway between the activation of TrkB receptors by BDNF and TRPC3 channels requires phospholipase C, diacylglycerol and IP3 receptors leading to increase in dendritic spine density (Amaral and Pozzo-Miller 2007; Li et al. 1999).

4.4 BDNF/TrkB Control of Dendritic Spines Modulated by Glucocorticoid and Mineralocorticoid Receptors

4.4.1 Glucocorticoid and Mineralocorticoid Modulation of ERK and GTPase Pathways for Control of Dendritic Spines: The GR-ERK-BDNF-Synaptic Proteins

An increase in synaptic proteins in developing cortical neurons due to BDNF is reduced by dexamethasone via suppression of ERK signaling through src homology-2 domain-containing phosphatase 2 (Shp2) and TrkB interaction (Fig. 4.8) (Kumamaru et al. 2008, 2011; Numakawa et al. 2010, 2013). Corticosterone administered for 1 h increases spine density of hippocampal pyramidal neurons, whereas GR antagonists and activation blockade of kinases PKA, PKC, P13K and ERK decrease the corticosterone-dependent spine development (Komatsuzaki et al. 2012). These effects after only a relatively short period of administration indicate that surface membrane GR are likely to be involved rather than cytoplasmic GR that have the opposite effect on spine density.

4.5 BDNF/TrkB Control of Dendritic Spines Modulated by Corticotropin-Releasing Factor (Hormone)

As described above, a canonical role for the glucocorticoids in stress-induced changes of dendritic spines is now well established. However, recent studies using forebrain-specific knockout of CRFR1 suggest that, even in the presence of a functional glucocorticoid system, the absence of CRFR attenuates the detrimental effects of chronic stress on dendritic arborization and the spines (Chen et al. 2012b). It appears therefore that CRF, besides GR, have a role to play in determining dendritic spine growth and regression. This is supported by the observations that stress in early life leads to overexpression of CRF in the hippocampus, which is accompanied by a decrease in dendritic spine density in CA3 (but not CA1) and interruption of learning and memory, whereas if there is a decrease in the expression of CRF, then dendritic spine density is restored (Wang et al. 2011). However, these effects of CRF do not hold for the whole brain as short periods of restraint stress, leading to increased CRF in the cerebellum, are accompanied by an increase rather than a decrease in the density of dendritic spines there through a mechanism in which the elevated CRF acts on CRFR1 to decrease RhoA and hence elevate the growth of spines (Gounko et al. 2013).

4.5.1 Corticotropin-Releasing Factor (Hormone) Modulation of Protein Lipase C (PLC)/Small GTPase Pathway for Control of Dendritic Spines

Considered here is the NMDA receptor control of F-actin cytoskeleton in the synaptic spine and the modulation of NMDA by CRFRs and by GR. The synaptic spine head grows as a consequence of the formation and elongation of actin filaments, namely, F-actin that pushes against the spine head plasmalemma. F-actin grows by incorporating monomeric actin subunits (G-actin) with bound ATP at the so-called barbed end of the F-actin. Older actin subunits are removed from the pointed end of F-actin in a treadmilling process. The older subunits are identified as a consequence of a process in which ATP-actin slowly hydrolyzes to ADP-actin. An F-actin severing protein, ADF/cofilin, has high affinity for ADP-actin, so promoting disassembly of the older pointed ends of the actin filaments. Profilin catalyses the exchange of ADP for ATP on these ADP-actin monomers. In this way ATP-actin monomers are made available for addition to the barbed ends of F-actin, so completing the treadmilling process. Filament growth is blocked by capping their barbed ends, for instance, with gelsolin. The filament branching network is enabled by arp2/3, which both caps the pointed end of F-actin and acts as a template for the assembly of new filaments. This process of spine growth has been described in detail and modelled (Bennett et al. 2011).

Synaptic spine growth and regression are under direct control of the NMDA receptors in the spine head so that the spines do not grow if NMDA receptors are knocked out (Alvarez et al. 2007). The first consequence of NMDA receptor activation involves its interaction with the cytoskeletal-associated protein postsynaptic density-95 (PSD-95), mediated by the binding of conserved motifs in the carboxy-terminal tails of the NR2 subunit of NMDA with the PDZ domains in PSD-95, which recruits Rho GTPases and protein kinases (the PDZ domain is a common structural domain of approximately 80 amino acids found in signaling proteins) (Pollard et al. 2000). The Rho GTPases then interact with proteins of the WASP family that activate the Arp2/3 complex. This leads to nucleation of actin filaments from a pool of F-actin bound to actin-regulatory molecules, such as profilin. Such filaments grow at a fixed 70-degree angle from the side of pre-existing actin filaments. As these grow at their barbed ends, they push the membrane forward. The new filaments are capped rapidly. Arp2/3 complexes are incorporated into the filament so that new Arp2/3 complex must be continually supplied if the network of filaments is to keep growing. A second consequence of NMDA receptor activation involves an increase in intracellular calcium through an influx of calcium ions, leading to modulation of calcium/calmodulin-dependent protein kinases (CaMKs) (Ciani and Salinas 2008).

A calcium/calmodulin-dependent kinase kinase complexed with calcium/calmodulin-dependent kinase I (the CaMKII/CaMKI complex) then phosphorylates a guanine nucleotide exchange factor, bPIX, leading to Rac/PAK/LIMK1 activation that is responsible for controlling the extent of the pool of unpolymerized monomers bound to profilin through the action of cofilin (Saneyoshi et al. 2008). In this way CaMK regulates the pool of ATP-actin necessary for spine enlargement. Spines are maintained if a stable F-actin pool is retained at the base of the spine head, whereas spines regress if this pool of F-actin is released into the dendritic shaft (Honkura et al. 2008). If spines regress there is elevated phagocytosis involving microglia, but this activity is not a cause of the loss of spines (Schiefer et al. 1999). Regulation of NMDA receptors through activation of GRs and CRFRs may then have direct consequences for spine growth and regression through both the CaMKK and the PSD95/Rho GTPase pathways.

Acute cortisone application (30 min) to GR2 on the plasma membrane, or only intracellular to the membrane (Johnson et al. 2005), decreases calcium influx through NMDA receptor channels of CA1 hippocampal pyramidal neurons (Sato et al. 2004), probably through a protein kinase-mediated phosphorylation of the NR2 subunit of NMDA (Liu et al. 2007). Acute CRF acting on CRFR2 also decreases calcium currents through the NMDA channels on amygdala neurons (Liu et al. 2004). CRF acts on CRFR2 to activate protein lipase C that potentiates the release of calcium through IP3 channels of internal stores and protein kinase C, leading to phosphorylation of the NMDA channel and depression of its activity (Riegel and Williams 2008; Sheng et al. 2008). Calcium through the NMDA channel activates calcium–calmodulin kinase and thereby the cascade of events that results in the upregulation of a Rho GTPase (Rac1) and hence cofilin (Nakayama et al. 2000; Swinny and Valentino 2006). Cofilin severs and depolymerizes ADP-actin filaments in the synaptic spine, thereby enlarging the pool of F-actin. If this pool is released into the dendritic shaft, then the spines retract with eventual loss of the dendrite (Brunson et al. 2001; Chen et al. 2008). Blocking CRFRs can block spine removal and stabilize them (Chen et al. 2004).

4.5.2 Corticotropin-Releasing Factor (Hormone) Modulation of Tissue Plasminogen Activator (tPA) Pathway for Control of Dendritic Spines

Although pro-BDNF binds specifically to P75 and primarily promotes cell death and long-term depression, mature BDNF binds more readily to TrkB, promoting long-term potentiation and cell survival. Therefore, the balance between pro-BDNF and

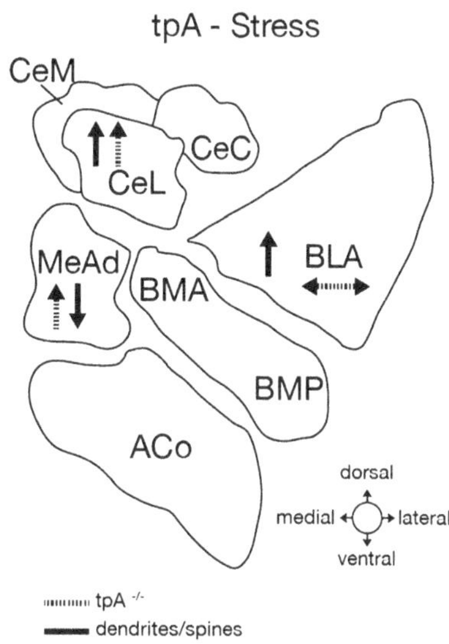

Fig. 4.10 Schematic of the effects of stress on tissue plasminogen activator (tpA) in the amygdala. Definition of the nuclei given in the legend to Fig. 4.7. Changes in dendrites and spines as in Fig. 4.7. The tpA changes are from Pawlak et al. (2003)

mature BDNF, on the one hand, and the P75 and TrkB receptors, on the other hand, is of critical importance in determining the functional characteristics of the BDNF signal. There is a positive feedback loop between BDNF and tissue plasminogen activator (tpA), as BDNF increases tpA gene transcription with the tpA so produced then cleaving the pro-form of BDNF to BDNF, by activating the extracellular protease plasmin (Daniel et al. 2007; Pang et al. 2004). Thus there is an increase in BDNF protein in the hippocampus in tpA overexpressed mice (Bahi and Dreyer 2012). tpA then plays a central role in determining the balance between pro-BDNF and mature BDNF in the synaptic cleft and is critical to the functional outcome of this BDNF signal (Fig. 4.9). tpA is present in the hippocampus, amygdala, hypothalamus and cerebellum (Skrzypiec et al. 2008). It is released from neurons upon their excitation, a process that is upregulated by CRF acting on CRFR1 (Fig. 4.9) (Matys et al. 2004). Stress causes an upregulation of CRF acting on CRFR1 that then facilitates tpA release during depolarization in the medial amygdala (Pawlak et al. 2003) (Fig. 4.10). Prenatal stress is accompanied by a decrease in the activity of the expression of tpA leading to increases in pro-BDNF and decreases in the mature form of BDNF (Yeh et al. 2012).

4.5.2.1 tpA and Dendritic Spines

tpA mediates stress-induced structural and functional rearrangements in the limbic system. In tpA-deficient mice, there are few biochemical, structural and behavioural signatures normally associated with stress (Bennur et al. 2007). In wild-type mice, chronic stress causes significant reductions in medial amygdala (MeAd) spine density and in CA1 hippocampus but enhances spine density in the neighbouring basolateral amygdala (BLA) (Bennur et al. 2007) (Fig. 4.10). tPA−/− mice exhibit significant attenuation of stress-induced spine regression in the MeAd and in CA1 hippocampus, but BLA spinogenesis is not affected (Bennur et al. 2007) (Fig. 4.10).

tpA may use plasminogen-dependent or independent mechanisms, such as, activation of NMDA receptors. If tpA activates NMDA receptors or plasminogen, then there is an increase in intracellular calcium, upregulation of phosphorylation of ERK1/2, activation of CREB and transcription of genes (Fig. 4.9) (Skrzypiec et al. 2008). Changes in spine regression in tPA−/− mice may be due to the fact that ERK1/2 is not phosphorylated in these mice (Bennur et al. 2007; Skrzypiec et al. 2008).

4.5.2.2 CRF Control of Spine Formation and Regression

CRF receptors are found in the limbic system, and in the central nucleus of the amygdala (CeA) as well as having been identified in dendrites and dendritic spines. Thirty-five percent of CRFR-labelled dendrites contain CRF immunoreactivity in the cytoplasm (Treweek et al. 2009). CRF is released at synapses during stress with severe stress giving rise to CRF levels associated with spine retraction through actin cytoskeletal collapse, following signaling through the actin-regulating small GTPase RhoA, downstream of the CRFR1 that reside within the spine head (Chen et al. 2012b, 2013). On the other hand, in the locus coeruleus, CRF peptides regulate dendritic arborization by upregulating Rac1 and downregulating RhoA (Swinny and Valentino 2006). In the cerebellum, CRF downregulates RhoA (Chen et al. 2012b).

CRF leads to changes in PKA and hence MAPK–Rac1 that regulates actin and the growth of dendritic spine of neurons in the locus coeruleus (Swinny and Valentino 2006). Application of CRF (and urocortin) to cerebellar slice cultures increases the density of spines on Purkinje cells via CRFR1 and CRFR2 leading to RhoA downregulation and spine growth. After 1 h of restraint stress, there is an increase in Purkinje cell spine density in the cerebellum, which is in contrast to chronic stress that decreases spine numbers in hippocampus and prefrontal cortex yet increases spine density in the BLA (Magarinos et al. 2011).

Mice can be separated on behavioural grounds into ‘PTSD-like’ and ‘resilient’ phenotypes (Lebow et al. 2012). PTSD-like phenotypes have attenuated corticosterone levels and upregulation of BNST (bed nucleus of the stria terminals)—CRF receptor 2 mRNA levels and BNST de-acetylation enzyme HDAC5. BDNF-CRF receptor 2-specific lentiviral knockdown reduces susceptibility to PTSD-like behaviour.

4.6 Cannabinoid (Receptor CB1) Modulation of BDNF Gene Transcription and BDNF/TrkB Control of Dendritic Spines

Endocannabinoids (eCBs) modulate the levels of BDNF in certain parts of the brain. Thus endocannabinoid receptor 1 (eCBR1) knockout mice have a decreased BDNF level in the hippocampus but not in the amygdala or neocortex. This may be specific to BDNF as neither nerve growth factor nor neurotrophic factor-3 is affected in these knockout mice (Aso et al. 2008). The principal cannabinoids at synapses are anandamide (AEA) and 2-arachidonoylglycerol. Metabolism of AEA is by fatty acid amide hydrolase (FAAH) (Taylor et al. 2011).

Normally there is a negative control exerted by eCBRs on the release of transmitter that include glutamate, GABA and catecholamines consequent on the release of cannabinoids from the dendritic spines (Haring et al. 2012). The effect of chronic stress on the amygdala is to increase FAAH resulting in a decrease of AEA with the consequences that there is an increase of glutamate release leading to activation of NMDA receptors and enhancement of dendritic arborization and spine density (Hill et al. 2013; Taylor et al. 2011). One of the effects of chronic stress in the hippocampus is to decrease eCBRs on GABAergic neurons, so impairing the normal negative control by these receptors of GABA release (Haring et al. 2012). Indeed even relatively mild stress can decrease eCBR1 by significant amounts (about 50%) in the hippocampus (Reich et al. 2009).

4.7 Serotonin Transporter (SERT) Modulation of BDNF Gene Transcription and BDNF/TrkB Control of Dendritic Spines

Modification of serotonin reuptake transport, with inhibitors such as fluoxetine, augments BDNF exon I mRNA levels in the BLA as well as in the hippocampus (Karpova et al. 2011). This augmentation is lost and replaced by a decrease in BDNF levels if the mice are homozygous for the BDNF Val66Met SNP (Bath et al. 2012).

A better outcome is obtained for erasing fear memories in PTSD subjects than using D-cycloserine if a combination is used of extinction training with chronic fluoxetine treatment that augments BDNF exon I mRNA (Karpova et al. 2011).

4.8 Conclusion

The following points are suggested by the present review on identifying the changes in dendritic spine synapses in neural networks under stress, the mechanisms that drive these and how these networks can be reinstated to normality.

4.8.1 Dendritic Spines and BDNF

Activation of BDNF leads to the sprouting of dendrites in many areas of the brain, such as CA1 in the hippocampus (Grande et al. 2010; Tyler and Pozzo-Miller 2003). As glucocorticoids decrease BDNF expression, they decrease dendritic spine density in these areas (Rothman and Mattson 2013). Thus activation of both GR and MR with corticosterone leads to an increase in dendritic spine turnover on pyramidal neurons in these areas (Liston and Gan 2011). In other areas of the brain, glucocorticoids do not have this effect on dendritic spines. Thus stress, normally accompanied by elevation of glucocorticoids in the BLA, leads to an increase in the dendrites and their spines on pyramidal neurons there as well as of BDNF in these neurons, and both are still present some 21 days after relief from the stress. A single restraint stress event also leads to increases in spine density that is maintained for at least 10 days in the BLA together with upregulation of BDNF mRNA (Lakshminarasimhan and Chattarji 2012; Oztan et al. 2011) (Fig. 4.10). These correlations between the changes in spine growth and density on the one hand and of BDNF levels on the other, up or down, suggest that the principal factor in determining spines in neural networks is the synthesis of BDNF (Vigers et al. 2012). If this is the case then that is where research should be concentrated.

4.8.2 BDNF, Glucocorticoid and Mineralocorticoid Receptors

Although it is argued above that changes in the expression of BDNF genes are likely to be the principal means of controlling dendritic spine growth and density, the question arises as to whether differential distributions of GR and MR might at least help explain why stress changes spine growth up in some regions but down in others (Figs. 4.7 and 4.10). It has been noted that activating the GR suppresses the action of BDNF through interruption of a number of pathways determining gene transcriptions that give products that modify, for example, actin polymerization in dendritic spines (see Figs. 4.8 and 4.9). One of these pathways is the mitogen-activated protein kinase/extracellular signal-regulated kinase ½ (MAPK/ERK). Another is through phospholipase C-gamma (PLCgamma) (Kawashima et al. 2010), and another is blocking the interaction between TrkB and src homology2 domains containing tyrosine phosphatase (Shp2) (Fig. 4.8). There is yet insufficient evidence to claim that these pathways are different from one neuron type or brain area to another that would support a claim of either changing the effects of BDNF on dendritic spines via the action of GR or of GR differently changing the expression of the BDNF genes.

MR are by far at the highest density in the hippocampus, whereas GRs are found at about the same density throughout the brain (Reul et al. 2000). Although activation of GR in most brain regions, like the hippocampus, reduces BDNF levels, activation of MR increases them (Adzic et al. 2009). Here then is a basis for differentiating brain regions like the BLA and CeM, in which stress gives rise to an increase in dendritic spines, from the MeAd, the CA1 hippocampus and the prefrontal cortex where spine density and growth decrease following stress as they do following corticosterone (Gourley et al. 2012) (Fig. 4.7). This hormone in the rat and cortisol in humans are released in a pulsatile pattern that defines the normal circadian rhythm. The frequency of the pulses is increased under states of chronic stress. Alterations in the pattern of glucocorticoid rhythms have different effects on MR and GR binding to DNA and offer a mechanism for tissue-specific responses to altered glucocorticoid dynamics. At the nadir levels of corticosterone, there is only MR binding to DNA, but at peak stress levels, there is both GR and MR binding. BDNF mRNA is enhanced in the frontal lobes but unaffected in the hippocampus in mice heterozygous for the GR. Such mice, following an episode of acute stress, also show no sign of changes in the cleaving of pro-BDNF to mature BDNF in the hippocampus, as normally occurs in wild-type mice (Molteni et al. 2010). Whether these differences could explain the differential spine growth and regression in different regions is not known.

4.8.3 Post-traumatic Stress Disorder

PTSD is characterized by the failure to restrict the memory of the fearful event to the environment in which the event was experienced. In males that have experienced trauma in combat, 62–72% of the variance in PTSD can be attributed to genetic influences and the remainder to quite specific environmental factors (Koenen et al. 2008; Sartor et al. 2011). How do these gene x environment mechanisms operate? The formation of fearful memories in the hippocampus involves GR activating the CaMKIIalpha–CREB-dependent pathway, which involves activation of GRs and NMDARs leading to epigenetic changes in the expression of BDNF (Chen et al. 2012a) (Fig. 4.4). Fearful memories triggered by environmental experiences largely unrelated to the original fearful event are associated with polymorphisms in the PKCgamma gene (de Quervain et al. 2012).

Extinction of a fear memory, such as of the negative effects of opiate withdrawal, involves increases of BDNF mRNA and protein in the ventromedial prefrontal cortex, through the action of CREB at histone H3 of the BDNF exon I transcript promoter with acetylation of the histone. This could be enhanced before extinction training with histone deacetylase inhibitors such as trichostatin A or inhibitors such as U0126 of ERK (Fig. 4.4) (Wang et al. 2012).

Major risk factors for PTSD are low levels of cortisol in the blood immediately after the trauma occasion and, before the trauma, in peripheral blood mononuclear cells, the presence of high GR numbers, low FKBP5 expression and high levels of GILZ mRNA (Fig. 4.11) (Aardal-Eriksson et al. 1998; van Zuiden et al. 2013). All of these risk factors are involved in the action of cytoplasmic GR in modulating gene transduction, including most likely that for the BDNF gene, as well as regulating the capacity for BDNF itself to act (Figs. 4.2, 4.6, 4.8 and 4.11). This emphasis on GR in PTSD is enforced by the observations that there is an association between two polymorphisms in the GR gene (N363S and BclI) and PTSD (Bachmann et al. 2005; Hauer et al. 2011) as there is between that of FKBP5 and GILZ on the one hand (Figs. 4.2 and 4.11) and the capacity of GR to modulate gene function on the other (Fig. 4.4) (Gillespie et al. 2009).

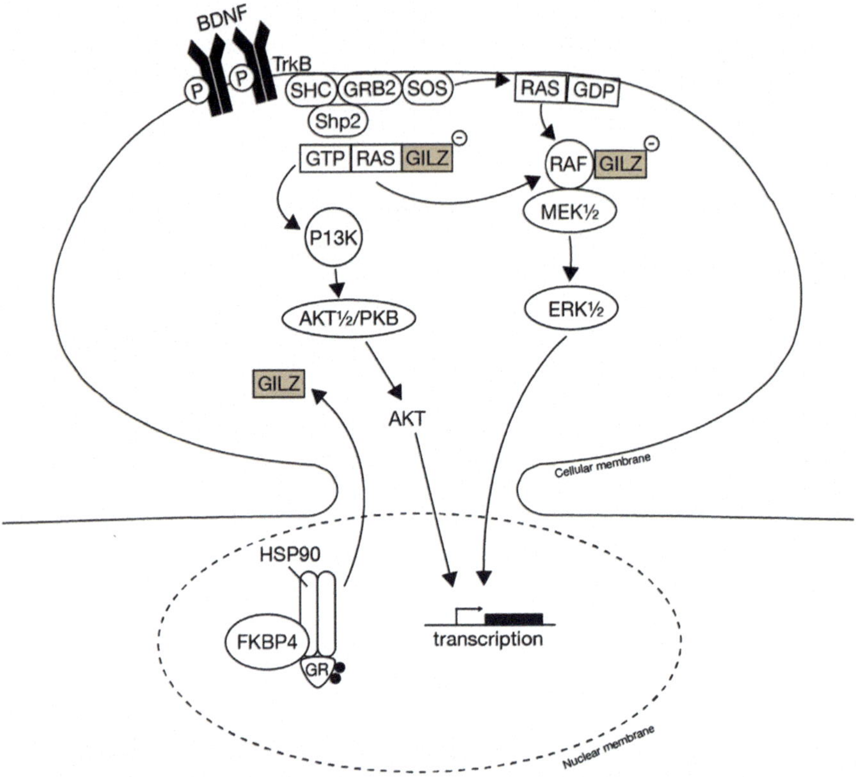

Fig. 4.11 Schematic illustration of the role of GILZ as a mediator of GR activity. The SHC/GRB2/SOS complex converts RAS into its active GTP-bound form leading to the activation of ERK1/2 and AKT1/2/PKB pathways. GILZ, induced by GR, directly interacts with RAS and RAF to determine the activation of ERK1/2 and AKT1/2/PKB pathways, leading to inhibition of transcription normally stimulated by BDNF/TrkB (from Bennett and Lagopoulos 2014)

References

Aardal-Eriksson E, Karlberg BE, Holm AC (1998) Salivary cortisol—an alternative to serum cortisol determinations in dynamic function tests. Clin Chem Lab Med 36(4):215–222. https://doi.org/10.1515/CCLM.1998.037

Abuhatzira L, Makedonski K, Kaufman Y, Razin A, Shemer R (2007) MeCP2 deficiency in the brain decreases BDNF levels by REST/CoREST-mediated repression and increases TRKB production. Epigenetics 2(4):214–222

Adzic M, Djordjevic J, Djordjevic A, Niciforovic A, Demonacos C, Radojcic M, Krstic-Demonacos M (2009) Acute or chronic stress induce cell compartment-specific phosphorylation of glucocorticoid receptor and alter its transcriptional activity in Wistar rat brain. J Endocrinol 202(1):87–97. https://doi.org/10.1677/JOE-08-0509

Aid T, Kazantseva A, Piirsoo M, Palm K, Timmusk T (2007) Mouse and rat BDNF gene structure and expression revisited. J Neurosci Res 85(3):525–535. https://doi.org/10.1002/jnr.21139

Alonso M, Medina JH, Pozzo-Miller L (2004) ERK1/2 activation is necessary for BDNF to increase dendritic spine density in hippocampal CA1 pyramidal neurons. Learn Mem 11(2):172–178. https://doi.org/10.1101/lm.67804

Alvarez VA, Ridenour DA, Sabatini BL (2007) Distinct structural and ionotropic roles of NMDA receptors in controlling spine and synapse stability. J Neurosci 27(28):7365–7376. https://doi.org/10.1523/JNEUROSCI.0956-07.2007

Amaral MD, Pozzo-Miller L (2007) TRPC3 channels are necessary for brain-derived neurotrophic factor to activate a nonselective cationic current and to induce dendritic spine formation. J Neurosci 27(19):5179–5189. https://doi.org/10.1523/JNEUROSCI.5499-06.2007

An JJ, Gharami K, Liao GY, Woo NH, Lau AG, Vanevski F et al (2008) Distinct role of long 3′ UTR BDNF mRNA in spine morphology and synaptic plasticity in hippocampal neurons. Cell 134(1):175–187. https://doi.org/10.1016/j.cell.2008.05.045

Aso E, Ozaita A, Valdizan EM, Ledent C, Pazos A, Maldonado R, Valverde O (2008) BDNF impairment in the hippocampus is related to enhanced despair behavior in CB1 knockout mice. J Neurochem 105(2):565–572. https://doi.org/10.1111/j.1471-4159.2007.05149.x

Attwood BK, Bourgognon JM, Patel S, Mucha M, Schiavon E, Skrzypiec AE et al (2011) Neuropsin cleaves EphB2 in the amygdala to control anxiety. Nature 473(7347):372–375. https://doi.org/10.1038/nature09938

Bachmann AW, Sedgley TL, Jackson RV, Gibson JN, Young RM, Torpy DJ (2005) Glucocorticoid receptor polymorphisms and post-traumatic stress disorder. Psychoneuroendocrinology 30 (3):297–306. https://doi.org/10.1016/j.psyneuen.2004.08.006

Bahi A, Dreyer JL (2012) Hippocampus-specific deletion of tissue plasminogen activator "tPA" in adult mice impairs depression- and anxiety-like behaviors. Eur Neuropsychopharmacol 22 (9):672–682. https://doi.org/10.1016/j.euroneuro.2012.01.008

Bariohay B, Tardivel C, Pio J, Jean A, Felix B (2008) BDNF-TrkB signaling interacts with the GABAergic system to inhibit rhythmic swallowing in the rat. Am J Physiol Regul Integr Comp Physiol 295(4):R1050–R1059. https://doi.org/10.1152/ajpregu.90407.2008

Bath KG, Jing DQ, Dincheva I, Neeb CC, Pattwell SS, Chao MV et al (2012) BDNF Val66Met impairs fluoxetine-induced enhancement of adult hippocampus plasticity. Neuropsychopharmacology 37(5):1297–1304. https://doi.org/10.1038/npp.2011.318

Bennett M (2009) Positive and negative symptoms in schizophrenia: the NMDA receptor hypofunction hypothesis, neuregulin/ErbB4 and synapse regression. Aust N Z J Psychiatry 43 (8):711–721. https://doi.org/10.1080/00048670903001943

Bennett MR, Farnell L, Gibson WG (2011) A model of NMDA receptor control of F-actin treadmilling in synaptic spines and their growth. Bull Math Biol 73(9):2109–2131. https://doi.org/10.1007/s11538-010-9614-4

Bennett MR, Lagopoulos J (2014) Stress and trauma: BDNF control of dendritic-spine formation and regression. Prog Neurobiol 112:80–99. https://doi.org/10.1016/j.pneurobio.2013.10.005

Bennur S, Shankaranarayana Rao BS, Pawlak R, Strickland S, McEwen BS, Chattarji S (2007) Stress-induced spine loss in the medial amygdala is mediated by tissue-plasminogen activator. Neuroscience 144(1):8–16. https://doi.org/10.1016/j.neuroscience.2006.08.075

Binder EB (2009) The role of FKBP5, a co-chaperone of the glucocorticoid receptor in the pathogenesis and therapy of affective and anxiety disorders. Psychoneuroendocrinology 34 (Suppl 1):S186–S195. https://doi.org/10.1016/j.psyneuen.2009.05.021

Binder EB, Bradley RG, Liu W, Epstein MP, Deveau TC, Mercer KB et al (2008) Association of FKBP5 polymorphisms and childhood abuse with risk of posttraumatic stress disorder symptoms in adults. JAMA 299(11):1291–1305. https://doi.org/10.1001/jama.299.11.1291

Bishop JF, Mueller GP, Mouradian MM (1994) Alternate 5′ exons in the rat brain-derived neurotrophic factor gene: differential patterns of expression across brain regions. Brain Res Mol Brain Res 26(1–2):225–232

Boscarino JA, Erlich PM, Hoffman SN, Rukstalis M, Stewart WF (2011) Association of FKBP5, COMT and CHRNA5 polymorphisms with PTSD among outpatients at risk for PTSD. Psychiatry Res 188(1):173–174. https://doi.org/10.1016/j.psychres.2011.03.002

Boulle F, van den Hove DL, Jakob SB, Rutten BP, Hamon M, van Os J et al (2012) Epigenetic regulation of the BDNF gene: implications for psychiatric disorders. Mol Psychiatry 17 (6):584–596. https://doi.org/10.1038/mp.2011.107

Bramham CR, Wells DG (2007) Dendritic mRNA: transport, translation and function. Nat Rev Neurosci 8(10):776–789. https://doi.org/10.1038/nrn2150

Branchi I, Karpova NN, D'Andrea I, Castren E, Alleva E (2011) Epigenetic modifications induced by early enrichment are associated with changes in timing of induction of BDNF expression. Neurosci Lett 495(3):168–172. https://doi.org/10.1016/j.neulet.2011.03.038

Bredy TW, Wu H, Crego C, Zellhoefer J, Sun YE, Barad M (2007) Histone modifications around individual BDNF gene promoters in prefrontal cortex are associated with extinction of conditioned fear. Learn Mem 14(4):268–276. https://doi.org/10.1101/lm.500907

Brunson KL, Khan N, Eghbal-Ahmadi M, Baram TZ (2001) Corticotropin (ACTH) acts directly on amygdala neurons to down-regulate corticotropin-releasing hormone gene expression. Ann Neurol 49(3):304–312

Carvalho AL, Caldeira MV, Santos SD, Duarte CB (2008) Role of the brain-derived neurotrophic factor at glutamatergic synapses. Br J Pharmacol 153(Suppl 1):S310–S324. https://doi.org/10.1038/sj.bjp.0707509

Chen Y, Bender RA, Brunson KL, Pomper JK, Grigoriadis DE, Wurst W, Baram TZ (2004) Modulation of dendritic differentiation by corticotropin-releasing factor in the developing hippocampus. Proc Natl Acad Sci USA 101(44):15782–15787. https://doi.org/10.1073/pnas.0403975101

Chen TJ, Gehler S, Shaw AE, Bamburg JR, Letourneau PC (2006) Cdc42 participates in the regulation of ADF/cofilin and retinal growth cone filopodia by brain derived neurotrophic factor. J Neurobiol 66(2):103–114. https://doi.org/10.1002/neu.20204

Chen Y, Dube CM, Rice CJ, Baram TZ (2008) Rapid loss of dendritic spines after stress involves derangement of spine dynamics by corticotropin-releasing hormone. J Neurosci 28 (11):2903–2911. https://doi.org/10.1523/JNEUROSCI.0225-08.2008

Chen DY, Bambah-Mukku D, Pollonini G, Alberini CM (2012a) Glucocorticoid receptors recruit the CaMKIIalpha-BDNF-CREB pathways to mediate memory consolidation. Nat Neurosci 15 (12):1707–1714. https://doi.org/10.1038/nn.3266

Chen X, Li Y, Li S, Kirouac GJ (2012b) Early fear as a predictor of avoidance in a rat model of post-traumatic stress disorder. Behav Brain Res 226(1):112–117. https://doi.org/10.1016/j.bbr.2011.09.004

Chen Y, Kramar EA, Chen LY, Babayan AH, Andres AL, Gall CM et al (2013) Impairment of synaptic plasticity by the stress mediator CRH involves selective destruction of thin dendritic spines via RhoA signaling. Mol Psychiatry 18(4):485–496. https://doi.org/10.1038/mp.2012.17

Chiba S, Numakawa T, Ninomiya M, Richards MC, Wakabayashi C, Kunugi H (2012) Chronic restraint stress causes anxiety- and depression-like behaviors, downregulates glucocorticoid receptor expression, and attenuates glutamate release induced by brain-derived neurotrophic factor in the prefrontal cortex. Prog Neuro-Psychopharmacol Biol Psychiatry 39(1):112–119. https://doi.org/10.1016/j.pnpbp.2012.05.018

Ciani L, Salinas PC (2008) From neuronal activity to the actin cytoskeleton: a role for CaMKKs and betaPIX in spine morphogenesis. Neuron 57(1):3–4. https://doi.org/10.1016/j.neuron.2007.12.017

D'Amore DE, Tracy BA, Parikh V (2013) Exogenous BDNF facilitates strategy set-shifting by modulating glutamate dynamics in the dorsal striatum. Neuropharmacology 75:312–323. https://doi.org/10.1016/j.neuropharm.2013.07.033

Daniel PB, Lux W, Samson AL, Schleuning WD, Niego B, Weiss TW et al (2007) Two conserved regions within the tissue-type plasminogen activator gene promoter mediate regulation by brain-derived neurotrophic factor. FEBS J 274(9):2411–2423. https://doi.org/10.1111/j.1742-4658.2007.05777.x

de Quervain DJ, Kolassa IT, Ackermann S, Aerni A, Boesiger P, Demougin P et al (2012) PKCalpha is genetically linked to memory capacity in healthy subjects and to risk for posttraumatic stress disorder in genocide survivors. Proc Natl Acad Sci USA 109(22):8746–8751. https://doi.org/10.1073/pnas.1200857109

Djordjevic A, Adzic M, Djordjevic J, Radojcic MB (2009) Stress type dependence of expression and cytoplasmic-nuclear partitioning of glucocorticoid receptor, hsp90 and hsp70 in Wistar rat brain. Neuropsychobiology 59(4):213–221. https://doi.org/10.1159/000223733

Dwivedi Y (2009) Brain-derived neurotrophic factor: role in depression and suicide. Neuropsychiatr Dis Treat 5:433–449

Dwivedi Y, Rizavi HS, Conley RR, Roberts RC, Tamminga CA, Pandey GN (2003) Altered gene expression of brain-derived neurotrophic factor and receptor tyrosine kinase B in postmortem brain of suicide subjects. Arch Gen Psychiatry 60(8):804–815. https://doi.org/10.1001/archpsyc.60.8.804

Dwivedi Y, Rizavi HS, Pandey GN (2006) Antidepressants reverse corticosterone-mediated decrease in brain-derived neurotrophic factor expression: differential regulation of specific exons by antidepressants and corticosterone. Neuroscience 139(3):1017–1029. https://doi.org/10.1016/j.neuroscience.2005.12.058

Elia LP, Yamamoto M, Zang K, Reichardt LF (2006) p120 catenin regulates dendritic spine and synapse development through Rho-family GTPases and cadherins. Neuron 51(1):43–56. https://doi.org/10.1016/j.neuron.2006.05.018

Fuchikami M, Morinobu S, Segawa M, Okamoto Y, Yamawaki S, Ozaki N et al (2011) DNA methylation profiles of the brain-derived neurotrophic factor (BDNF) gene as a potent diagnostic biomarker in major depression. PLoS One 6(8):e23881. https://doi.org/10.1371/journal.pone.0023881

Fung SJ, Xi M, Zhang J, Torterolo P, Sampogna S, Morales FR, Chase MH (2011) Projection neurons from the central nucleus of the amygdala to the nucleus pontis oralis. J Neurosci Res 89(3):429–436. https://doi.org/10.1002/jnr.22554

Gillespie CF, Phifer J, Bradley B, Ressler KJ (2009) Risk and resilience: genetic and environmental influences on development of the stress response. Depress Anxiety 26(11):984–992. https://doi.org/10.1002/da.20605

Gounko NV, Swinny JD, Kalicharan D, Jafari S, Corteen N, Seifi M (2013) Corticotropin-releasing factor and urocortin regulate spine and synapse formation: structural basis for stress-induced neuronal remodeling and pathology. Mol Psychiatry 18(1):86–92. https://doi.org/10.1038/mp.2012.43

Gourley SL, Swanson AM, Jacobs AM, Howell JL, Mo M, Dileone RJ (2012) Action control is mediated by prefrontal BDNF and glucocorticoid receptor binding. Proc Natl Acad Sci USA 109(50):20714–20719. https://doi.org/10.1073/pnas.1208342109

Govek EE, Newey SE, Van Aelst L (2005) The role of the Rho GTPases in neuronal development. Genes Dev 19(1):1–49. https://doi.org/10.1101/gad.1256405

Grande I, Fries GR, Kunz M, Kapczinski F (2010) The role of BDNF as a mediator of neuroplasticity in bipolar disorder. Psychiatry Investig 7(4):243–250. https://doi.org/10.4306/pi.2010.7.4.243

Guidotti G, Calabrese F, Anacker C, Racagni G, Pariante CM, Riva MA (2013) Glucocorticoid receptor and FKBP5 expression is altered following exposure to chronic stress: modulation by antidepressant treatment. Neuropsychopharmacology 38(4):616–627. https://doi.org/10.1038/npp.2012.225

Gupta-Agarwal S, Franklin AV, Deramus T, Wheelock M, Davis RL, McMahon LL, Lubin FD (2012) G9a/GLP histone lysine dimethyltransferase complex activity in the hippocampus and the entorhinal cortex is required for gene activation and silencing during memory consolidation. J Neurosci 32(16):5440–5453. https://doi.org/10.1523/JNEUROSCI.0147-12.2012

Hale CF, Dietz KC, Varela JA, Wood CB, Zirlin BC, Leverich LS et al (2011) Essential role for vav Guanine nucleotide exchange factors in brain-derived neurotrophic factor-induced dendritic spine growth and synapse plasticity. J Neurosci 31(35):12426–12436. https://doi.org/10.1523/JNEUROSCI.0685-11.2011

Hansson AC, Sommer WH, Metsis M, Stromberg I, Agnati LF, Fuxe K (2006) Corticosterone actions on the hippocampal brain-derived neurotrophic factor expression are mediated by exon IV promoter. J Neuroendocrinol 18(2):104–114. https://doi.org/10.1111/j.1365-2826.2005.01390.x

Haring M, Guggenhuber S, Lutz B (2012) Neuronal populations mediating the effects of endocannabinoids on stress and emotionality. Neuroscience 204:145–158. https://doi.org/10.1016/j.neuroscience.2011.12.035
Harvey RJ, Morando L, Rasetti R, Strata P (2005) Spontaneous electrical activity and dendritic spine size in mature cerebellar Purkinje cells. Eur J Neurosci 21(7):1777–1784. https://doi.org/10.1111/j.1460-9568.2005.04010.x
Hauer D, Weis F, Papassotiropoulos A, Schmoeckel M, Beiras-Fernandez A, Lieke J et al (2011) Relationship of a common polymorphism of the glucocorticoid receptor gene to traumatic memories and posttraumatic stress disorder in patients after intensive care therapy. Crit Care Med 39(4):643–650. https://doi.org/10.1097/CCM.0b013e318206bae6
Hill MN, Hillard CJ, McEwen BS (2011) Alterations in corticolimbic dendritic morphology and emotional behavior in cannabinoid CB1 receptor-deficient mice parallel the effects of chronic stress. Cereb Cortex 21(9):2056–2064. https://doi.org/10.1093/cercor/bhq280
Hill MN, Bierer LM, Makotkine I, Golier JA, Galea S, McEwen BS et al (2013) Reductions in circulating endocannabinoid levels in individuals with post-traumatic stress disorder following exposure to the World Trade Center attacks. Psychoneuroendocrinology 38(12):2952–2961. https://doi.org/10.1016/j.psyneuen.2013.08.004
Honkura N, Matsuzaki M, Noguchi J, Ellis-Davies GC, Kasai H (2008) The subspine organization of actin fibers regulates the structure and plasticity of dendritic spines. Neuron 57(5):719–729. https://doi.org/10.1016/j.neuron.2008.01.013
Huang F, Chotiner JK, Steward O (2007) Actin polymerization and ERK phosphorylation are required for Arc/Arg3.1 mRNA targeting to activated synaptic sites on dendrites. J Neurosci 27 (34):9054–9067. https://doi.org/10.1523/JNEUROSCI.2410-07.2007
Huang Y, Ko H, Cheung ZH, Yung KK, Yao T, Wang JJ et al (2012) Dual actions of brain-derived neurotrophic factor on GABAergic transmission in cerebellar Purkinje neurons. Exp Neurol 233 (2):791–798. https://doi.org/10.1016/j.expneurol.2011.11.043
Jaaskelainen T, Makkonen H, Palvimo JJ (2011) Steroid up-regulation of FKBP51 and its role in hormone signaling. Curr Opin Pharmacol 11(4):326–331. https://doi.org/10.1016/j.coph.2011.04.006
Johnson LR, Farb C, Morrison JH, McEwen BS, LeDoux JE (2005) Localization of glucocorticoid receptors at postsynaptic membranes in the lateral amygdala. Neuroscience 136(1):289–299. https://doi.org/10.1016/j.neuroscience.2005.06.050
Johnson SA, Wang JF, Sun X, McEwen BS, Chattarji S, Young LT (2009) Lithium treatment prevents stress-induced dendritic remodeling in the rodent amygdala. Neuroscience 163 (1):34–39. https://doi.org/10.1016/j.neuroscience.2009.06.005
Kang H, Schuman EM (1996) A requirement for local protein synthesis in neurotrophin-induced hippocampal synaptic plasticity. Science 273(5280):1402–1406
Kaplan GB, Moore KA (2011) The use of cognitive enhancers in animal models of fear extinction. Pharmacol Biochem Behav 99(2):217–228. https://doi.org/10.1016/j.pbb.2011.01.009
Karpova NN, Pickenhagen A, Lindholm J, Tiraboschi E, Kulesskaya N, Agustsdottir A et al (2011) Fear erasure in mice requires synergy between antidepressant drugs and extinction training. Science 334(6063):1731–1734. https://doi.org/10.1126/science.1214592
Kawashima H, Numakawa T, Kumamaru E, Adachi N, Mizuno H, Ninomiya M et al (2010) Glucocorticoid attenuates brain-derived neurotrophic factor-dependent upregulation of glutamate receptors via the suppression of microRNA-132 expression. Neuroscience 165 (4):1301–1311. https://doi.org/10.1016/j.neuroscience.2009.11.057
Keller S, Sarchiapone M, Zarrilli F, Videtic A, Ferraro A, Carli V et al (2010) Increased BDNF promoter methylation in the Wernicke area of suicide subjects. Arch Gen Psychiatry 67 (3):258–267. https://doi.org/10.1001/archgenpsychiatry.2010.9
Klein ME, Lioy DT, Ma L, Impey S, Mandel G, Goodman RH (2007) Homeostatic regulation of MeCP2 expression by a CREB-induced microRNA. Nat Neurosci 10(12):1513–1514. https://doi.org/10.1038/nn2010

Klengel T, Mehta D, Anacker C, Rex-Haffner M, Pruessner JC, Pariante CM et al (2013) Allele-specific FKBP5 DNA demethylation mediates gene-childhood trauma interactions. Nat Neurosci 16(1):33–41. https://doi.org/10.1038/nn.3275

Koenen KC, Saxe G, Purcell S, Smoller JW, Bartholomew D, Miller A et al (2005) Polymorphisms in FKBP5 are associated with peritraumatic dissociation in medically injured children. Mol Psychiatry 10(12):1058–1059. https://doi.org/10.1038/sj.mp.4001727

Koenen KC, Fu QJ, Ertel K, Lyons MJ, Eisen SA, True WR et al (2008) Common genetic liability to major depression and posttraumatic stress disorder in men. J Affect Disord 105 (1–3):109–115. https://doi.org/10.1016/j.jad.2007.04.021

Komatsuzaki Y, Hatanaka Y, Murakami G, Mukai H, Hojo Y, Saito M et al (2012) Corticosterone induces rapid spinogenesis via synaptic glucocorticoid receptors and kinase networks in hippocampus. PLoS One 7(4):e34124. https://doi.org/10.1371/journal.pone.0034124

Kozlovsky N, Matar MA, Kaplan Z, Kotler M, Zohar J, Cohen H (2007) Long-term down-regulation of BDNF mRNA in rat hippocampal CA1 subregion correlates with PTSD-like behavioural stress response. Int J Neuropsychopharmacol 10(6):741–758. https://doi.org/10.1017/S1461145707007560

Kumamaru E, Numakawa T, Adachi N, Yagasaki Y, Izumi A, Niyaz M et al (2008) Glucocorticoid prevents brain-derived neurotrophic factor-mediated maturation of synaptic function in developing hippocampal neurons through reduction in the activity of mitogen-activated protein kinase. Mol Endocrinol 22(3):546–558. https://doi.org/10.1210/me.2007-0264

Kumamaru E, Numakawa T, Adachi N, Kunugi H (2011) Glucocorticoid suppresses BDNF-stimulated MAPK/ERK pathway via inhibiting interaction of Shp2 with TrkB. FEBS Lett 585 (20):3224–3228. https://doi.org/10.1016/j.febslet.2011.09.010

Lakshminarasimhan H, Chattarji S (2012) Stress leads to contrasting effects on the levels of brain derived neurotrophic factor in the hippocampus and amygdala. PLoS One 7(1):e30481. https://doi.org/10.1371/journal.pone.0030481

Lebow M, Neufeld-Cohen A, Kuperman Y, Tsoory M, Gil S, Chen A (2012) Susceptibility to PTSD-like behavior is mediated by corticotropin-releasing factor receptor type 2 levels in the bed nucleus of the stria terminalis. J Neurosci 32(20):6906–6916. https://doi.org/10.1523/JNEUROSCI.4012-11.2012

Li HS, Xu XZ, Montell C (1999) Activation of a TRPC3-dependent cation current through the neurotrophin BDNF. Neuron 24(1):261–273

Li Y, Calfa G, Inoue T, Amaral MD, Pozzo-Miller L (2010) Activity-dependent release of endogenous BDNF from mossy fibers evokes a TRPC3 current and Ca2+ elevations in CA3 pyramidal neurons. J Neurophysiol 103(5):2846–2856. https://doi.org/10.1152/jn.01140.2009

Lightman SL (2008) The neuroendocrinology of stress: a never ending story. J Neuroendocrinol 20 (6):880–884. https://doi.org/10.1111/j.1365-2826.2008.01711.x

Liston C, Gan WB (2011) Glucocorticoids are critical regulators of dendritic spine development and plasticity in vivo. Proc Natl Acad Sci USA 108(38):16074–16079. https://doi.org/10.1073/pnas.1110444108

Liu J, Yu B, Neugebauer V, Grigoriadis DE, Rivier J, Vale WW et al (2004) Corticotropin-releasing factor and Urocortin I modulate excitatory glutamatergic synaptic transmission. J Neurosci 24 (16):4020–4029. https://doi.org/10.1523/JNEUROSCI.5531-03.2004

Liu L, Wang C, Ni X, Sun J (2007) A rapid inhibition of NMDA receptor current by corticosterone in cultured hippocampal neurons. Neurosci Lett 420(3):245–250. https://doi.org/10.1016/j.neulet.2007.05.003

Lubin FD (2011) Epigenetic gene regulation in the adult mammalian brain: multiple roles in memory formation. Neurobiol Learn Mem 96(1):68–78. https://doi.org/10.1016/j.nlm.2011.03.001

Lubin FD, Roth TL, Sweatt JD (2008) Epigenetic regulation of BDNF gene transcription in the consolidation of fear memory. J Neurosci 28(42):10576–10586. https://doi.org/10.1523/JNEUROSCI.1786-08.2008

Maddox SA, Schafe GE (2011) Epigenetic alterations in the lateral amygdala are required for reconsolidation of a Pavlovian fear memory. Learn Mem 18(9):579–593. https://doi.org/10.1101/lm.2243411

Magarinos AM, Li CJ, Gal Toth J, Bath KG, Jing D, Lee FS, McEwen BS (2011) Effect of brain-derived neurotrophic factor haploinsufficiency on stress-induced remodeling of hippocampal neurons. Hippocampus 21(3):253–264. https://doi.org/10.1002/hipo.20744

Matys T, Pawlak R, Matys E, Pavlides C, McEwen BS, Strickland S (2004) Tissue plasminogen activator promotes the effects of corticotropin-releasing factor on the amygdala and anxiety-like behavior. Proc Natl Acad Sci USA 101(46):16345–16350. https://doi.org/10.1073/pnas.0407355101

McGowan PO, Sasaki A, D'Alessio AC, Dymov S, Labonte B, Szyf M et al (2009) Epigenetic regulation of the glucocorticoid receptor in human brain associates with childhood abuse. Nat Neurosci 12(3):342–348. https://doi.org/10.1038/nn.2270

Mehta D, Gonik M, Klengel T, Rex-Haffner M, Menke A, Rubel J et al (2011) Using polymorphisms in FKBP5 to define biologically distinct subtypes of posttraumatic stress disorder: evidence from endocrine and gene expression studies. Arch Gen Psychiatry 68(9):901–910. https://doi.org/10.1001/archgenpsychiatry.2011.50

Meijer OC (2002) Coregulator proteins and corticosteroid action in the brain. J Neuroendocrinol 14(6):499–505

Merighi A, Bardoni R, Salio C, Lossi L, Ferrini F, Prandini M et al (2008) Presynaptic functional trkB receptors mediate the release of excitatory neurotransmitters from primary afferent terminals in lamina II (substantia gelatinosa) of postnatal rat spinal cord. Dev Neurobiol 68(4):457–475. https://doi.org/10.1002/dneu.20605

Mitra R, Ferguson D, Sapolsky RM (2009) Mineralocorticoid receptor overexpression in basolateral amygdala reduces corticosterone secretion and anxiety. Biol Psychiatry 66(7):686–690. https://doi.org/10.1016/j.biopsych.2009.04.016

Molteni R, Calabrese F, Chourbaji S, Brandwein C, Racagni G, Gass P, Riva MA (2010) Depression-prone mice with reduced glucocorticoid receptor expression display an altered stress-dependent regulation of brain-derived neurotrophic factor and activity-regulated cyto-skeleton-associated protein. J Psychopharmacol 24(4):595–603. https://doi.org/10.1177/0269881108099815

Nagappan G, Zaitsev E, Senatorov VV Jr, Yang J, Hempstead BL, Lu B (2009) Control of extracellular cleavage of ProBDNF by high frequency neuronal activity. Proc Natl Acad Sci USA 106(4):1267–1272. https://doi.org/10.1073/pnas.0807322106

Nair A, Vadodaria KC, Banerjee SB, Benekareddy M, Dias BG, Duman RS, Vaidya VA (2007) Stressor-specific regulation of distinct brain-derived neurotrophic factor transcripts and cyclic AMP response element-binding protein expression in the postnatal and adult rat hippocampus. Neuropsychopharmacology 32(7):1504–1519. https://doi.org/10.1038/sj.npp.1301276

Nakayama AY, Harms MB, Luo L (2000) Small GTPases Rac and Rho in the maintenance of dendritic spines and branches in hippocampal pyramidal neurons. J Neurosci 20(14):5329–5338

Numakawa T, Suzuki S, Kumamaru E, Adachi N, Richards M, Kunugi H (2010) BDNF function and intracellular signaling in neurons. Histol Histopathol 25(2):237–258. https://doi.org/10.14670/HH-25.237

Numakawa T, Adachi N, Richards M, Chiba S, Kunugi H (2013) Brain-derived neurotrophic factor and glucocorticoids: reciprocal influence on the central nervous system. Neuroscience 239:157–172. https://doi.org/10.1016/j.neuroscience.2012.09.073

Ohbuchi T, Yokoyama T, Saito T, Hashimoto H, Suzuki H, Otsubo H et al (2009) Brain-derived neurotrophic factor inhibits spontaneous inhibitory postsynaptic currents in the rat supraoptic nucleus. Brain Res 1258:34–42. https://doi.org/10.1016/j.brainres.2008.12.057

Ou LC, Gean PW (2007) Transcriptional regulation of brain-derived neurotrophic factor in the amygdala during consolidation of fear memory. Mol Pharmacol 72(2):350–358. https://doi.org/10.1124/mol.107.034934

Oztan O, Aydin C, Isgor C (2011) Chronic variable physical stress during the peripubertal-juvenile period causes differential depressive and anxiogenic effects in the novelty-seeking phenotype: functional implications for hippocampal and amygdalar brain-derived neurotrophic factor and the mossy fibre plasticity. Neuroscience 192:334–344. https://doi.org/10.1016/j.neuroscience.2011.06.077

Pang PT, Teng HK, Zaitsev E, Woo NT, Sakata K, Zhen S et al (2004) Cleavage of proBDNF by tPA/plasmin is essential for long-term hippocampal plasticity. Science 306(5695):487–491. https://doi.org/10.1126/science.1100135

Pawlak R, Magarinos AM, Melchor J, McEwen B, Strickland S (2003) Tissue plasminogen activator in the amygdala is critical for stress-induced anxiety-like behavior. Nat Neurosci 6 (2):168–174. https://doi.org/10.1038/nn998

Pitman RK, Rasmusson AM, Koenen KC, Shin LM, Orr SP, Gilbertson MW et al (2012) Biological studies of post-traumatic stress disorder. Nat Rev Neurosci 13(11):769–787. https://doi.org/10.1038/nrn3339

Pollard TD, Blanchoin L, Mullins RD (2000) Molecular mechanisms controlling actin filament dynamics in nonmuscle cells. Annu Rev Biophys Biomol Struct 29:545–576. https://doi.org/10.1146/annurev.biophys.29.1.545

Porcher C, Hatchett C, Longbottom RE, McAinch K, Sihra TS, Moss SJ et al (2011) Positive feedback regulation between gamma-aminobutyric acid type A (GABA(A)) receptor signaling and brain-derived neurotrophic factor (BDNF) release in developing neurons. J Biol Chem 286 (24):21667–21677. https://doi.org/10.1074/jbc.M110.201582

Rabiner CA, Mains RE, Eipper BA (2005) Kalirin: a dual Rho guanine nucleotide exchange factor that is so much more than the sum of its many parts. Neuroscientist 11(2):148–160. https://doi.org/10.1177/1073858404271250

Rattiner LM, Davis M, Ressler KJ (2004) Differential regulation of brain-derived neurotrophic factor transcripts during the consolidation of fear learning. Learn Mem 11(6):727–731

Reich CG, Taylor ME, McCarthy MM (2009) Differential effects of chronic unpredictable stress on hippocampal CB1 receptors in male and female rats. Behav Brain Res 203(2):264–269. https://doi.org/10.1016/j.bbr.2009.05.013

Reul JM, Chandramohan Y (2007) Epigenetic mechanisms in stress-related memory formation. Psychoneuroendocrinology 32(Suppl 1):S21–S25. https://doi.org/10.1016/j.psyneuen.2007.03.016

Reul JM, Gesing A, Droste S, Stec IS, Weber A, Bachmann C et al (2000) The brain mineralocorticoid receptor: greedy for ligand, mysterious in function. Eur J Pharmacol 405(1–3):235–249

Riegel AC, Williams JT (2008) CRF facilitates calcium release from intracellular stores in midbrain dopamine neurons. Neuron 57(4):559–570. https://doi.org/10.1016/j.neuron.2007.12.029

Rodenas-Ruano A, Chavez AE, Cossio MJ, Castillo PE, Zukin RS (2012) REST-dependent epigenetic remodeling promotes the developmental switch in synaptic NMDA receptors. Nat Neurosci 15(10):1382–1390. https://doi.org/10.1038/nn.3214

Roth TL, Sweatt JD (2011) Epigenetic marking of the BDNF gene by early-life adverse experiences. Horm Behav 59(3):315–320. https://doi.org/10.1016/j.yhbeh.2010.05.005

Roth TL, Lubin FD, Funk AJ, Sweatt JD (2009) Lasting epigenetic influence of early-life adversity on the BDNF gene. Biol Psychiatry 65(9):760–769. https://doi.org/10.1016/j.biopsych.2008.11.028

Roth TL, Zoladz PR, Sweatt JD, Diamond DM (2011) Epigenetic modification of hippocampal Bdnf DNA in adult rats in an animal model of post-traumatic stress disorder. J Psychiatr Res 45 (7):919–926. https://doi.org/10.1016/j.jpsychires.2011.01.013

Rothman SM, Mattson MP (2013) Activity-dependent, stress-responsive BDNF signaling and the quest for optimal brain health and resilience throughout the lifespan. Neuroscience 239:228–240. https://doi.org/10.1016/j.neuroscience.2012.10.014

Sala C, Piech V, Wilson NR, Passafaro M, Liu G, Sheng M (2001) Regulation of dendritic spine morphology and synaptic function by Shank and Homer. Neuron 31(1):115–130

Saneyoshi T, Wayman G, Fortin D, Davare M, Hoshi N, Nozaki N et al (2008) Activity-dependent synaptogenesis: regulation by a CaM-kinase kinase/CaM-kinase I/betaPIX signaling complex. Neuron 57(1):94–107. https://doi.org/10.1016/j.neuron.2007.11.016

Santos AR, Comprido D, Duarte CB (2010) Regulation of local translation at the synapse by BDNF. Prog Neurobiol 92(4):505–516. https://doi.org/10.1016/j.pneurobio.2010.08.004

Sartor CE, McCutcheon VV, Pommer NE, Nelson EC, Grant JD, Duncan AE et al (2011) Common genetic and environmental contributions to post-traumatic stress disorder and alcohol dependence in young women. Psychol Med 41(7):1497–1505. https://doi.org/10.1017/S0033291710002072

Sato S, Osanai H, Monma T, Harada T, Hirano A, Saito M, Kawato S (2004) Acute effect of corticosterone on N-methyl-D-aspartate receptor-mediated Ca2+ elevation in mouse hippocampal slices. Biochem Biophys Res Commun 321(2):510–513. https://doi.org/10.1016/j.bbrc.2004.06.168

Schaaf MJ, De Kloet ER, Vreugdenhil E (2000) Corticosterone effects on BDNF expression in the hippocampus. Implications for memory formation. Stress 3(3):201–208

Scharf SH, Liebl C, Binder EB, Schmidt MV, Muller MB (2011) Expression and regulation of the Fkbp5 gene in the adult mouse brain. PLoS One 6(2):e16883. https://doi.org/10.1371/journal.pone.0016883

Schiefer J, Kampe K, Dodt HU, Zieglgansberger W, Kreutzberg GW (1999) Microglial motility in the rat facial nucleus following peripheral axotomy. J Neurocytol 28(6):439–453

Schratt GM, Nigh EA, Chen WG, Hu L, Greenberg ME (2004) BDNF regulates the translation of a select group of mRNAs by a mammalian target of rapamycin-phosphatidylinositol 3-kinase-dependent pathway during neuronal development. J Neurosci 24(33):7366–7377. https://doi.org/10.1523/JNEUROSCI.1739-04.2004

Sharma RP, Tun N, Grayson DR (2008) Depolarization induces downregulation of DNMT1 and DNMT3a in primary cortical cultures. Epigenetics 3(2):74–80

Shen W, Wu B, Zhang Z, Dou Y, Rao ZR, Chen YR, Duan S (2006) Activity-induced rapid synaptic maturation mediated by presynaptic cdc42 signaling. Neuron 50(3):401–414. https://doi.org/10.1016/j.neuron.2006.03.017

Sheng H, Zhang Y, Sun J, Gao L, Ma B, Lu J, Ni X (2008) Corticotropin-releasing hormone (CRH) depresses n-methyl-D-aspartate receptor-mediated current in cultured rat hippocampal neurons via CRH receptor type 1. Endocrinology 149(3):1389–1398. https://doi.org/10.1210/en.2007-1378

Skrzypiec AE, Buczko W, Pawlak R (2008) Tissue plasminogen activator in the amygdala: a new role for an old protease. J Physiol Pharmacol 59(Suppl 8):135–146

Smith MA, Makino S, Kvetnansky R, Post RM (1995) Stress and glucocorticoids affect the expression of brain-derived neurotrophic factor and neurotrophin-3 mRNAs in the hippocampus. J Neurosci 15(3 Pt 1):1768–1777

So AY, Chaivorapol C, Bolton EC, Li H, Yamamoto KR (2007) Determinants of cell- and gene-specific transcriptional regulation by the glucocorticoid receptor. PLoS Genet 3(6):e94. https://doi.org/10.1371/journal.pgen.0030094

Spijker AT, van Rossum EF (2012) Glucocorticoid sensitivity in mood disorders. Neuroendocrinology 95(3):179–186. https://doi.org/10.1159/000329846

Stranahan AM, Arumugam TV, Mattson MP (2011) Lowering corticosterone levels reinstates hippocampal brain-derived neurotropic factor and Trkb expression without influencing deficits in hypothalamic brain-derived neurotropic factor expression in leptin receptor-deficient mice. Neuroendocrinology 93(1):58–64. https://doi.org/10.1159/000322808

Suri D, Vaidya VA (2013) Glucocorticoid regulation of brain-derived neurotrophic factor: relevance to hippocampal structural and functional plasticity. Neuroscience 239:196–213. https://doi.org/10.1016/j.neuroscience.2012.08.065

Swinny JD, Valentino RJ (2006) Corticotropin-releasing factor promotes growth of brain norepinephrine neuronal processes through Rho GTPase regulators of the actin cytoskeleton in rat. Eur J Neurosci 24(9):2481–2490. https://doi.org/10.1111/j.1460-9568.2006.05129.x

Takei S, Morinobu S, Yamamoto S, Fuchikami M, Matsumoto T, Yamawaki S (2011) Enhanced hippocampal BDNF/TrkB signaling in response to fear conditioning in an animal model of posttraumatic stress disorder. J Psychiatr Res 45(4):460–468. https://doi.org/10.1016/j.jpsychires.2010.08.009

Tanaka J, Horiike Y, Matsuzaki M, Miyazaki T, Ellis-Davies GC, Kasai H (2008) Protein synthesis and neurotrophin-dependent structural plasticity of single dendritic spines. Science 319 (5870):1683–1687. https://doi.org/10.1126/science.1152864

Tashiro A, Yuste R (2004) Regulation of dendritic spine motility and stability by Rac1 and Rho kinase: evidence for two forms of spine motility. Mol Cell Neurosci 26(3):429–440. https://doi.org/10.1016/j.mcn.2004.04.001

Tashiro A, Yuste R (2008) Role of Rho GTPases in the morphogenesis and motility of dendritic spines. Methods Enzymol 439:285–302. https://doi.org/10.1016/S0076-6879(07)00421-1

Tashiro A, Minden A, Yuste R (2000) Regulation of dendritic spine morphology by the rho family of small GTPases: antagonistic roles of Rac and Rho. Cereb Cortex 10(10):927–938

Taylor AH, Finney M, Lam PM, Konje JC (2011) Modulation of the endocannabinoid system in viable and non-viable first trimester pregnancies by pregnancy-related hormones. Reprod Biol Endocrinol 9:152. https://doi.org/10.1186/1477-7827-9-152

Timmusk T, Palm K, Metsis M, Reintam T, Paalme V, Saarma M, Persson H (1993) Multiple promoters direct tissue-specific expression of the rat BDNF gene. Neuron 10(3):475–489

Treweek JB, Jaferi A, Colago EE, Zhou P, Pickel VM (2009) Electron microscopic localization of corticotropin-releasing factor (CRF) and CRF receptor in rat and mouse central nucleus of the amygdala. J Comp Neurol 512(3):323–335. https://doi.org/10.1002/cne.21884

Trollope AF, Gutierrez-Mecinas M, Mifsud KR, Collins A, Saunderson EA, Reul JM (2012) Stress, epigenetic control of gene expression and memory formation. Exp Neurol 233(1):3–11. https://doi.org/10.1016/j.expneurol.2011.03.022

Tsai SJ, Hong CJ, Liou YJ (2011) Recent molecular genetic studies and methodological issues in suicide research. Prog Neuro-Psychopharmacol Biol Psychiatry 35(4):809–817. https://doi.org/10.1016/j.pnpbp.2010.10.014

Tsankova NM, Berton O, Renthal W, Kumar A, Neve RL, Nestler EJ (2006) Sustained hippocampal chromatin regulation in a mouse model of depression and antidepressant action. Nat Neurosci 9(4):519–525. https://doi.org/10.1038/nn1659

Turner JD, Alt SR, Cao L, Vernocchi S, Trifonova S, Battello N, Muller CP (2010) Transcriptional control of the glucocorticoid receptor: CpG islands, epigenetics and more. Biochem Pharmacol 80(12):1860–1868. https://doi.org/10.1016/j.bcp.2010.06.037

Tyler WJ, Pozzo-Miller L (2003) Miniature synaptic transmission and BDNF modulate dendritic spine growth and form in rat CA1 neurones. J Physiol 553(Pt 2):497–509. https://doi.org/10.1113/jphysiol.2003.052639

Vadodaria KC, Brakebusch C, Suter U, Jessberger S (2013) Stage-specific functions of the small Rho GTPases Cdc42 and Rac1 for adult hippocampal neurogenesis. J Neurosci 33 (3):1179–1189. https://doi.org/10.1523/JNEUROSCI.2103-12.2013

van der Laan S, Meijer OC (2008) Pharmacology of glucocorticoids: beyond receptors. Eur J Pharmacol 585(2–3):483–491. https://doi.org/10.1016/j.ejphar.2008.01.060

van Winkel R, van Nierop M, Myin-Germeys I, van Os J (2013) Childhood trauma as a cause of psychosis: linking genes, psychology, and biology. Can J Psychiatr 58(1):44–51. https://doi.org/10.1177/070674371305800109

van Zuiden M, Kavelaars A, Geuze E, Olff M, Heijnen CJ (2013) Predicting PTSD: pre-existing vulnerabilities in glucocorticoid-signaling and implications for preventive interventions. Brain Behav Immun 30:12–21. https://doi.org/10.1016/j.bbi.2012.08.015

Vigers AJ, Amin DS, Talley-Farnham T, Gorski JA, Xu B, Jones KR (2012) Sustained expression of brain-derived neurotrophic factor is required for maintenance of dendritic spines and normal behavior. Neuroscience 212:1–18. https://doi.org/10.1016/j.neuroscience.2012.03.031

Vyas A, Mitra R, Shankaranarayana Rao BS, Chattarji S (2002) Chronic stress induces contrasting patterns of dendritic remodeling in hippocampal and amygdaloid neurons. J Neurosci 22 (15):6810–6818. https://doi.org/10.1523/JNEUROSCI.22-15-06810.2002
Vyas A, Bernal S, Chattarji S (2003) Effects of chronic stress on dendritic arborization in the central and extended amygdala. Brain Res 965(1–2):290–294
Wang JC, Derynck MK, Nonaka DF, Khodabakhsh DB, Haqq C, Yamamoto KR (2004) Chromatin immunoprecipitation (ChIP) scanning identifies primary glucocorticoid receptor target genes. Proc Natl Acad Sci USA 101(44):15603–15608. https://doi.org/10.1073/pnas.0407008101
Wang XD, Rammes G, Kraev I, Wolf M, Liebl C, Scharf SH et al (2011) Forebrain CRF (1) modulates early-life stress-programmed cognitive deficits. J Neurosci 31 (38):13625–13634. https://doi.org/10.1523/JNEUROSCI.2259-11.2011
Wang WS, Kang S, Liu WT, Li M, Liu Y, Yu C et al (2012) Extinction of aversive memories associated with morphine withdrawal requires ERK-mediated epigenetic regulation of brain-derived neurotrophic factor transcription in the rat ventromedial prefrontal cortex. J Neurosci 32 (40):13763–13775. https://doi.org/10.1523/JNEUROSCI.1991-12.2012
Waterhouse EG, Xu B (2009) New insights into the role of brain-derived neurotrophic factor in synaptic plasticity. Mol Cell Neurosci 42(2):81–89. https://doi.org/10.1016/j.mcn.2009.06.009
Wayman GA, Davare M, Ando H, Fortin D, Varlamova O, Cheng HY et al (2008) An activity-regulated microRNA controls dendritic plasticity by down-regulating p250GAP. Proc Natl Acad Sci USA 105(26):9093–9098. https://doi.org/10.1073/pnas.0803072105
Xie Z, Cahill ME, Penzes P (2010) Kalirin loss results in cortical morphological alterations. Mol Cell Neurosci 43(1):81–89. https://doi.org/10.1016/j.mcn.2009.09.006
Yeh CM, Huang CC, Hsu KS (2012) Prenatal stress alters hippocampal synaptic plasticity in young rat offspring through preventing the proteolytic conversion of pro-brain-derived neurotrophic factor (BDNF) to mature BDNF. J Physiol 590(4):991–1010. https://doi.org/10.1113/jphysiol.2011.222042
Yehuda R, Flory JD, Pratchett LC, Buxbaum J, Ising M, Holsboer F (2010) Putative biological mechanisms for the association between early life adversity and the subsequent development of PTSD. Psychopharmacology 212(3):405–417. https://doi.org/10.1007/s00213-010-1969-6
Yin Y, Edelman GM, Vanderklish PW (2002) The brain-derived neurotrophic factor enhances synthesis of Arc in synaptoneurosomes. Proc Natl Acad Sci USA 99(4):2368–2373. https://doi.org/10.1073/pnas.042693699
Yudt MR, Cidlowski JA (2002) The glucocorticoid receptor: coding a diversity of proteins and responses through a single gene. Mol Endocrinol 16(8):1719–1726. https://doi.org/10.1210/me.2002-0106
Zai CC, Manchia M, De Luca V, Tiwari AK, Chowdhury NI, Zai GC et al (2012) The brain-derived neurotrophic factor gene in suicidal behaviour: a meta-analysis. Int J Neuropsychopharmacol 15 (8):1037–1042. https://doi.org/10.1017/S1461145711001313
Zhang Z, Fan J, Ren Y, Zhou W, Yin G (2013) The release of glutamate from cortical neurons regulated by BDNF via the TrkB/Src/PLC-gamma1 pathway. J Cell Biochem 114(1):144–151. https://doi.org/10.1002/jcb.24311
Zocchi L, Sassone-Corsi P (2012) SIRT1-mediated deacetylation of MeCP2 contributes to BDNF expression. Epigenetics 7(7):695–700. https://doi.org/10.4161/epi.20733

Modulation of the Core Neural Network in Stress: The Role of Endocannabinoids and LTD

5

5.1 The Role of Endocannabinoids

Both clinical and animal studies point to the importance of endocannabiniods in extinction (Sect. 5.1.1). Single nucleotide polymorphisms in pathways controlling endocannabinoids have profound effects on the level of these in the amygdala in response to stress (Sect. 5.1.2). The core network controlling the responses to stress in the amygdala is of major interest in this review, particularly in the context of extinction, for which we now have considerable insights from animal studies (Sect. 5.1.3). These, together with clinical observations, indicate that endocannabinoids and BDNF may act together in a feedback pathway that sustains the level of the former and therefore the extent and duration of extinction (Sect. 5.1.4).

5.1.1 Clinical Observations on Endocannabinoids and Extinction

Baseline serum levels of the endocannabinoids anandamide (AEA) and/or 2-arachidonoylglycerol (2-AG) are relatively low in those that experience stress associated with motion sickness (Chouker et al. 2010) and postoperative deliriums (Hauer et al. 2012) as well as psycho-social stress (Dlugos et al. 2012). Post-traumatic stress disorder (PTSD) is also accompanied by low levels of AEA in serum (Neumeister et al. 2013, 2015; but see Hauer et al. 2013) and in hair (Wilker et al. 2014), as well as of 2-AG (Hill et al. 2013a, b). Furthermore, low serum levels of 2-AG are found in those that have been stressed through prolonged isolation (Yi et al. 2016). As baseline circulating levels of endocannabinoids, such as AEA, are negatively correlated with stress and anxiety, and following stressful events are transiently elevated in healthy subjects (Dlugos et al. 2012), then it might be expected that facilitating an increase in

Reprinted from Behav Brain Res. Bennett MR, Arnold J, Hatton SN, Lagopoulos J. Regulation of fear extinction by long-term depression: The roles of endocannabinoids and brain derived neurotrophic factor. Vol. 19:148–164. Copyright (2017) With permission from Elsevier.

M. Bennett, J. Lagopoulos, *Stress, Trauma and Synaptic Plasticity*,
https://doi.org/10.1007/978-3-319-91116-8_5

baseline endocannabinoids or their receptor (CB1R) agonists might significantly decrease stressful responses. There is some evidence for this as the synthetic CB1R Yiagonist Nabilone, when taken orally by those suffering from PTSD, significantly reduces nightmares and insomnia in at least a cohort of these patients (Cameron et al. 2014; Fraser 2009; Jetly et al. 2015). Indeed, Δ^9-tetrahydroclannabinol (Δ^9-THC), the principal active component of cannabis, gives significant improvements in insomnia and the frequency of nightmares (Roitman et al. 2014). Although the number of patients in these trials varied from 3 to 147, one study followed a double-blind, randomized procedure which showed significant results even for those patients that had previously been non-responsive to traditional therapies (Jetly et al. 2015). Set against these positive effects of cannabinoids are the observations that high systemic doses of CB1R agonists are ineffective or even impair extinction in animal experiments (Chhatwal et al. 2006; Pamplona et al. 2006, 2008) as does chronic administration of low doses (Lin et al. 2008). Indeed, meta-analyses of the effects of CB1R agent rimonabant in clinical trials indicate an increased risk of poor psychiatric outcomes (Christensen et al. 2007; Rucker et al. 2007).

5.1.2 Genetic Observations Implicating Endocannabinoids in Extinction

Additional evidence that endocannabinoids have a major role in regulating responses to stress comes from genetics. The endocannabinoids are regulated by the enzyme fatty acid amide hydrolase (FAAH), for which carriers of the single nucleotide polymorphism (SNP) 385A allele (rs324420) possess increased AEA (Sipe et al. 2010) and so reduced amygdala activity (Hariri et al. 2009). They are also less stress reactive than those without the SNP (Gunduz-Cinar et al. 2013). Furthermore, these A allele carriers show reduced anxiety, enhanced fear extinction and reduced FAAH expression (Dincheva et al. 2015).

5.1.3 The Core Neural Network in Endocannabinoid-Mediated Extinction

This review is partly concerned with identifying the core neural network through which endocannabinoids such as AEA and 2-arachidonoylglycerol (2-AG) exert their influence to ameliorate stress through the mechanism of extinction (Bennett et al. 2016). The critical role that endocannabinoids released in the amygdala play in extinction was discovered in 2002 (Marsicano et al. 2002), and various suggestions have been made since that time of its source and targets of action (Lafenetre et al. 2007; Ramikie and Patel 2012). This has led to various hypotheses concerning how the action of endocannabinoids can be integrated into the circuitry of the amygdala relevant to fear extinction (Anastasio 2013) (for earlier reviews of this topic, see Ganon-Elazar and Akirav 2013; Lutz 2007; Gunduz-Cinar et al. 2013; Ruehle et al. 2012) and in particular their possible role as enhancers of extinction (de Bitencourt et al. 2013; Fitzgerald et al. 2014). Here, we take advantage of recent findings that clarify this circuitry and lay emphasis on inhibitory synapses of principal neurons in the basal nucleus as the main site of endocannabinoid action in extinction (for recent reviews of this subject, see Gunduz-Cinar et al. 2013; Lafenetre et al. 2007).

5.1.4 The Maintenance of Extinction by Endocannabinoids and BDNF

Given the role of endocannabinoids in extinction, it is important to develop plausible hypotheses as to how the release of these could be sustained so as to maintain extinction over long times once it has been initiated. The interaction between the production of endocannabinoids and brain-derived neurotrophic factor (BDNF) provides a possible solution to this problem. BDNF plays an important role in brain function associated with trauma and stress, particularly through its control of the shape and size of synaptic spines (for a recent review, see Bennett and Lagopoulos 2014). BDNF is elevated by the cannabinoid Δ^9-THC (D'Souza et al. 2009), and both plasma levels of AEA and BDNF are elevated during and after exercise stress, showing a temporal correlation suggesting that endocannabinoids are causally responsible for the elevated BDNF (Heyman et al. 2012). BDNF is also under control by glucocorticoids, which are themselves central to the stress response (for recent reviews, see Bennett et al. 2016; Bennett and Lagopoulos 2014) and are upregulated by endocannabinoids (Cameron et al. 2014; Hill et al. 2013a, b). Polymorphisms in the glucocorticoid receptor regulatory protein FKBP5, such as that for the risk alleles 1360780(T), lead to decreases in the production of BDNF and of mushroom synaptic spines (Young et al. 2015). PTSD carriers of this allele were at increased risk of symptom relapse, following narrative exposure therapy, an exposure-based short-term therapy, compared with non-carriers (Wilker et al. 2014). In this work, we also consider plausible cellular and network mechanisms that relate the action of BDNF and glucocorticoids to that of endocannabinoids and in particular whether a plausible mechanism exists whereby an autocrine loop of BDN production, such as that described for neuronal survival (Marini et al. 2008), could lead to a sustained release of endocannabinoids necessary to maintain extinction once it is initiated.

5.2 Functional Amygdala Networks in Acquisition and Extinction

5.2.1 Lateral Amygdala (LA)

All available evidence points to the lateral amygdala as the site of laying down of the CS-US memory as well as a major site of subsequent plasticity changes to the CS-US memory following extinction (Helmstetter and Bellgowan 1994; Maren 1996, 1999; Muller et al. 1997; Wilensky et al. 2000). Glutamatergic neurons in the LA receive inputs from the medial division of the geniculate (MGm) that are long-term potentiated (LTP) following the acquisition of conditioned fear (Blair et al. 2001; Davis 1992; McKernan and Shinnick-Gallagher 1997; Quirk et al. 1997; Tsvetkov et al. 2002) or following direct stimulation of the MGm input in conjunction with the US (Kwon and

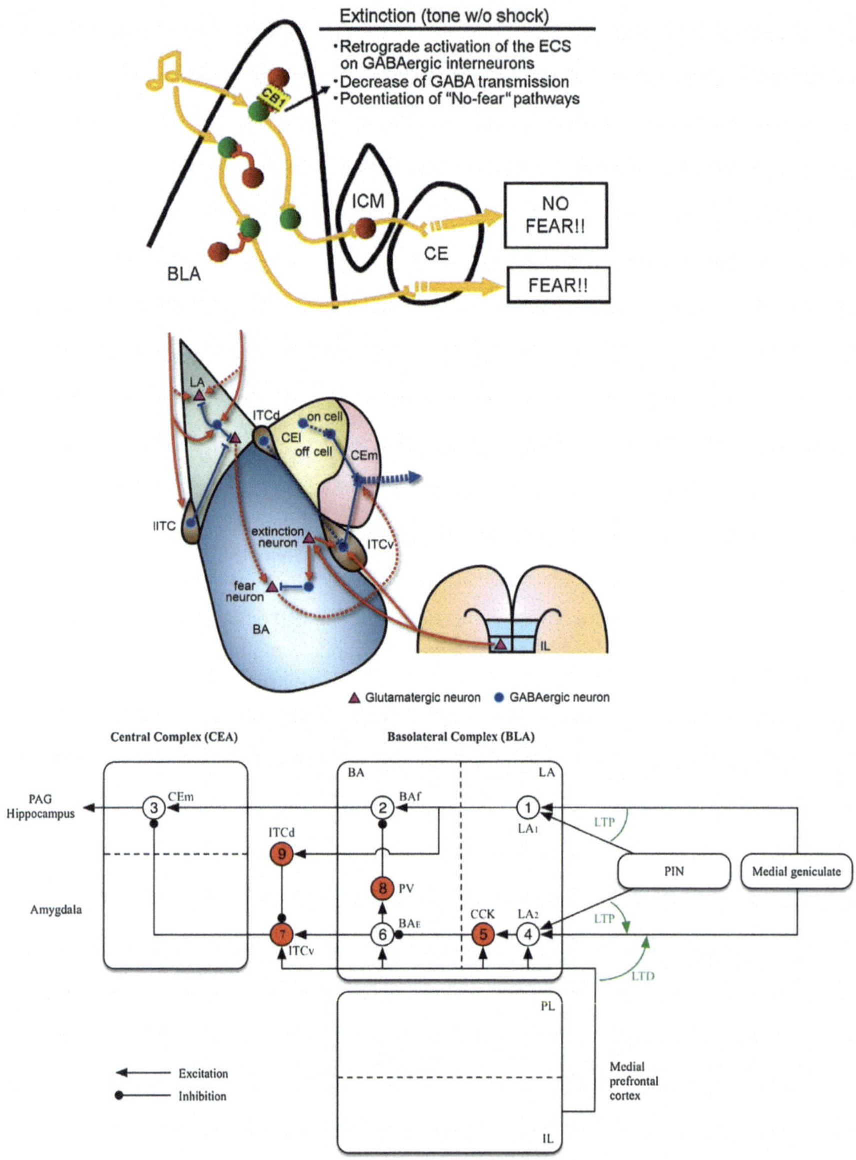

Fig. 5.1 A model of the amygdala network engaged in extinction: the role of endocannabinoids (from Bennett et al. 2017)

Choi 2009). This potentiated activity is projected from the LA to the basal (BA) and central (CEm) nuclei to initiate aversive behaviour (Davis and Whalen 2001; LeDoux 2000). On the other hand, the glutamatergic neurons in the LA fall into two classes (LA1 and LA2) in relation to extinction. One (LA2) shows a decrease in response to the CS after extinction and the other (LA1) does not (Fig. 5.1c; An et al. 2012). The LA1 neurons are found in the ventral regions of the dorsal LA and the LA2 neurons in the dorsal tip of the dorsal LA (Herry et al. 2008; Repa et al. 2001).

The LA1 neurons do not show changes after the original CS-US association, for example after reconditioning, whereas the LA2 neurons do, reflecting the reconditioned CS-US association, with a high degree of potentiation compared with the original CS-US (An et al. 2012). The original CS-US is not negated by subsequent extinction and the establishment of a separate extinction memory (Pavlov 1927; Patterson et al. 2001).

Stimulation of the MGm auditory input to the LA at low frequencies to induce long-term depression (LTD) at synapses in the LA inactivates the original CS-US memory, possibly by reducing the potentiated response of LA2 neurons to the original CS-US memory, without affecting the CS-US memory retained by LA1 neurons. This can be restored by stimulation of the MGm auditory input at high frequencies to induce LTP (Nabavi et al. 2014) possibly by restoring the relatively high efficacy of MGm synapses on LA2 neurons (Fig. 5.1c).

5.2.2 Basal Amygdala (BA) and Infralimbic (IL) Cortex

Extinction requires enhanced activity in the projection from the neurons in the infralimbic (IL) medial prefrontal cortex to neurons in the basolateral complex (BLA) and the intercalated cells (ITC) of the amygdala (Fig. 5.1; Holmes et al. 2012; Milad et al. 2004; Santini et al. 2004). If then the IL is lesioned before extinction training, then extinction is curtailed post-training (Quirk et al. 2000), whereas if IL is stimulated, extinction is facilitated (Bukalo et al. 2014, 2015). However, there is evidence of a decrease in the efficacy of transmission from the IL to excitatory neurons in the BLA and no change in efficacy of transmission from the IL to inhibitory interneurons in either the BLA or ITC (Fig. 5.1c; (Cho et al. 2013), with all of these neurons receiving direct monosynaptic glutamatergic projection from the IL (Cho et al. 2013).

The question arises as to how the IL, which comes into play during extinction, interacts with LA2 to exert its inhibitory influence on the output of the central nucleus (CEm; Fig. 5.1c) (Haubensak et al. 2010; Maren and Quirk 2004; Pape and Pare 2010). This process probably occurs as a consequence at least in part of depotentiation of LA2 neuron inputs of MGm synapses (Fig. 5.1c), concomitant with an increase in the activity of inhibitory interneurons in the BA and ITC (Fig. 5.1a, c) that also receive direct IL inputs (Fig. 5.1c) (Chhatwal et al. 2005; Hong et al. 2009; Likhtik et al. 2005; Lin et al. 2009; Milad et al. 2004; Rosenkranz et al. 2003; Sotres-Bayon et al. 2006). According to this scheme then extinction requires both a loss of memory of the original CS-US, occurring through at least depotentiation at the input of MGm projections on LA2 neurons (Fig. 5.1c), and an increase in the activity of inhibitory neurons, perhaps both in the BA and ITC (Fig. 5.1c), all of which receive a direct input from IL (Fig. 5.1c). So the new extinction memory and the old CS-US memory occur at LA2 and LA1 respectively in this scheme (Fig. 5.1c) (Medina et al. 2002). Furthermore, the observation that extinction can inhibit at least part of the original CS-US memory which can be restored by subsequent reconditioning (Bouton and King 1983; Myers and Davis 2007; Quirk 2006) and that this component of the original CS-US memory dominates the LA population of glutamatergic neurons as determined using either early gene expression methods (Hall et al. 2001; Reijmers et al. 2007) or field potential recordings (Rogan et al. 1997) suggests that it is the synapses on LA2 neurons in the scheme of Fig. 5.1c that mediate these forms of plasticity, leaving those on LA1 representing part of the original CS-US.

5.3 Classical LTP and LTD in the Functional Networks Mediating Acquisition and Extinction

The importance of LTP and LTD in the functioning of the amygdala network in acquisition and extinction highlights the importance of the molecular events that sustain these, which are considered in this section. In Figs. 5.2 and 5.3 which are used to summarize the mechanisms involved, we have superimposed the endocannabinoid synthesis pathways for economy of figures. These pathways are considered in a later section (Sect. 5.4).

5.3.1 Mechanism of Classical Early LTP and LTD at Synaptic Spines

The extent of the increase in calcium concentration in the spine following the binding of *N*-methyl-D-aspartate receptors (NMDAR) on the spine head by glutamate determines the molecular mechanisms that are set in train leading to either LTP or LTD (Bear et al. 1987; He et al. 2014; Lisman 1989; Yang et al. 1999). If the calcium concentration is sufficiently high, such as that which accompanies impulses at 100–200 Hz over a second (Citri and Malenka 2008), then protein kinases lead to phosphorylation that may increase both alpha-amino-3-hydroxy-5-methyl-4-isoxazolepropionic acid receptor (AMPAR) unitary channel conductance as well as the density of these receptors (Fig. 5.2). Alternatively if the calcium concentration increase is low (such as that which accompanies low frequency impulses at 1 Hz for 900 s), then protein phosphatases are activated which lead to dephosphorylations that both decrease the unitary AMPAR channel conductance and decrease the number of AMPAR receptors through accelerating AMPAR endocytosis from the spine head (Fig. 5.2). These early phases of LTP and LTD may be succeeded by late phases that last for many days and require gene expressions and are associated with structural changes in the spines (Engert and Bonhoeffer 1999).

The sequence of molecular events leading to LTP is as follows: first, at the time of glutamate binding to the NMDAR, magnesium is removed from the open channel (Nowak et al. 1984) by considerable depolarization across the channel-containing membrane (Lisman et al. 2002); second, the rise in intracellular calcium leads to the release of calcium calmodulin (CaM) bound to neurogranin (Baudier et al. 1991; LaPorte et al. 1980), and in the presence of high spine-head calcium the formation of CaMCa, which leads to both the formation of calcium calmodulin kinase II

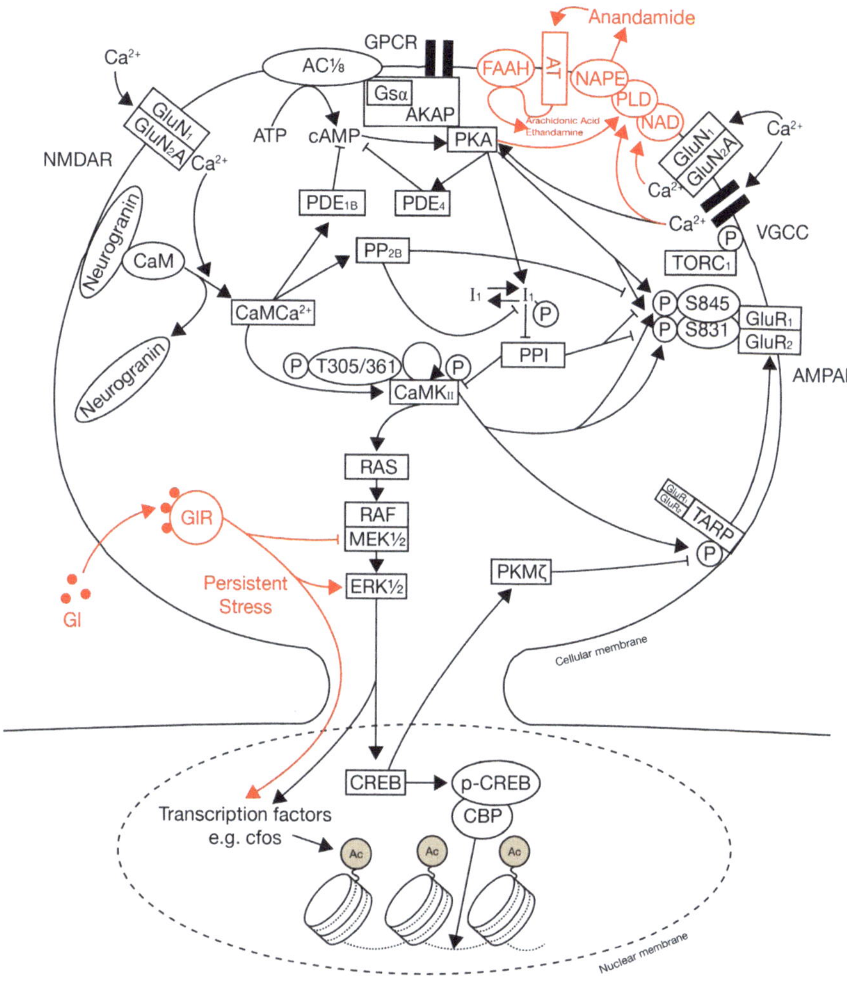

Fig. 5.2 Long-term potentiation (LTP), long-term depression (LTD) and endocannabinoids in the extinction network (from Bennett et al. 2017)

(CaMKII), constitutively active through autophosphorylation (Shonesy et al. 2014), and activation of adenylate cyclase and hence the production of high levels of cAMP (Fig. 5.2). This subsequently leads to the regulatory subunit of PKA releasing the catalytic subunit, which in turn phosphorylates serine 845 (S845) of the AMPAR GluR1 subunit to promote the movement of AMPAR into an extra-synaptic site away from PSD-95, the receptor anchorage region at the postsynaptic membrane in the spine head (Banke et al. 2000). The phosphorylation of S845 is not only modulated by PKA but also calcineurin (PP2b) and protein phosphatase (PP1) (Fig. 5.2; Beattie et al. 2000; Man et al. 2007). The binding partner for the AMPAR to PSD-95 in the spine head is the transmembrane AMPAR regulation

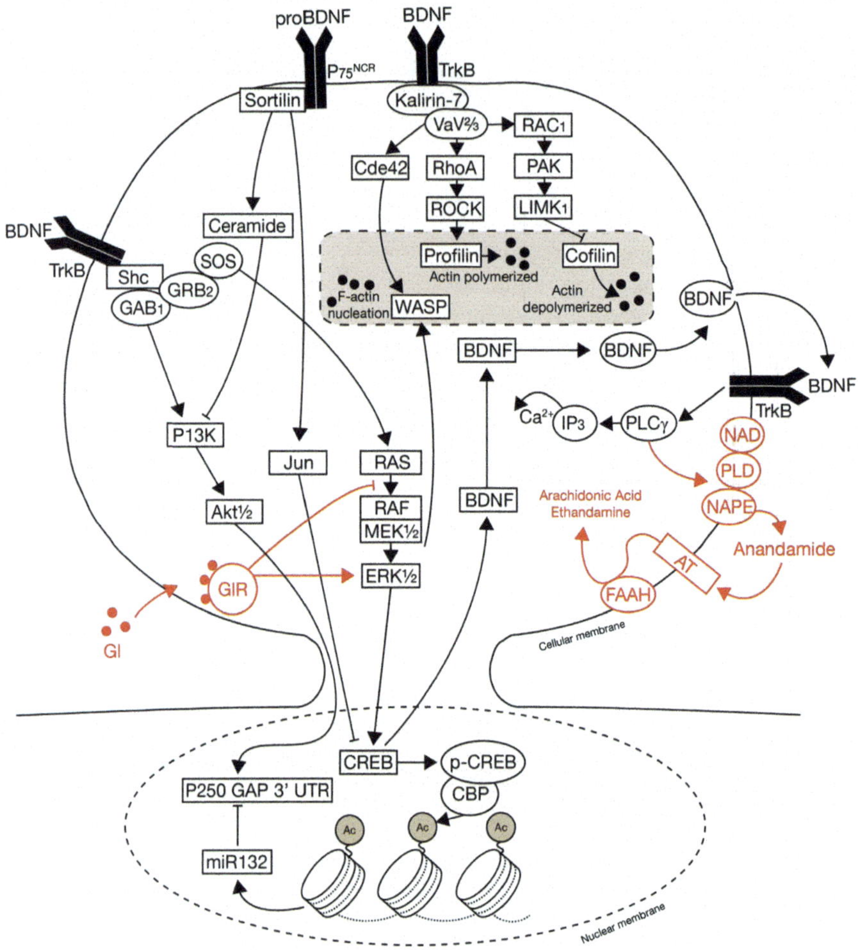

Fig. 5.3 The role of BDNF and endocannabinoids such as AEA in the extinction network (from Bennett et al. 2017)

protein (TARP; Banke et al. 2000). TARP binds AMPAR through its cytoplasmic tail (Chen et al. 2000; Tomita et al. 2003, 2005) and is a principal site of AMPAR and PSD-95 anchorage regulation through its phosphorylation, for example by CaMKII (Hayashi et al. 2000; Sumioka et al. 2010). In addition to CaMKII increasing the number of AMPAR at PSD-95, it also increases the unitary conductance of these receptors by phosphorylating residue serine 831 (S831) of the AMPAR GluR1 subunit (Fig. 5.2) (Barria et al. 1997; Kristensen et al. 2011).

The early events accompanying LTP are in contrast to those accompanying LTD under conditions of low CaMKII complex, for example through phosphorylation of T305/306 of CaMKII that is favoured under conditions of the sustained low-calcium levels during LTD inducing patterns of stimulation (Coultrap and Bayer 2012). Under these conditions, the extra-synaptic AMPAR, away from PSD-95 in the spine head, are removed by a process of endocytosis conditional on the dephosphorylation of S845 by phosphatases PP2b (Fig. 5.2) (Mulkey et al. 1993) with a concomitant decrease in S845 phosphorylation. PP2b also activates PP1 by dephosphorylation of inhibitor I1 (Mulkey et al. 1994), leading to decreases in CaMKII phosphorylation of S831 and hence in the AMPAR unitary conductance (Fig. 5.2).

Another critical source of control over AMPAR in the PSD-95 concerns the anchoring of protein kinase A (PKA) at the PSD-95 in the spine head by the A-kinase anchoring protein (AKAP; Fig. 5.2; Sanderson and Dell'Acqua 2011; Carr et al. 1992). AKAP is removed from the spines during LTD (Gomez et al. 2002), following actin polymerization by PP2b and cleavage of PIP2 by phospholipase C (Gomez et al. 2002).

In summary, there are three principal determinants of whether a synaptic spine participates in LTP or LTD: the relative calcium sensitivity of CaM binding proteins, the actions of phosphatases and kinases and AKAP localization and regulation of cAMP (Fig. 5.2).

5.3.2 Mechanism of Late Classical LTP and LTD at Synaptic Spines

5.3.2.1 Changes in Receptor Distributions

Synaptic spines fall into four main morphological groups: small stubby (Ia), large stubby (Ib), mushroom (II) and thin filipodia-like with a small head (III). These different shapes have important implications for the changes in intercellular calcium concentrations that accompany transmission and which give rise to LTP or LTD. In the LA, glutamatergic neurons may receive synapses on their small dendritic spines (type Ia) from medial prefrontal cortex neurons such as IL neurons and large dendritic spines (Ib and II) from thalamic afferents such as MGm, with both spine types sharing the same dendritic tree (Humeau et al. 2005). This distribution of synaptic boutons and spine morphological types is consistent with the generation of LTP at the thalamic synapses. Indeed, there is a large increase in GluR1- and GluR2-AMPAR with only few GluR2b at these spines during fear acquisition in a CS-US (Nedelescu et al. 2010; Radley et al. 2007). This corresponds with AMPAR at spines during acquisition being of the calcium-impermeable stable type (Hong et al. 2013; Maren 2005). NMDAR on synaptic spines in the LA possess both GluN1 and GluN2A or GluN2B subunits (Muller et al. 2009), and these control the endocytosis of new AMPAR during induction of LTP at the thalamic inputs on LA neurons (namely LA1, Fig. 5.1), with selective antagonists to GluN2A or GluN2B in NMDAR blocking this AMPAR/NMDAR-dependent endocytosis at the spines (Yu et al. 2010).

NMDAR in the LA are implicated in both UC-CS fear acquisition and extinction, as each of these are greatly impaired by NMDAR blockers (Falls et al. 1992; Kim et al. 2007; Miserendino et al. 1990; Sotres-Bayon et al. 2006). Fear acquisition is also decreased by interfering with cyclic AMP-dependent protein kinase (PKA) and macromolecular synthesis, such as that of BDNF (Herry et al. 2006; Schafe et al. 2001). This is to be expected given that these pathways are necessary for induction, expression and maintenance of LTP and LTD (Figs. 5.2 and 5.3). Fear extinction is impaired by blocking ventral medial prefrontal cortex IL protein synthesis (Santini et al. 2004), MAP kinase (MAPK) (Hugues et al. 2006) as well as antagonizing NMDAR (Burgos-Robles et al. 2007).

5.3.2.2 Changes in Spine Shape

As noted above, spines come in four principal shapes, small and large stubby, mushroom and thin (filipodia-like with a thin head) (Humeau et al. 2005). During LTP, the size of some of these spine shapes increases (Bourne and Harris 2008; Lisman and Raghavachari 2006: Tanaka et al. 2008), often accompanied by the formation of entirely new spines (Maletic-Savatic et al. 1999), with synaptic boutons forming on many of these (Kwon and Sabatini 2011; Nagerl et al. 2007; Zito et al. 2009). The shape and size of these spines is correlated with the efficacy of the synapses (Bellot et al. 2014) as in LTP (Panja and Bramham 2014). The changes in spine shape and growth involve changes to the dynamics of actin in the spine head (Fig. 5.3) (Bourne et al. 2007; Bramham 2008; Fischer et al. 1998; Murakoshi and Yasuda 2012; Okamoto et al. 2004), which occur during induction and maintenance of LTP (Fukazawa et al. 2003; Kelly and Lynch 2000; Krucker et al. 2000) and require protein synthesis (Tanaka et al. 2008). In particular, the synthesis of BDNF and its subsequent action on TrkB receptors in the membrane promote spine growth (Fig. 5.3), (Tanaka et al. 2008), especially that of stubby spines (Tyler and Pozzo-Miller 2003), thereby determining the density and distribution of different spine types. BDNF effects spine density through activation of the extracellular signal-regulated kinase (ERK)1/2 pathway, and so increasing WASP induced nucleation of F-actin at Arp2/3 (Fig. 5.4) (Alonso et al. 2004; Finkbeiner et al. 1997; Lonze and Ginty 2002; Rex et al. 2007) as well as through the Rho-GTPase-Rac1-LIMK1 pathway (Fig. 5.3), (Yin et al. 2002) inhibiting cofilin/ADB responsible for F-actin depolymerization (Meng et al. 2002) through actin severing protein cofilin (Fig. 5.3), (Bramham and Wells 2007). The increase in the growth of spine heads during maintenance of LTP also occurs through BDNF promotion of actin polymerization also (Meng et al. 2002; Sala et al. 2001). Actin filaments in the spine head are responsible for both exerting forces on the membrane-containing AMPAR and the transport of AMPAR in vesicles into specific spine regions. Accordingly these mechanisms are under the control of BDNF with AMPAR proteins interacting with the actin cytoskeleton being identified for GluR1 as MyoVI/SAP97, RIL/alpha-actinin-2, MyoVa and band 4.1. For the GluR2 subunit of AMPAR, the associated proteins are Pick1/Arp2/3 (Dixon et al. 2009). Note that the transport of PSD95 and probably PKMeta into spines is dependent on incursions of microtubules from the dendritic shaft into the spine, under the control of BDNF (Hu et al. 2011).

Myosin IIb motors exert mechanical effects on the spine actin cytoskeleton following synaptic activity. By binding to and contracting actin filaments (Ryu et al. 2006) myosin IIb motors the shape of the spines, especially the mushroom-

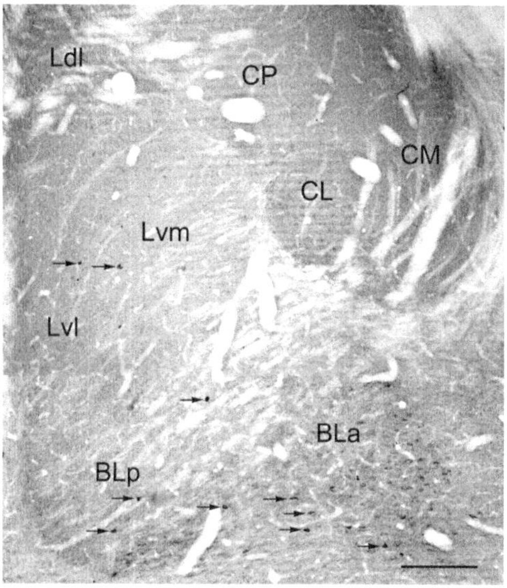

Fig. 5.4 Photomontage of CB1 cannabinoid receptor immunoreactivity in the basolateral and central amygdalar nuclei in a coronal section (bregma level 33.14 of the atlas by Paxinos and Watson Paxinos and Watson, 1986. Note that there are a small number of intensely stained CB1+ non-pyramidal neurons (arrows) in the nuclei of the basolateral nuclear complex (Ldl, dorsolateral subdivision of the lateral nucleus; Lvl, ventrolateral subdivision of the lateral nucleus; Lvm, ventromedial subdivision of the lateral nucleus; BLa, anterior subdivision of the basolateral nucleus; BLp, posterior subdivision of the basolateral nucleus). In addition, there are numerous lightly stained pyramidal cells in the ventromedial (lower right) portions of the BLa and BLp. Other abbreviations: CP, caudate-putamen; CL, lateral subdivision of the central amygdalar nucleus; CM, medial subdivision of the central amygdalar nucleus. Scale bar = 200 μm (from McDonald and Mascagni 2001)

shape spines (Rex et al. 2010). This non-muscle myosin IIb determines the transition of the thin filipodia-like spines to mushroom shape as well as the formation of the filipodia-like spine protrusions from the dendritic shaft in the first place (Hodges et al. 2011). Both alpha CaMKII mRNA and polyribosomes increase significantly during induction of LTP (Bourne et al. 2007; Havik et al. 2003; Ostroff et al. 2002) due to motor protein myosin Va (Kneussel and Wagner 2013; Yoshimura et al. 2006).

proBDNF bound to its receptor p75NTR depresses spine growth, particularly mushroom spines as well as depressing spine density (Fig. 5.3) (Chapleau and Pozzo-Miller 2012; Koshimizu et al. 2009). The importance of proBDNF in long-term depression and potentiation is shown by the observation that LTP decreases and LTD increases in proBDNF knock-in mice (Woo et al. 2005; Yang et al. 2014), whereas proBDNF +/− mice are unable to maintain LTD following theta burst stimulation (Korte et al. 1995). It is interesting in this regard that acute stress impairs extinction and enhances mushroom spine density in the BLA (Maroun et al. 2013; Maren and Holmes 2016).

5.4 Endocannabinoids in Functioning Networks in the Basolateral Amygdala

The endocannabinoids (Di Marzo et al. 1998) and their receptor CB1R (Matsuda et al. 1990) are implicated in synaptic plasticity in many areas of the brain (Katona et al. 2001; Marsicano et al. 2002) and are known to modulate behavioural memory (De Oliveira Alvares et al. 2008; Hampson and Deadwyler 1998; Reibaud et al. 1999), including spatial memory (Varvel et al. 2005).

5.4.1 Evidence that Endocannabinoids Have a Critical Role in Extinction

The BLA possesses exceptionally high concentrations of CB1Rs, particularly on non-pyramidal neurons in the basolateral nuclei (BA), with also a relatively large number of pyramidal neurons in the BA possessing low concentrations of CB1R (Katona et al. 2001; Marsicano et al. 2002). These receptors, most likely on the non-pyramidal neurons, mediate synaptic plasticity at GABAergic synapses in the BLA (Marsicano et al. 2002). Endocannabinoids, such as anandamide, are synthesized and released from the membrane following increases in spine intracellular calcium, mostly through activation of voltage-gated channels following depolarization, as well as increases in PKA following metabotropic receptor activation (Fig. 5.2). Although CB2Rs occur in the amygdala, and small elevations in CB2R mRNA occur following stress, there is no quantitative measure of their distribution (Gong et al. 2006); furthermore, CB2Rs do not mediate contextual extinction whereas CB1Rs do (Ganon-Elazar and Akirav 2013).

5.4.1.1 Acquisition

Acquisition of a conditioned fear response is dependent on the LA (see Sect. 5.1.1), which as part of the BLA has an extensive CB1R/endocannabinoid system. Since injection of the CB1R agonist WIN 55, 212-2 (WIN) or the antagonist AM251 into the BLA has no effect on acquisition, endocannabinoids are not considered to be involved in acquisition (Kuhnert et al. 2013).

5.4.1.2 Extinction

Classical fear conditioning, involving acquisition and extinction, is modulated by monoamines, opioids, neuropeptide Y, oxytocin, glucocorticoids and endocannabinoids (for a recent review, see Stockhorst and Antov 2015). Endocannabinoids are particularly important in the process of extinction (Marsicano et al. 2002; Pamplona et al. 2008; Terranova et al. 1996). Levels of endocannabinoids such as anandamide (N arachidonoylethanolamine; AEA) and 2-arachidonoylglycerol (2-AG) increase significantly in the BLA during extinction procedures (Marsicano et al. 2002), with even greater increases observed following systemic administration of AM3506, an inhibitor of the membrane-bound enzyme FAAH that normally degrades AEA (Gunduz-Cinar et al. 2013). The importance of cannabinoids in extinction is emphasized by the extent that it is impaired in CB1R null mutant mice (Kamprath et al. 2006), leaving acquisition unaffected (Marsicano et al. 2002). It is also impaired following administration of systemic CB1R antagonists prior to extinction training (Chhatwal et al. 2005). Indeed, the normal stress-induced interruption of extinction is prevented by prior injection of a CB1R agonist (WIN) into the BLA (Ganon-Elazar and Akirav 2013; Pamplona et al. 2006) without affecting conditioned fear (Pamplona et al. 2006, 2008).

Extinction is also facilitated by enhancing the activation of CB1R in the infralimbic medial prefrontal cortex by injecting QIN or using FAAH inhibitors which increase synaptic AEA concentrations (Lin et al. 2009). On the other hand, extinction is diminished following injections of the CB1R antagonist AM251 into the IL (Kuhnert et al. 2013). Injection of CB1R antagonists into CA1 blocks contextual fear extinction, whereas injection of AEA antagonizes the consolidation of contextual fear (De Oliveira Alvares et al. 2008).

5.4.2 Mechanisms of Action of Endocannabinoids

Activated CB1R on both glutamatergic and GABAergic terminals decrease the release of these transmitters (Fig. 5.5); (Ramikie and Patel 2012), so that deletion of CB1R on only glutamatergic cells in conditional knockout mice, and so decrease in the release of glutamate from these mediated by the mechanisms of depolarization suppression of excitation (DSE) and classical LTD, gives enhanced LTP (Fig. 5.5); on the other hand, CB1R deletion on only GABAergic cells decreases LTP, as expected given the enhanced release of GABA following loss of long-term depression of inhibition (LTDi; Monory et al. 2015). However, CB1R knockout mice show increased LTP (Bohme et al. 2000; Marsicano et al. 2002) in both CA1 and amygdala, as well as an increase in dendritic spine density (Hill et al. 2011), presumably because the enhanced glutamatergic transmission overcomes the enhanced GABA transmission.

Activation of CB1R on GABAergic nerve terminals inhibits adenyl cyclase, decreases cAMP levels and reduces PKA activity, with in addition increased intraterminal Ca2+ activating calcineurin; together these enhance dephosphorylation of proteins necessary for transmitter release, leading to LTDi of transmission in the amygdala (Fig. 5.5) (Chevaleyre et al. 2007) (for reviews, see Kano 2014; Ohno-Shosaku et al. 2012). Evidence suggests that AEA is the principal endocannabinoid mediating LTDi during high-frequency impulse firing (Lerner and Kreitzer 2012), with such firing required to give persistent activation of CB1Rs in order to initiate LTDi in somatosensory cortex (Oberheim et al. 2009).

The hypothesis that there is a tonic inhibition of extinction neurons in the BLA, due to GABAergic neurons, which is relieved during LTDi accompanying extinction is supported by the observation that FAAH inhibitors as well as FAAH KO mice

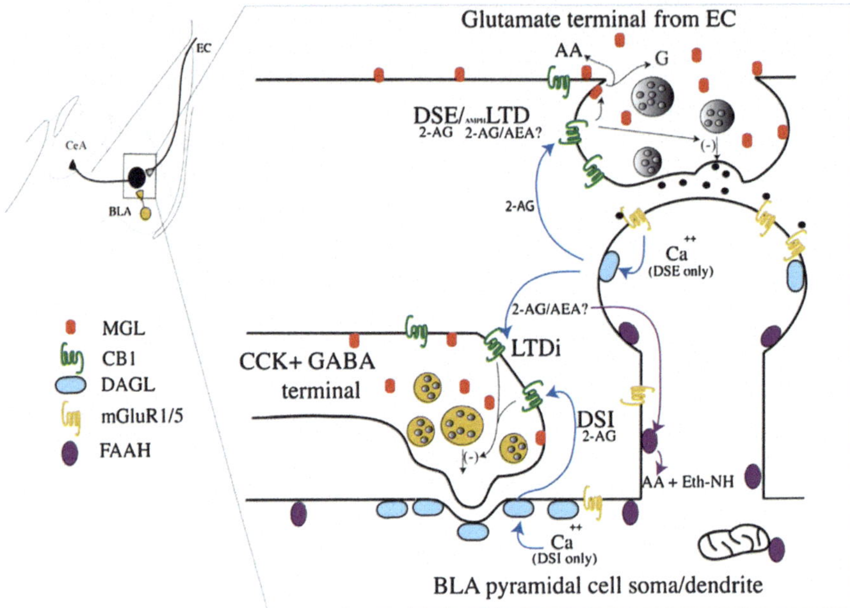

Fig. 5.5 Schematic diagram depicting the regional heterogeneity in the expression of the molecular components of EC signalling between different amygdala subnuclei. The localization of $DAGL_a$, CB1 receptors, FAAH and MGL are depicted based on data summarized in the text. Left top inset: diagram of the amygdala with simplified cellular organization. BLA projection neuron in black, GABAergic interneuron in yellow and glutamatergic inputs in grey. Boxed area is expanded in the main figure. Synaptic localization of EC signalling components in a schematized BLA neuron receiving synaptic inputs from a GABAergic interneuron and glutamatergic input from outside the amygdala, e.g. infralimbic cortex. Known forms of EC-mediated synaptic signalling at these synapses are overlaid, and specific EC ligands noted where data is available (from Ramikie and Patel 2012). (For interpretation of the references to colour in this figure legend, the reader is referred to the web version of this article.)

enhance the LTDi of GABAergic terminals during extinction (Gunduz-Cinar et al. 2013; Marsicano et al. 2002). Whether extinction is accompanied by DSE of excitatory glutamatergic inputs to pyramidal neurons mediated by endocannabinoids is unclear (Fig. 5.5), especially as these inputs appear to only possess CB1R at low density (Fig. 5.4).

5.4.3 Endocannabinoid Modulation by Glucocorticoids

The facilitation of fear extinction by glucocorticoids is mediated by endocannabinoids (Bitencourt et al. 2014). The consolidation of memories for emotionally arousing experiences is enhanced by endocannabinoids through a mechanism triggered by glucocorticoids (see Fig. 5.3; Campolongo et al. 2009). Elevation of a glucocorticoid in the amygdala increases AEA (see Fig. 5.3; Hill et al.

2010). The action of glucocorticoids in the amygdala in consolidating emotionally arousing experiences is mediated by endocannabinoids (Atsak et al. 2015).

The extinction of conditioned fear is enhanced by glucocorticoids and diminished by glucocorticoid receptor (GlR) antagonists in animals (Barrett and Gonzalez-Lima 2004; Blundell et al. 2011; Clay et al. 2011; de Quervain et al. 2011; Ninomiya et al. 2010; Yang et al. 2006, 2007), and cortisol is effective in improving extinction-based cognitive therapies in human patients (Aerni et al. 2004; de Quervain et al. 2011). These observations are consistent with the idea that glucocorticoids probably have their action through the release of endocannabinoids from the postsynaptic membranes (Fig. 5.2) (McEwen et al. 2015). This may be effected through the known action of glucocorticoids to elevate protein lipase D (PLD) in some cells (Kim et al. 2000), as PLD acts on the membrane lipid precursor *N*-arachidonoyl-phosphatidylethanolamine to release AEA from the membrane into the extracellular space (Fig. 5.2). On the other hand, activated GlR, for example with cortisol, have variable effects on CaMKII stimulation of CREB and therefore on BDNF synthesis (Fig. 5.2).

The level of glucocorticoids in the amygdala and elsewhere is indirectly under the control of levels of endocannabinoids in the BLA. This occurs as the endocannabinoids set the excitability of the BLA input to the CEm output of the amygdala, which in turn projects to the hypothalamus in the hypothalamic–pituitary–adrenal (HPA) axis (Di et al. 2003; Freund et al. 2003; Gorzalka et al. 2008; Patel et al. 2004; Steiner and Wotjak 2008).

5.4.4 Endocannabinoid Modulation by BDNF

Endogenous cannabinoids such as AEA and 2-AG act on CB1R to activate both MAPK and ERK, downstream from MAPK, at least in CA1 and CA3 (Fig. 5.3) (Derkinderen et al. 2001, 2003; Karanian et al. 2005; Winder et al. 1999). In addition, there is upregulation of BDNF mRNA (Kang and Schuman 1995; Patterson et al. 2001; Ying et al. 2002), possibly through ERK activation of immediate early genes and/or CREB (Fig. 5.2), thereby enhancing the maintenance of LTP (Collins et al. 1995; Impey et al. 1999; Misner and Sullivan 1999; Terranova et al. 1995). Importantly, endocannabinoids are released from somatosensory neurons by BDNF (at least from layer 2/3 neurons) (Zhao et al. 2015), through a PLC/gamma-dependant pathway (Fig. 5.3) that can also upregulate CB1R in at least cerebellar neurons (Maison et al. 2009). As newly synthesized BDNF in a neuron can be packaged into endosomes and subsequently released from these into the extracellular space by impulses (Wong et al. 2015), it is possible that such endogenously released BDNF acting on TrkB receptors on the same neuron can lead to further BDNF synthesis, thus maintaining a level of extracellular BDNF sufficient to trigger continual synthesis and release of cannabinoids (Fig. 5.3). Such a mechanism might be responsible for the known dependence of the retention of extinction on BDNF in the BLA (Andero and Ressler 2012; Chhatwal et al. 2006; Heldt et al. 2007).

5.5 Functional Amygdala Networks that Mediate Extinction: The Roles of LTDi and Endocannabinoids at Extinction Neurons in the Basolateral Amygdala

5.5.1 Summary of Key Observations

(a) CB1R null mutants are severely impaired in extinction of a CS (fear tone) and it appears that the activation of CB1R is required at the moment of fear memory extinction for this impairment to occur (Kamprath et al. 2006).

(b) ECs and CB1R are critically involved in LTDi at GABAergic synapses in the BLA as evidenced by their role in determining the amplitude of inhibitory postsynaptic currents (IPSCs).

(c) As CB1R are only found on GABA/CCK interneurons in the BLA in high density in the vicinity of principal excitatory pyramidal neurons (extinction neurons, BAe) (Collins and Pare 1999: McDonald and Pearson 1989; Pare and Gaudreau 1996; Rainnie et al. 1991; Takagi et al. 1981) (see Fig. 5.4 and especially Katona et al. 2001; McDonald and Mascagni 2001), it appears likely that the very large increase in AEA found in the BLA during extinction has its origins from the BAe neurons.

In the BLA, cannabinoid receptor agonists modulate both excitatory and inhibitory synaptic transmission via CB1R presynaptic receptors (Azad et al. 2003; Domenici et al. 2006). Boosting the extent of endocannabinoid release with depolarization of the source neuron (Shin et al. 2010) or enhancement of precursors in endocannabinoid releasing neurons (Shonesy et al. 2014) inhibits local release of glutamate from nerve terminals. The question of whether there is an endogenous suppression of glutamate release during extinction, given the relatively low density of CB1R at these sites (Fig. 5.4), still remains unanswered, even given that such transmission is blocked by unknown concentrations of exogenous excitants of CB1R.

(d) The GABA/CCK interneurons are likely to be firing tonically, as they do in other parts of the brain (Hausser and Clark 1997), thereby exerting a tonic suppression of BAe activity.

(e) Depolarization, together with activation of mGLuR (using the metabotropic agonist DHPG), induces a short-term depression of inhibitory postsynaptic currents (IPSCs) due to ECs, namely LTDi (at least in hippocampal CA1; Sheinin et al. 2008). It is probably this mechanism that leads to disinhibition of the BAe neurons during extinction and so their excitation via the input from the IL (see Fig. 5.1). However, as noted, this is transient and presumably occurs for such a short period that it cannot provide an explanation for the fact that extinction typically lasts for weeks or months (see (f) below).

(f) ECs are released from the BAe by a combination of depolarization due to the action of glutamate released from the IL terminals acting on NMDA receptors together with that of GABA released from the GABA/CCK interneurons onto GPCR to produce protein lipase C (Fig. 5.2). ECs can also be mobilized in the BAe through the impulse-dependent production of BDNF that through a diacylglycerol (DGL)-dependent pathway stimulates EC production (Zhao et al. 2015) (Fig. 5.3). This provides a much longer time course production of ECs and therefore of extinction. In this regard, ECs themselves can initiate the release of BDNF, by activating on a pool of ERK (at least in the hippocampus) which leads to immediate early gene products like BDNF mRNA; these in turn can then give rise to the release of further ECs as noted above (Fig. 5.3), in a positive closed loop (autocrine) process that could greatly extend the time of release of ECs. (Derkinderen et al. 2001; Patterson et al. 2001; Winder et al. 1999; Ying et al. 2002).

(g) The synaptic strength of IL synapses on the GABA interneurons does not change during extinction, so providing a strong drive to the tonic inhibition of the GABA/CCK synapses. On the other hand, some weakening of the IL synapses on the principal pyramidal excitatory neurons may occur (Cho et al. 2013). It has been argued that this will lead to a general increase in inhibition in the BLA and a decrease in excitation.

(h) Heterosynaptic inhibition of auditory (MGm) input to the BLA (LA) occurs with priming stimulation of medial prefrontal cortex (IL) (Cho et al. 2013). This cannot involve the release of ECs, as there are no CB1R at this site in the BLA (see point c above). Presumably then it involves the classical LTD processes (see Sect. 5.1). It may be that this involves incorporation of GluR2/3 into the AMPAR, a subunit that is known to lead to long-term weakening of presynaptic function at LA synapses (Kim et al. 2015).

5.5.2 A Model of the Amygdala in Extinction

1. Considering the acquisition circuit (UC-CS conditioning; see Fig. 5.1c; Sect. 5.2.1). MGm projects synapses onto LA glutamatergic neurons, in particular on large spines, which are potentiated (LTP) during fear/pain CS-US; the molecular mechanisms involved are those for classical LTP. The existence of two LA neurons types requires a division of the classical projection patterns which are as follows: the LA1 neurons project directly to the BA fear neurons (BAf) and the dorsal ITC inhibitory neurons as in the classical fear network; the LA2 neurons project to inhibitory neurons in the BA and thence to the extinction neurons (BAe) that in turn project to the ventral ITC neurons (ITCv). Thus, the LA1 pathway directly excites the CEm output of the central amygdala, which is also excited by the ITC dorsal neurons (ITCd) through disinhibition of the ITCv neurons. The LA2 pathway also excites the CEm during acquisition by lowering the firing of inhibitory neurons in ITCv that project to CEm.
2. Considering the extinction circuit (see Fig. 5.1c and Sect. 5.2). There are two classes of LA glutamatergic neurons: one which is relatively unaffected by extinction of the CS-US fear acquisition potentiation (LA1) and the other which is LA2. LA2 neurons alone receive input from IL cortex and concurrently with repeated exposure to the CS leads to depotentiation as a consequence of classical LTP at its MGm synapses. This leaves the memory of the UC-CS in the

form of potentiated CS synapses on LA1 unaffected but that on LA2 depotentiated, with the net effect that the BAe are disinhibited, so allowing BAe to drive via inhibitory PV neurons, a decrease in BAf firing and an increase in ITCv firing, both leading to a decrease in CEm output. This network then separates out the acquisition (LA1) and extinction (LA2) memories in the LA; it does not involve extinction of the LA1 memory that is then retained, but does leave open the question of the IL projection to BLA outside of the input on LA2 in the LA. Very importantly, it is now known that IL has direct monosynaptic projections to both excitatory and inhibitory neurons in the BA and ITC (Fig. 5.1c) and that there is tonic firing of ITC neurons (Pare et al. 2004; Haubensak et al. 2010). Given the former, if the IL is firing it will excite not only LA2 neurons but also CCK inhibitory interneurons in BA, the BAe neurons and the ITCv neurons (Fig. 5.1c)

3. Considering the processes involved in LTP and LTD at LA synaptic spines (see Sect. 5.2). (1) The relative calcium sensitivity of the calcium/calmodulin (CaM)-binding protein CaMKII and hence on the calcium entering the spine head through GluN2A containing NMDA receptors in the case of LTP and GluN2B containing NMDA receptors in the case of LTD. (2) The shape/dimensions of the spine as a determinant of the transient calcium concentration for a given calcium entry through the NMDA channel. (3) The action of various phosphatases and kinases. (4) AKAP localization and therefore capacity to regulate cAMP and AMPAR exocytosis (Fig. 5.2). (5) The extent of activation of the PP2B path to enhance AMPAR endocytosis (Fig. 5.2).
4. The production of BDNF in the polyribosomes (under control of CaMKII–MAPK–Erk pathway) and its subsequent action on Trk1 receptors to lead (via RAC1 to PAK to LIM) to regulation of actin polymerization as do myosin IIb motors (Fig. 5.3 and Sect. 5.2). BDNF is packaged and released from endosomes, with such BDNF being available to act in an autocrine manner on the neurons own Trk1 receptors, thus sustaining its own release. Also the production of proBDNF (via Sortilin to Ceramide to P13K to Akt1/2 pathway) to decrease spine size (e.g. from mushroom to thin spines; Fig. 5.3).
5. During extinction IL neurons produce a classical LTD of the input from MGm, so lowering the capacity of the CS fear to utilize the LA2 pathway (Fig. 5.1 and Sects. 5.1 and 5.2).
6. ECs are only released in the BLA during extinction, and without their action extinction is greatly restricted. ECs have no role to play in acquisition (see Sect. 5.3).
7. ECs, such as AEA, are synthesized from neurons in the BLA through a classical calcium/diacylglycerol (DAG) pathway and released to act presynaptically where they inhibit the secretion of GABA (Figs. 5.1c and 5.5; see Sect. 5.3). CB1R are only found in high concentrations on the GABA/CCK containing inhibitory neurons in the BA, so it is only the terminals of these that can have their GABA release reduced by the release of ECs from postsynaptic neurons, which in this case are the BAe neurons, so producing LTDi.

8. The inhibitory GABA/CCK neurons fail to inhibit the BAe neurons during extinction, as the excitatory input to these neurons from the IL cortex leads to the release of ECs that acts presynaptically to block GABA release. Thus, GABA/CCK neurons have significant repression of their transmitter release through an LTPi mechanism using ECs as well as depression of the medial geniculate input to LA2 through a classical LTD mechanism (see point 5 above). The net effect of this is increased excitation of the BAe neurons, their drive on the ITCv neurons and so inhibition of the CEm output neurons (Fig. 5.1c and Sect. 5.3).
9. As BDNF acts in an autocrine manner on its release from endosomes into the extracellular space, and BDNF enhances the synthesis and release of anandamide, then a pathway can be established during extinction that leads to long-term elevation of AEA in the BLA. This has the potential to contribute to a sustained decrease of GABA/CCK neuron inhibition of BAe neurons and so enhanced activity of ITC neurons and their depression of the CEm output neurons of the amygdala.

5.5.3 Caveats Concerning Postulated Roles for Endocannabinoids and Brain-Derived Neurotrophic Factor in Extinction

1. Fear extinction gives a net depression of conditioning-induced potentiation of cortical input (IL) synapses into the LA, suggesting that extinction of a conditional fear involves a "distinct" form of depotentiation at these synapses dependent on both NMDARs and group II mGluR (these are the collective triggers for release of ECs (Hong et al. 2009).
2. Claims that extinction involves activation of mGluR2/3 causing long-term weakening of the enhanced presynaptic function acquired during conditioned fear in the LA (Kim et al. 2015). This seems to claim that extinction is a weakening of acquisition and in LA, contrary to many other observations.
3. Agonists at mGluR1 synapses reduce synaptic depotentiation only at thalamic input synapses in the LA where potentiation is consolidated (acquisition), but not at non-consolidated synapses. These consolidated synapses can become labile following memory reactivation in which they are insensitive to mGluR1 agonists.
4. It is unclear at present whether the LA2 neurons project directly to the GABA/CCK interneurons in BLA as in Fig. 5.1c, or directly to the BAe neurons leaving the input to the GABA/CCK neurons unaccounted for (compare Fig. 5.1a with 5.1c). In either case this does not substantially affect the hypothesis regarding the origin and sites of action of ECs in BLA.
5. Finally, it remains unclear whether there is a decrease in glutametergic synaptic transmission to pyramidal neurons in the BLA during extinction, mediated by the EC activation of DSE.

5.6 Conclusion

A review of the relevant literature suggests the following answers to the questions posed in the introduction, namely which are the principal cells in the amygdala that release ECs and what is their main site(s) of action during extinction? Recent observations suggest that the principal cells involved in EC synthesis and release during extinction are the BAe neurons in the basal nucleus. The main site of action of the ECs are the adjacent GABA/CCK interneurons, whose normal role of blocking the BAe neurons is greatly diminished through LTDi. The molecular pathways leading to the release of ECs from dendritic spines of the BAe neurons, together with the positive effects exerted on these pathways by concomitantly released BDNF, are also noted. These pathways suggest that the long-term effects of extinction trials may be sustained by a positive feedback (autocrine) loop in the production and release of BDNF from BAe neurons that in turn sustains the continual synthesis and release of ECs from the neurons. Further experimental work is required to justify these conclusions from the literature.

(A) Working hypothesis of extinction processing in the amygdala. Schematic representation of the amygdala, with reference to subnuclei. During conditioning, the simultaneous presentation of the tone (CS) and the shock (US) potentiates "fear" circuits. During extinction, non-reinforced and prolonged tone presentation causes sustained stimulation of "no-fear" pathways, which is "disinhibited" through ECS (endocannabinoid)-mediated inhibition of GABAergic transmission (BLA, basolateral amygdala; CE, central nucleus of amygdala; ICM, intercalated cell masses. Green, glutamatergic pathways; red, GABAergic neurons (from Lafenetre et al. 2007).
(B) Model of amygdala circuitry during fear conditioning and extinction. The model contains both known and hypothetical neural connections. Solid and dashed lines indicate strengthened and weakened connections, respectively, during the learning processes. During fear extinction, enhanced neural activity in LA is reduced by increased suppression from local inhibitory circuits and depotentiation of glutamatergic synapses. Decreased connectivity between extinction-resistant neurons in LA and fear neurons in BA could also contribute to fear extinction. Furthermore, IL cortex sends glutamatergic projections to ITCv and possibly extinction neurons of BA. These IL projections may participate in suppression of BA fear neurons via local inhibitory circuits (from Lee et al. 2013).
(C) In acquisition the US stimulus (e.g. a foot shock with the associated pain signals relayed through the posterior intralaminar nucleus of the thalamus; PIN) occurs at a time that overlaps with the CS (e.g. a sound relayed through the medial division of the medial geniculate body; MGm), with each of these delivering impulses at synapses on two different sets of lateral amygdala (LA) neurons: the LA1 neurons (numbered 1) found in the ventral LA and the LA2 neurons (4) in the dorsal LA. The MGm synapses on LA1 and LA2 are long-term potentiated (LTP) by the simultaneous inputs from PIN. When MGm alone is now excited

the output of the amygdala is as follows: LA1 projects to the fear neurons in the basal amygdala (BA), namely BAf (2), which in turn project to the output neurons of the central amygdala (CEA), namely CEm (3), which have control over the periaqueductal grey (PAG) and the hypothalamus (Hyp). LA1 neurons, in addition to projecting to BAf, also project to inhibitory interneurons (GABA) in the dorsal intercalated cells region (ITCd) (9) of the amygdala, which in turn project to inhibitory neurons in the ventral ITC region (ITCv) (7); these synapse on the output neurons, CEm. So this LA1 pathway produces disinhibition of ITCv neurons and hence relieves the CEm of an inhibitory input, thus allowing its excitation via BAf to continue unabated.

In acquisition, the LA2 neurons (4) similarly acquire a potentiated MGm input from simultaneous firing with the PIN input, and when MGm alone now fires it has the following effects on CEm. LA2 excites inhibitory GABA/CCK neurons (5) in the BLA which in turn inhibit the BA extinction neurons, namely BAe (6). Consequently, this no longer excites ITCv neurons nor GABA/PV neurons (8) that project to BAf neurons, with the net effect that the enhanced MGm input to LA2 (the CS sound stimulus) has no effect on the amygdala output via CEm.

In extinction, there is repeated delivery of the MGm (the CS sound stimulus) in the absence of the PIN input (the US pain stimulus). The effect of the LA1 neurons on the CEm neurons remains the same, as that for the fully potentiated MGm-LA1 synapses at the end of acquisition. However, the effect of the LA2 neurons is now quite different, for now there is an excitatory input to these neurons from the infralimbic (IL) medial prefrontal cortex to all the neurons in the LA2 to CEm pathway (that is to neurons LA2, GABA/CCK, BAe and ITCv) during the process of extinction. The simultaneous firing of IL and MGm leads to long-term depression of the MGm input (LTD), so that it is depressed below its previously potentiated values, failing to excite the GABA/CCK neurons and leading to disinhibition of ITCv neurons and continuing excitement of CEm through the LA1 pathway. It would then be anticipated that the IL input to the GABA/CCK neurons would be able to drive these to inhibit BAe, but instead this pathway fails because of the depolarization of the BAe neurons together with the release of GABA onto metabotropic receptors that leads to the release of endocannabinoids from BAe onto presynaptic cannabinoid receptors (CB1R) found uniquely on the GABA/CCK terminals, leading to significant reduction in GABA release; this is the process of LTDi (see Fig. 5.5). Thus, the GABA/CCK terminals decrease their secretory activity because of both a decreased input from LA2 neurons and the release of endocannabinoids (e.g. AEA) from BAe. The result is that BAe excitation from IL continues in the absence of an inhibitory input, allowing BAe to excite ITCv, which also receives an excitatory input directly from IL, collectively providing a powerful inhibitory input to CEm and hence extinction of this amygdala output. (Note also that the uninhibited activity of BAe, excited by IL, enhances the inhibitory input from GABA/PV neurons (8) on BAf. Thus, this decreases the excitation of CEm from the LA1 pathway in extinction, without affecting the acquisition "memory" that resides in the unaffected LTP MGm synapse on LA1 neurons.)

5.6.1 LTP Occurs Consequent on High-Frequency Stimulation (100–200 Hz for 1 S)

This figure shows the pathways for LTP induction and maintenance. During LTP there is a concomitant increase in intracellular calcium concentration in the synaptic spine, as a consequence of activation of voltage-gated calcium (Ca2+) channels (VGCC) and/or activation of *N*-methyl-D-aspartate receptors (NMDAR) through the action of glutamate. This occurs in conjunction with activation of the catalytic subunit of protein kinase A (PKA) subsequent to the activation of metabotropic receptors by transmitters such as glutamate, GABA and dopamine. On the one hand, the increase in Ca2+ concentration gives rise to the release of calmodulin (CaM) from neurogranin at the membrane with the subsequent formation of calcium–calmodulin (CaMCa2+) and the autophosphorylating calcium–calmodulin kinase II (CaMKII). This remains constitutively active long after the Ca2+ concentration declines. CaMKII is the major determinant of the appearance of alpha-amino-3-hydroxy-5-methyl-4-isoxazolepropionic acid receptors (AMPAR) into the membrane with their binding protein TARP, as a consequence of phosphorylation of TARP, as well as phosphorylation of serine S831 of the GluR1 subunit of the AMPAR. PKA also phosphorylates S831 and S845 of the GluR1, thereby increasing its contribution to the unitary conductance of AMPAR. Important regulatory mechanisms modulating the principal pathways are inhibition of cAMP via PKA by phosphodiesterase 4 (PDE4) as well as by PDE1b, stimulated by CaMCa2+. This latter also stimulates protein phosphatase 2B (PP2B or calcineurin) that inhibits the phosphorylation of serine S845 on the GluR1 subunit of AMPAR as well as stimulating the phosphorylation of protein phosphatase inhibitor 1 (I1). This in turn inhibits protein phosphatase 1 (PP1) and relieves its dephosphorylation of serines S831 and S845 on GluR1 of AMPAR. Thus, the main effects leading to LTP are the depolarization and binding of glutamate at NMDAR leading to sufficient calcium influx to promote the autophosphorylating CaMKII that maintains through TARP the localization of AMPAR at PSD 95 beneath the postsynaptic membrane and the phosphorylation of its subunits, providing both an increase in the number of receptors at the membrane as well as their unit conductance.

Note that CaMKII also stimulates the activity of CREB and its products BDNF and PKMzeta via a RAS-RAF-MEK1/2-ERK1/2 pathway. This BDNF may be made available to act on the TrkB membrane receptors and so has the potential to enhance synaptic spine growth through nucleation of F-actin. Furthermore, PKMzeta can block the endocytosis of AMPAR (see below), so enhancing transmission, an action which can possibly be maintained over a long time scale, so contributing to LTP (Kwapis and Helmstetter 2014). Neurons in LA with elevated CREB have relatively large numbers of synaptic spines that through their enhanced NMDA receptors (Marie et al. 2005) gives them a competitive advantage among LA neurons in developing LTP (Han et al. 2007) and being involved in the LA "memory" in acquisition (Kim et al. 2014; Morrison et al. 2016). Acronyms are AC, adenylate cyclase; AKAP, A-kinase anchor protein.

5.6.2 LTD Occurs as a Consequence of Low-Frequency Stimulation (1 Hz for 900 s)

This stimulation may produce sufficient depolarization to remove magnesium from the pore of NMDAR and allow the influx of calcium ions into the synaptic spine. As for LTP, this stimulation triggers CaMKII which does not then switch to a sufficient state of autophosphorylation to maintain phosphorylation of TARP and hence placement and anchorage of AMPAR at the membrane and their binding at the postsynaptic density-95 (PSD-95). This is probably due to phosphorylation of T305/306 of CaMKII maintaining it clamped in a low state of phosphorylation. Additionally, PP1 exerts an inhibitory influence on CaMKII which can be relieved by the action of phosphorylated I1, depending on its regulation by PKA and PP2B. Both PP1 and PP2b maintain the serines S831 and S845 on the GluR1 subunit of AMPAR dephosphorylated and therefore AMPAR in a relatively low conductance state. Under these conditions, CaMKII can also phosphorylate S567 on GluR1 that reduces the conductance of AMPAR. Most importantly, these AMPAR are now removed from the membrane by the process of endocytosis that is inhibited by PKMzeta which is produced via CREB following its activation via CaMKII-RAS-RAF-MEK1/2-ERK1/2.

5.6.3 The Synthesis, Release and Uptake of Endocannabinoids

Endocannabinoids, such as AEA, are released via a membrane lipid precursor *N*-arachidonoyl-phosphatidylethanolamine (NAPE), by the action of phospholipase D (PLD) activated by depolarization of G-protein coupled receptor stimulation via glutamate or GABA or dopamine. NAPE biosynthesis is catalysed by the membrane enzyme *N*-acetyltransferase (NAT) activated by Ca2+ and cAMP. The sources of Ca2+ increase in the spines are both depolarization-activated channels (VGCC) and the action of glutamate on NMDAR, together with activation of metabotropic receptors (acted on by glutamate or GABA or dopamine or acetylcholine). After the release of an endocannabinoid such as anandamide, an active transporter is engaged (an AEA transporter, AN, is shown, but there is little information on this transporter). This uptake is followed by degradation by enzymatic cleavage to arachidonic acid and ethanolamide by the membrane-bound enzyme fatty acid amidohydrolase (FAAH).

The following pathways are indicated from the BDNF plasmalemma receptor TrkB activation to the control of actin nucleation and spine growth (read clockwise around the spine head).

1. TrkB acting on GRB1/2 protein through its SH2 domain activates the RAS-RAF-MEK1/2-ERK1/2 pathway to WASP nucleation of G-actin around Arp2/3 (and in addition, through CREB, provides newly synthesized BDNF). In addition, GRB1 through its SH2 domain leads to activation of P13K and so Akt1/2, which through CREB also has the potential to provide newly synthesized BDNF.

2. Activation of Kalirin-7/Vav2/3 leading to activation of RAC1/PAK to excite LIMK and so inhibits cofilin depolymerization of F-actin; in addition, a pathway exists from Kalirin-7/Vav2/3 to RhoA-ROCK which also excites LIMK.
3. BDNF acting on TrkB receptors can also lead to enhanced PLCgamma and the release of Ca2+ from IP3-stimulated calcium stores, increasing CaMKII.

Newly synthesized BDNF can be retained in endosomes for the stimulated release of BDNF from the membrane. This may then act on TrkB receptors to stimulate the synthesis and release of endocannabinoids, such as anandamide, via PLCgamma.

A rapid action of glucocorticoids (Gl), which do not reach the nucleus, involves these binding to a G-protein coupled glucocorticoid receptor (GlR) and then activation of a signalling pathway (red arrow) which may or may not promote the synthesis of BDNF and hence of an endocannabinoid (in this case anandamide) and its release through the pathway described in the last paragraph.

5.6.4 proBDNF Control of Synaptic Spine Regression

The neurotrophin proBDNFacts on P75NCR/Sortilin receptors in the membrane through two different pathways reduce the expression of CREB and hence the synthesis of BDNF. One of these pathways proceeds through Ceramide to inhibit P13K and hence reduces the serine/threonine-protein kinases Akt1/2 with consequent decrease in activity of CREB. The other pathway leads directly through the transcription factor Jun to CREB to have the same effect on lowering BDNF (and also the GluN2B subunit of NMDAR).

References

Aerni A, Traber R, Hock C, Roozendaal B, Schelling G, Papassotiropoulos A et al (2004) Low-dose cortisol for symptoms of posttraumatic stress disorder. Am J Psychiatry 161(8):1488–1490. https://doi.org/10.1176/appi.ajp.161.8.1488

Alonso M, Medina JH, Pozzo-Miller L (2004) ERK1/2 activation is necessary for BDNF to increase dendritic spine density in hippocampal CA1 pyramidal neurons. Learn Mem 11(2):172–178. https://doi.org/10.1101/lm.67804

An B, Hong I, Choi S (2012) Long-term neural correlates of reversible fear learning in the lateral amygdala. J Neurosci 32(47):16845–16856. https://doi.org/10.1523/JNEUROSCI.3017-12.2012

Anastasio TJ (2013) Computational search for hypotheses concerning the endocannabinoid contribution to the extinction of fear conditioning. Front Comput Neurosci 7:74. https://doi.org/10.3389/fncom.2013.00074

Andero R, Ressler KJ (2012) Fear extinction and BDNF: translating animal models of PTSD to the clinic. Genes Brain Behav 11(5):503–512. https://doi.org/10.1111/j.1601-183X.2012.00801.x

Atsak P, Hauer D, Campolongo P, Schelling G, Fornari RV, Roozendaal B (2015) Endocannabinoid signaling within the basolateral amygdala integrates multiple stress hormone effects on memory consolidation. Neuropsychopharmacology 40(6):1485–1494. https://doi.org/10.1038/npp.2014.334

Azad SC, Eder M, Marsicano G, Lutz B, Zieglgänsberger W, Rammes G (2003) Activation of the cannabinoid receptor type 1 decreases glutamatergic and GABAergic synaptic transmission in the lateral amygdala of the mouse. Learn Mem 10(2):116–128. https://doi.org/10.1101/lm.53303

Banke TG, Bowie D, Lee H, Huganir RL, Schousboe A, Traynelis SF (2000) Control of GluR1 AMPA receptor function by cAMP-dependent protein kinase. J Neurosci 20(1):89–102

Barrett D, Gonzalez-Lima F (2004) Behavioral effects of metyrapone on Pavlovian extinction. Neurosci Lett 371(2–3):91–96. https://doi.org/10.1016/j.neulet.2004.08.046

Barria A, Derkach V, Soderling T (1997) Identification of the Ca2+/calmodulin-dependent protein kinase II regulatory phosphorylation site in the alpha-amino-3-hydroxyl-5-methyl-4-isoxazole-propionate-type glutamate receptor. J Biol Chem 272(52):32727–32730

Baudier J, Deloulme JC, Van Dorsselaer A, Black D, Matthes HW (1991) Purification and characterization of a brain-specific protein kinase C substrate, neurogranin (p17). Identification of a consensus amino acid sequence between neurogranin and neuromodulin (GAP43) that corresponds to the protein kinase C phosphorylation site and the calmodulin-binding domain. J Biol Chem 266(1):229–237

Bear MF, Cooper LN, Ebner FF (1987) A physiological basis for a theory of synapse modification. Science 237(4810):42–48

Beattie EC, Carroll RC, Yu X, Morishita W, Yasuda H, von Zastrow M, Malenka RC (2000) Regulation of AMPA receptor endocytosis by a signaling mechanism shared with LTD. Nat Neurosci 3(12):1291–1300. https://doi.org/10.1038/81823

Bellot A, Guivernau B, Tajes M, Bosch-Morato M, Valls-Comamala V, Munoz FJ (2014) The structure and function of actin cytoskeleton in mature glutamatergic dendritic spines. Brain Res 1573:1–16. https://doi.org/10.1016/j.brainres.2014.05.024

Bennett MR, Lagopoulos J (2014) Stress and trauma: BDNF control of dendritic-spine formation and regression. Prog Neurobiol 112:80–99. https://doi.org/10.1016/j.pneurobio.2013.10.005

Bennett MR, Hatton SN, Lagopoulos J (2016) Stress, trauma and PTSD: translational insights into the core synaptic circuitry and its modulation. Brain Struct Funct 221(5):2401–2426. https://doi.org/10.1007/s00429-015-1056-1

Bennett MR, Arnold J, Hatton SN, Lagopoulos J (2017) Regulation of fear extinction by long-term depression: the roles of endocannabinoids and brain derived neurotrophic factor. Behav Brain Res 319:148–164. https://doi.org/10.1016/j.bbr.2016.11.029

Bitencourt RM, Pamplona FA, Takahashi RN (2014) Corticosteroid-endocannabinoid loop supports decrease of fear-conditioned response in rats. Eur Neuropsychopharmacol 24(7):1091–1102. https://doi.org/10.1016/j.euroneuro.2014.01.010

Blair HT, Schafe GE, Bauer EP, Rodrigues SM, LeDoux JE (2001) Synaptic plasticity in the lateral amygdala: a cellular hypothesis of fear conditioning. Learn Mem 8(5):229–242. https://doi.org/10.1101/lm.30901

Blundell J, Blaiss CA, Lagace DC, Eisch AJ, Powell CM (2011) Block of glucocorticoid synthesis during re-activation inhibits extinction of an established fear memory. Neurobiol Learn Mem 95(4):453–460. https://doi.org/10.1016/j.nlm.2011.02.006

Bohme GA, Laville M, Ledent C, Parmentier M, Imperato A (2000) Enhanced long-term potentiation in mice lacking cannabinoid CB1 receptors. Neuroscience 95(1):5–7

Bourne JN, Harris KM (2008) Balancing structure and function at hippocampal dendritic spines. Annu Rev Neurosci 31:47–67. https://doi.org/10.1146/annurev.neuro.31.060407.125646

Bourne JN, Sorra KE, Hurlburt J, Harris KM (2007) Polyribosomes are increased in spines of CA1 dendrites 2 h after the induction of LTP in mature rat hippocampal slices. Hippocampus 17(1):1–4. https://doi.org/10.1002/hipo.20238

Bouton ME, King DA (1983) Contextual control of the extinction of conditioned fear: tests for the associative value of the context. J Exp Psychol Anim Behav Process 9(3):248–265

Bramham CR (2008) Local protein synthesis, actin dynamics, and LTP consolidation. Curr Opin Neurobiol 18(5):524–531. https://doi.org/10.1016/j.conb.2008.09.013

Bramham CR, Wells DG (2007) Dendritic mRNA: transport, translation and function. Nat Rev Neurosci 8(10):776–789. https://doi.org/10.1038/nrn2150

Bukalo O, Pinard CR, Holmes A (2014) Mechanisms to medicines: elucidating neural and molecular substrates of fear extinction to identify novel treatments for anxiety disorders. Br J Pharmacol 171(20):4690–4718. https://doi.org/10.1111/bph.12779

Bukalo O, Pinard CR, Silverstein S, Brehm C, Hartley ND, Whittle N et al (2015) Prefrontal inputs to the amygdala instruct fear extinction memory formation. Sci Adv 1(6):e1500251. https://doi.org/10.1126/sciadv.1500251

Burgos-Robles A, Vidal-Gonzalez I, Santini E, Quirk GJ (2007) Consolidation of fear extinction requires NMDA receptor-dependent bursting in the ventromedial prefrontal cortex. Neuron 53 (6):871–880. https://doi.org/10.1016/j.neuron.2007.02.021

Cameron C, Watson D, Robinson J (2014) Use of a synthetic cannabinoid in a correctional population for posttraumatic stress disorder-related insomnia and nightmares, chronic pain, harm reduction, and other indications: a retrospective evaluation. J Clin Psychopharmacol 34 (5):559–564. https://doi.org/10.1097/JCP.0000000000000180

Campolongo P, Roozendaal B, Trezza V, Hauer D, Schelling G, McGaugh JL, Cuomo V (2009) Endocannabinoids in the rat basolateral amygdala enhance memory consolidation and enable glucocorticoid modulation of memory. Proc Natl Acad Sci USA 106(12):4888–4893. https://doi.org/10.1073/pnas.0900835106

Carr DW, Stofko-Hahn RE, Fraser ID, Cone RD, Scott JD (1992) Localization of the cAMP-dependent protein kinase to the postsynaptic densities by A-kinase anchoring proteins. Characterization of AKAP 79. J Biol Chem 267(24):16816–16823

Chapleau CA, Pozzo-Miller L (2012) Divergent roles of p75NTR and Trk receptors in BDNF's effects on dendritic spine density and morphology. Neural Plast 2012:578057. https://doi.org/10.1155/2012/578057

Chen L, Chetkovich DM, Petralia RS, Sweeney NT, Kawasaki Y, Wenthold RJ et al (2000) Stargazin regulates synaptic targeting of AMPA receptors by two distinct mechanisms. Nature 408(6815):936–943. https://doi.org/10.1038/35050030

Chevaleyre V, Heifets BD, Kaeser PS, Sudhof TC, Castillo PE (2007) Endocannabinoid-mediated long-term plasticity requires cAMP/PKA signaling and RIM1alpha. Neuron 54(5):801–812. https://doi.org/10.1016/j.neuron.2007.05.020

Chhatwal JP, Davis M, Maguschak KA, Ressler KJ (2005) Enhancing cannabinoid neurotransmission augments the extinction of conditioned fear. Neuropsychopharmacology 30(3):516–524. https://doi.org/10.1038/sj.npp.1300655

Chhatwal JP, Stanek-Rattiner L, Davis M, Ressler KJ (2006) Amygdala BDNF signaling is required for consolidation but not encoding of extinction. Nat Neurosci 9(7):870–872. https://doi.org/10.1038/nn1718

Cho JH, Deisseroth K, Bolshakov VY (2013) Synaptic encoding of fear extinction in mPFC-amygdala circuits. Neuron 80(6):1491–1507. https://doi.org/10.1016/j.neuron.2013.09.025

Chouker A, Kaufmann I, Kreth S, Hauer D, Feuerecker M, Thieme D et al (2010) Motion sickness, stress and the endocannabinoid system. PLoS One 5(5):e10752. https://doi.org/10.1371/journal.pone.0010752

Christensen R, Kristensen PK, Bartels EM, Bliddal H, Astrup A (2007) Efficacy and safety of the weight-loss drug rimonabant: a meta-analysis of randomised trials. Lancet 370 (9600):1706–1713. https://doi.org/10.1016/s0140-6736(07)61721-8

Citri A, Malenka RC (2008) Synaptic plasticity: multiple forms, functions, and mechanisms. Neuropsychopharmacology 33(1):18–41. https://doi.org/10.1038/sj.npp.1301559

Clay R, Hebert M, Gill G, Stapleton LA, Pridham A, Coady M et al (2011) Glucocorticoids are required for extinction of predator stress-induced hyperarousal. Neurobiol Learn Mem 96 (2):367–377. https://doi.org/10.1016/j.nlm.2011.06.012

Collins DR, Pare D (1999) Spontaneous and evoked activity of intercalated amygdala neurons. Eur J Neurosci 11(10):3441–3448

Collins DR, Pertwee RG, Davies SN (1995) Prevention by the cannabinoid antagonist, SR141716A, of cannabinoid-mediated blockade of long-term potentiation in the rat hippocampal slice. Br J Pharmacol 115(6):869–870
Coultrap SJ, Bayer KU (2012) CaMKII regulation in information processing and storage. Trends Neurosci 35(10):607–618. https://doi.org/10.1016/j.tins.2012.05.003
D'Souza DC, Pittman B, Perry E, Simen A (2009) Preliminary evidence of cannabinoid effects on brain-derived neurotrophic factor (BDNF) levels in humans. Psychopharmacology 202 (4):569–578. https://doi.org/10.1007/s00213-008-1333-2
Davis M (1992) The role of the amygdala in fear and anxiety. Annu Rev Neurosci 15:353–375. https://doi.org/10.1146/annurev.neuro.15.1.353
Davis M, Whalen PJ (2001) The amygdala: vigilance and emotion. Mol Psychiatry 6(1):13–34
de Bitencourt RM, Pamplona FA, Takahashi RN (2013) A current overview of cannabinoids and glucocorticoids in facilitating extinction of aversive memories: potential extinction enhancers. Neuropharmacology 64:389–395. https://doi.org/10.1016/j.neuropharm.2012.05.039
De Oliveira Alvares L, Genro BP, Diehl F, Quillfeldt JA (2008) Differential role of the hippocampal endocannabinoid system in the memory consolidation and retrieval mechanisms. Neurobiol Learn Mem 90(1):1–9. https://doi.org/10.1016/j.nlm.2008.01.009
de Quervain DJ, Bentz D, Michael T, Bolt OC, Wiederhold BK, Margraf J, Wilhelm FH (2011) Glucocorticoids enhance extinction-based psychotherapy. Proc Natl Acad Sci USA 108 (16):6621–6625. https://doi.org/10.1073/pnas.1018214108
Derkinderen P, Ledent C, Parmentier M, Girault JA (2001) Cannabinoids activate p38 mitogen-activated protein kinases through CB1 receptors in hippocampus. J Neurochem 77(3):957–960
Derkinderen P, Valjent E, Toutant M, Corvol J-C, Enslen H, Ledent C et al (2003) Regulation of extracellular signal-regulated kinase by cannabinoids in hippocampus. J Neurosci 23(6):2371
Di Marzo V, Melck D, Bisogno T, De Petrocellis L (1998) Endocannabinoids: endogenous cannabinoid receptor ligands with neuromodulatory action. Trends Neurosci 21(12):521–528
Di S, Malcher-Lopes R, Halmos KC, Tasker JG (2003) Nongenomic glucocorticoid inhibition via endocannabinoid release in the hypothalamus: a fast feedback mechanism. J Neurosci 23 (12):4850–4857
Dincheva I, Drysdale AT, Hartley CA, Johnson DC, Jing D, King EC et al (2015) FAAH genetic variation enhances fronto-amygdala function in mouse and human. Nat Commun 6:6395. https://doi.org/10.1038/ncomms7395
Dixon RM, Mellor JR, Hanley JG (2009) PICK1-mediated glutamate receptor subunit 2 (GluR2) trafficking contributes to cell death in oxygen/glucose-deprived hippocampal neurons. J Biol Chem 284(21):14230–14235. https://doi.org/10.1074/jbc.M901203200
Dlugos A, Childs E, Stuhr KL, Hillard CJ, de Wit H (2012) Acute stress increases circulating anandamide and other N-acylethanolamines in healthy humans. Neuropsychopharmacology 37 (11):2416–2427. https://doi.org/10.1038/npp.2012.100
Domenici MR, Azad SC, Marsicano G, Schierloh A, Wotjak CT, Dodt HU et al (2006) Cannabinoid receptor type 1 located on presynaptic terminals of principal neurons in the forebrain controls glutamatergic synaptic transmission. J Neurosci 26(21):5794–5799. https://doi.org/10.1523/jneurosci.0372-06.2006
Engert F, Bonhoeffer T (1999) Dendritic spine changes associated with hippocampal long-term synaptic plasticity. Nature 399(6731):66–70. https://doi.org/10.1038/19978
Falls WA, Miserendino MJ, Davis M (1992) Extinction of fear-potentiated startle: blockade by infusion of an NMDA antagonist into the amygdala. J Neurosci 12(3):854–863
Finkbeiner S, Tavazoie SF, Maloratsky A, Jacobs KM, Harris KM, Greenberg ME (1997) CREB: a major mediator of neuronal neurotrophin responses. Neuron 19(5):1031–1047
Fischer M, Kaech S, Knutti D, Matus A (1998) Rapid actin-based plasticity in dendritic spines. Neuron 20(5):847–854
Fitzgerald PJ, Whittle N, Flynn SM, Graybeal C, Pinard CR, Gunduz-Cinar O et al (2014) Prefrontal single-unit firing associated with deficient extinction in mice. Neurobiol Learn Mem 113:69–81. https://doi.org/10.1016/j.nlm.2013.11.002

Fraser GA (2009) The use of a synthetic cannabinoid in the management of treatment-resistant nightmares in posttraumatic stress disorder (PTSD). CNS Neurosci Ther 15(1):84–88. https://doi.org/10.1111/j.1755-5949.2008.00071.x

Freund TF, Katona I, Piomelli D (2003) Role of endogenous cannabinoids in synaptic signaling. Physiol Rev 83(3):1017–1066. https://doi.org/10.1152/physrev.00004.2003

Fukazawa Y, Saitoh Y, Ozawa F, Ohta Y, Mizuno K, Inokuchi K (2003) Hippocampal LTP is accompanied by enhanced F-actin content within the dendritic spine that is essential for late LTP maintenance in vivo. Neuron 38(3):447–460

Ganon-Elazar E, Akirav I (2013) Cannabinoids and traumatic stress modulation of contextual fear extinction and GR expression in the amygdala-hippocampal-prefrontal circuit. Psychoneuroendocrinology 38(9):1675–1687. https://doi.org/10.1016/j.psyneuen.2013.01.014

Gomez LL, Alam S, Smith KE, Horne E, Dell'Acqua ML (2002) Regulation of A-kinase anchoring protein 79/150-cAMP-dependent protein kinase postsynaptic targeting by NMDA receptor activation of calcineurin and remodeling of dendritic actin. J Neurosci 22(16):7027–7044. https://doi.org/10.1523/JNEUROSCI.22-16-07027.2002

Gong JP, Onaivi ES, Ishiguro H, Liu QR, Tagliaferro PA, Brusco A, Uhl GR (2006) Cannabinoid CB2 receptors: immunohistochemical localization in rat brain. Brain Res 1071(1):10–23. https://doi.org/10.1016/j.brainres.2005.11.035

Gorzalka BB, Hill MN, Hillard CJ (2008) Regulation of endocannabinoid signaling by stress: implications for stress-related affective disorders. Neurosci Biobehav Rev 32(6):1152–1160. https://doi.org/10.1016/j.neubiorev.2008.03.004

Gunduz-Cinar O, MacPherson KP, Cinar R, Gamble-George J, Sugden K, Williams B et al (2013) Convergent translational evidence of a role for anandamide in amygdala-mediated fear extinction, threat processing and stress-reactivity. Mol Psychiatry 18(7):813–823. https://doi.org/10.1038/mp.2012.72

Hall J, Thomas KL, Everitt BJ (2001) Fear memory retrieval induces CREB phosphorylation and Fos expression within the amygdala. Eur J Neurosci 13(7):1453–1458

Hampson RE, Deadwyler SA (1998) Role of cannabinoid receptors in memory storage. Neurobiol Dis 5(6 Pt B):474–482. https://doi.org/10.1006/nbdi.1998.0223

Han JH, Kushner SA, Yiu AP, Cole CJ, Matynia A, Brown RA et al (2007) Neuronal competition and selection during memory formation. Science 316(5823):457–460. https://doi.org/10.1126/science.1139438

Hariri AR, Gorka A, Hyde LW, Kimak M, Halder I, Ducci F et al (2009) Divergent effects of genetic variation in endocannabinoid signaling on human threat- and reward-related brain function. Biol Psychiatry 66(1):9–16. https://doi.org/10.1016/j.biopsych.2008.10.047

Haubensak W, Kunwar PS, Cai H, Ciocchi S, Wall NR, Ponnusamy R et al (2010) Genetic dissection of an amygdala microcircuit that gates conditioned fear. Nature 468 (7321):270–276. https://doi.org/10.1038/nature09553

Hauer D, Weis F, Campolongo P, Schopp M, Beiras-Fernandez A, Strewe C et al (2012) Glucocorticoid-endocannabinoid interaction in cardiac surgical patients: relationship to early cognitive dysfunction and late depression. Rev Neurosci 23(5–6):681–690. https://doi.org/10.1515/revneuro-2012-0058

Hauer D, Schelling G, Gola H, Campolongo P, Morath J, Roozendaal B et al (2013) Plasma concentrations of endocannabinoids and related primary fatty acid amides in patients with post-traumatic stress disorder. PLoS One 8(5):e62741. https://doi.org/10.1371/journal.pone.0062741

Hausser M, Clark BA (1997) Tonic synaptic inhibition modulates neuronal output pattern and spatiotemporal synaptic integration. Neuron 19(3):665–678

Havik B, Rokke H, Bardsen K, Davanger S, Bramham CR (2003) Bursts of high-frequency stimulation trigger rapid delivery of pre-existing alpha-CaMKII mRNA to synapses: a mechanism in dendritic protein synthesis during long-term potentiation in adult awake rats. Eur J Neurosci 17(12):2679–2689

Hayashi Y, Shi SH, Esteban JA, Piccini A, Poncer JC, Malinow R (2000) Driving AMPA receptors into synapses by LTP and CaMKII: requirement for GluR1 and PDZ domain interaction. Science 287(5461):2262–2267

He Y, Kulasiri D, Samarasinghe S (2014) Systems biology of synaptic plasticity: a review on N-methyl-D-aspartate receptor mediated biochemical pathways and related mathematical models. Biosystems 122:7–18. https://doi.org/10.1016/j.biosystems.2014.06.005

Heldt SA, Stanek L, Chhatwal JP, Ressler KJ (2007) Hippocampus-specific deletion of BDNF in adult mice impairs spatial memory and extinction of aversive memories. Mol Psychiatry 12 (7):656–670. https://doi.org/10.1038/sj.mp.4001957

Helmstetter FJ, Bellgowan PS (1994) Effects of muscimol applied to the basolateral amygdala on acquisition and expression of contextual fear conditioning in rats. Behav Neurosci 108 (5):1005–1009

Herry C, Trifilieff P, Micheau J, Luthi A, Mons N (2006) Extinction of auditory fear conditioning requires MAPK/ERK activation in the basolateral amygdala. Eur J Neurosci 24(1):261–269. https://doi.org/10.1111/j.1460-9568.2006.04893.x

Herry C, Ciocchi S, Senn V, Demmou L, Muller C, Luthi A (2008) Switching on and off fear by distinct neuronal circuits. Nature 454(7204):600–606. https://doi.org/10.1038/nature07166

Heyman E, Gamelin FX, Goekint M, Piscitelli F, Roelands B, Leclair E et al (2012) Intense exercise increases circulating endocannabinoid and BDNF levels in humans—possible implications for reward and depression. Psychoneuroendocrinology 37(6):844–851. https://doi.org/10.1016/j.psyneuen.2011.09.017

Hill MN, Karatsoreos IN, Hillard CJ, McEwen BS (2010) Rapid elevations in limbic endocannabinoid content by glucocorticoid hormones in vivo. Psychoneuroendocrinology 35 (9):1333–1338. https://doi.org/10.1016/j.psyneuen.2010.03.005

Hill MN, Hillard CJ, McEwen BS (2011) Alterations in corticolimbic dendritic morphology and emotional behavior in cannabinoid CB1 receptor-deficient mice parallel the effects of chronic stress. Cereb Cortex 21(9):2056–2064. https://doi.org/10.1093/cercor/bhq280

Hill MN, Bierer LM, Makotkine I, Golier JA, Galea S, McEwen BS et al (2013a) Reductions in circulating endocannabinoid levels in individuals with post-traumatic stress disorder following exposure to the World Trade Center attacks. Psychoneuroendocrinology 38(12):2952–2961. https://doi.org/10.1016/j.psyneuen.2013.08.004

Hill MN, Kumar SA, Filipski SB, Iverson M, Stuhr KL, Keith JM et al (2013b) Disruption of fatty acid amide hydrolase activity prevents the effects of chronic stress on anxiety and amygdalar microstructure. Mol Psychiatry 18(10):1125–1135. https://doi.org/10.1038/mp.2012.90

Hodges JL, Newell-Litwa K, Asmussen H, Vicente-Manzanares M, Horwitz AR (2011) Myosin IIb activity and phosphorylation status determines dendritic spine and post-synaptic density morphology. PLoS One 6(8):e24149. https://doi.org/10.1371/journal.pone.0024149

Holmes A, Fitzgerald PJ, MacPherson KP, DeBrouse L, Colacicco G, Flynn SM et al (2012) Chronic alcohol remodels prefrontal neurons and disrupts NMDAR-mediated fear extinction encoding. Nat Neurosci 15(10):1359–1361. https://doi.org/10.1038/nn.3204

Hong I, Song B, Lee S, Kim J, Kim J, Choi S (2009) Extinction of cued fear memory involves a distinct form of depotentiation at cortical input synapses onto the lateral amygdala. Eur J Neurosci 30(11):2089–2099. https://doi.org/10.1111/j.1460-9568.2009.07004.x

Hong I, Kim J, Kim J, Lee S, Ko HG, Nader K et al (2013) AMPA receptor exchange underlies transient memory destabilization on retrieval. Proc Natl Acad Sci USA 110(20):8218–8223. https://doi.org/10.1073/pnas.1305235110

Hu X, Ballo L, Pietila L, Viesselmann C, Ballweg J, Lumbard D et al (2011) BDNF-induced increase of PSD-95 in dendritic spines requires dynamic microtubule invasions. J Neurosci 31 (43):15597–15603. https://doi.org/10.1523/JNEUROSCI.2445-11.2011

Hugues S, Chessel A, Lena I, Marsault R, Garcia R (2006) Prefrontal infusion of PD098059 immediately after fear extinction training blocks extinction-associated prefrontal synaptic plasticity and decreases prefrontal ERK2 phosphorylation. Synapse 60(4):280–287. https://doi.org/10.1002/syn.20291

Humeau Y, Herry C, Kemp N, Shaban H, Fourcaudot E, Bissiere S, Luthi A (2005) Dendritic spine heterogeneity determines afferent-specific Hebbian plasticity in the amygdala. Neuron 45 (1):119–131. https://doi.org/10.1016/j.neuron.2004.12.019

Impey S, Obrietan K, Storm DR (1999) Making new connections: role of ERK/MAP kinase signaling in neuronal plasticity. Neuron 23(1):11–14

Jetly R, Heber A, Fraser G, Boisvert D (2015) The efficacy of nabilone, a synthetic cannabinoid, in the treatment of PTSD-associated nightmares: a preliminary randomized, double-blind, placebo-controlled cross-over design study. Psychoneuroendocrinology 51:585–588. https://doi.org/10.1016/j.psyneuen.2014.11.002

Kamprath K, Marsicano G, Tang J, Monory K, Bisogno T, Di Marzo V et al (2006) Cannabinoid CB1 receptor mediates fear extinction via habituation-like processes. J Neurosci 26 (25):6677–6686. https://doi.org/10.1523/JNEUROSCI.0153-06.2006

Kang H, Schuman EM (1995) Long-lasting neurotrophin-induced enhancement of synaptic transmission in the adult hippocampus. Science 267(5204):1658–1662

Kano M (2014) Control of synaptic function by endocannabinoid-mediated retrograde signaling. Proc Jpn Acad Ser B Phys Biol Sci 90(7):235–250

Karanian DA, Brown QB, Makriyannis A, Bahr BA (2005) Blocking cannabinoid activation of FAK and ERK1/2 compromises synaptic integrity in hippocampus. Eur J Pharmacol 508 (1–3):47–56. https://doi.org/10.1016/j.ejphar.2004.12.009

Katona I, Rancz EA, Acsady L, Ledent C, Mackie K, Hajos N, Freund TF (2001) Distribution of CB1 cannabinoid receptors in the amygdala and their role in the control of GABAergic transmission. J Neurosci 21(23):9506–9518

Kelly A, Lynch MA (2000) Long-term potentiation in dentate gyrus of the rat is inhibited by the phosphoinositide 3-kinase inhibitor, wortmannin. Neuropharmacology 39(4):643–651

Kim WJ, Lee MJ, Park MA, Jung JS, Uhlinger DJ, Kwak JY (2000) Dexamethasone enhances phospholipase D activity in M-1 cells. Exp Mol Med 32(3):170–177. https://doi.org/10.1038/emm.2000.28

Kim J, Lee S, Park K, Hong I, Song B, Son G et al (2007) Amygdala depotentiation and fear extinction. Proc Natl Acad Sci USA 104(52):20955–20960. https://doi.org/10.1073/pnas.0710548105

Kim J, Kwon JT, Kim HS, Josselyn SA, Han JH (2014) Memory recall and modifications by activating neurons with elevated CREB. Nat Neurosci 17(1):65–72. https://doi.org/10.1038/nn.3592

Kim J, An B, Kim J, Park S, Park S, Hong I et al (2015) mGluR2/3 in the lateral amygdala is required for fear extinction: cortical input synapses onto the lateral amygdala as a target site of the mGluR2/3 action. Neuropsychopharmacology 40(13):2916–2928. https://doi.org/10.1038/npp.2015.145

Kneussel M, Wagner W (2013) Myosin motors at neuronal synapses: drivers of membrane transport and actin dynamics. Nat Rev Neurosci 14(4):233–247. https://doi.org/10.1038/nrn3445

Korte M, Carroll P, Wolf E, Brem G, Thoenen H, Bonhoeffer T (1995) Hippocampal long-term potentiation is impaired in mice lacking brain-derived neurotrophic factor. Proc Natl Acad Sci USA 92(19):8856–8860

Koshimizu H, Kiyosue K, Hara T, Hazama S, Suzuki S, Uegaki K et al (2009) Multiple functions of precursor BDNF to CNS neurons: negative regulation of neurite growth, spine formation and cell survival. Mol Brain 2:27. https://doi.org/10.1186/1756-6606-2-27

Kristensen AS, Jenkins MA, Banke TG, Schousboe A, Makino Y, Johnson RC et al (2011) Mechanism of Ca2+/calmodulin-dependent kinase II regulation of AMPA receptor gating. Nat Neurosci 14(6):727–735. https://doi.org/10.1038/nn.2804

Krucker T, Siggins GR, Halpain S (2000) Dynamic actin filaments are required for stable long-term potentiation (LTP) in area CA1 of the hippocampus. Proc Natl Acad Sci USA 97 (12):6856–6861. https://doi.org/10.1073/pnas.100139797

Kuhnert S, Meyer C, Koch M (2013) Involvement of cannabinoid receptors in the amygdala and prefrontal cortex of rats in fear learning, consolidation, retrieval and extinction. Behav Brain Res 250:274–284. https://doi.org/10.1016/j.bbr.2013.05.002

Kwapis JL, Helmstetter FJ (2014) Does PKM(zeta) maintain memory? Brain Res Bull 105:36–45. https://doi.org/10.1016/j.brainresbull.2013.09.005

Kwon JT, Choi JS (2009) Cornering the fear engram: long-term synaptic changes in the lateral nucleus of the amygdala after fear conditioning. J Neurosci 29(31):9700–9703. https://doi.org/10.1523/jneurosci.5928-08.2009

Kwon HB, Sabatini BL (2011) Glutamate induces de novo growth of functional spines in developing cortex. Nature 474(7349):100–104. https://doi.org/10.1038/nature09986

Lafenetre P, Chaouloff F, Marsicano G (2007) The endocannabinoid system in the processing of anxiety and fear and how CB1 receptors may modulate fear extinction. Pharmacol Res 56(5):367–381. https://doi.org/10.1016/j.phrs.2007.09.006

LaPorte DC, Wierman BM, Storm DR (1980) Calcium-induced exposure of a hydrophobic surface on calmodulin. Biochemistry 19(16):3814–3819

LeDoux JE (2000) Emotion circuits in the brain. Annu Rev Neurosci 23:155–184. https://doi.org/10.1146/annurev.neuro.23.1.155

Lee S, Kim S-J, Kwon O-B, Lee JH, Kim J-H (2013) Inhibitory networks of the amygdala for emotional memory. Front Neural Circuits 7:129. https://doi.org/10.3389/fncir.2013.00129

Lerner TN, Kreitzer AC (2012) RGS4 is required for dopaminergic control of striatal LTD and susceptibility to parkinsonian motor deficits. Neuron 73(2):347–359. https://doi.org/10.1016/j.neuron.2011.11.015

Likhtik E, Pelletier JG, Paz R, Pare D (2005) Prefrontal control of the amygdala. J Neurosci 25(32):7429–7437. https://doi.org/10.1523/JNEUROSCI.2314-05.2005

Lin HC, Mao SC, Chen PS, Gean PW (2008) Chronic cannabinoid administration in vivo compromises extinction of fear memory. Learn Mem 15(12):876–884. https://doi.org/10.1101/lm.1081908

Lin HC, Mao SC, Gean PW (2009) Block of gamma-aminobutyric acid-A receptor insertion in the amygdala impairs extinction of conditioned fear. Biol Psychiatry 66(7):665–673. https://doi.org/10.1016/j.biopsych.2009.04.003

Lisman J (1989) A mechanism for the Hebb and the anti-Hebb processes underlying learning and memory. Proc Natl Acad Sci USA 86(23):9574–9578

Lisman J, Raghavachari S (2006) A unified model of the presynaptic and postsynaptic changes during LTP at CA1 synapses. Sci STKE 2006(356):re11. https://doi.org/10.1126/stke.3562006re11

Lisman J, Schulman H, Cline H (2002) The molecular basis of CaMKII function in synaptic and behavioural memory. Nat Rev Neurosci 3(3):175–190. https://doi.org/10.1038/nrn753

Lonze BE, Ginty DD (2002) Function and regulation of CREB family transcription factors in the nervous system. Neuron 35(4):605–623

Lutz B (2007) The endocannabinoid system and extinction learning. Mol Neurobiol 36(1):92–101. https://doi.org/10.1007/s12035-007-8004-x

Maison P, Walker DJ, Walsh FS, Williams G, Doherty P (2009) BDNF regulates neuronal sensitivity to endocannabinoids. Neurosci Lett 467(2):90–94. https://doi.org/10.1016/j.neulet.2009.10.011

Maletic-Savatic M, Malinow R, Svoboda K (1999) Rapid dendritic morphogenesis in CA1 hippocampal dendrites induced by synaptic activity. Science 283(5409):1923–1927

Man HY, Sekine-Aizawa Y, Huganir RL (2007) Regulation of {alpha}-amino-3-hydroxy-5-methyl-4-isoxazolepropionic acid receptor trafficking through PKA phosphorylation of the Glu receptor 1 subunit. Proc Natl Acad Sci USA 104(9):3579–3584. https://doi.org/10.1073/pnas.0611698104

Maren S (1996) Synaptic transmission and plasticity in the amygdala. An emerging physiology of fear conditioning circuits. Mol Neurobiol 13(1):1–22. https://doi.org/10.1007/BF02740749

Maren S (1999) Long-term potentiation in the amygdala: a mechanism for emotional learning and memory. Trends Neurosci 22(12):561–567
Maren S (2005) Synaptic mechanisms of associative memory in the amygdala. Neuron 47 (6):783–786. https://doi.org/10.1016/j.neuron.2005.08.009
Maren S, Holmes A (2016) Stress and fear extinction. Neuropsychopharmacology 41(1):58–79. https://doi.org/10.1038/npp.2015.180
Maren S, Quirk GJ (2004) Neuronal signalling of fear memory. Nat Rev Neurosci 5(11):844–852. https://doi.org/10.1038/nrn1535
Marie H, Morishita W, Yu X, Calakos N, Malenka RC (2005) Generation of silent synapses by acute in vivo expression of CaMKIV and CREB. Neuron 45(5):741–752. https://doi.org/10.1016/j.neuron.2005.01.039
Marini AM, Jiang H, Pan H, Wu X, Lipsky RH (2008) Hormesis: a promising strategy to sustain endogenous neuronal survival pathways against neurodegenerative disorders. Ageing Res Rev 7 (1):21–33. https://doi.org/10.1016/j.arr.2007.07.003
Maroun M, Ioannides PJ, Bergman KL, Kavushansky A, Holmes A, Wellman CL (2013) Fear extinction deficits following acute stress associate with increased spine density and dendritic retraction in basolateral amygdala neurons. Eur J Neurosci 38(4):2611–2620. https://doi.org/10.1111/ejn.12259
Marsicano G, Wotjak CT, Azad SC, Bisogno T, Rammes G, Cascio MG et al (2002) The endogenous cannabinoid system controls extinction of aversive memories. Nature 418 (6897):530–534. https://doi.org/10.1038/nature00839
Matsuda LA, Lolait SJ, Brownstein MJ, Young AC, Bonner TI (1990) Structure of a cannabinoid receptor and functional expression of the cloned cDNA. Nature 346(6284):561–564. https://doi.org/10.1038/346561a0
McDonald AJ, Mascagni F (2001) Colocalization of calcium-binding proteins and GABA in neurons of the rat basolateral amygdala. Neuroscience 105(3):681–693
McDonald AJ, Pearson JC (1989) Coexistence of GABA and peptide immunoreactivity in non-pyramidal neurons of the basolateral amygdala. Neurosci Lett 100(1–3):53–58
McEwen BS, Bowles NP, Gray JD, Hill MN, Hunter RG, Karatsoreos IN, Nasca C (2015) Mechanisms of stress in the brain. Nat Neurosci 18(10):1353–1363. https://doi.org/10.1038/nn.4086
McKernan MG, Shinnick-Gallagher P (1997) Fear conditioning induces a lasting potentiation of synaptic currents in vitro. Nature 390(6660):607–611. https://doi.org/10.1038/37605
Medina JF, Repa JC, Mauk MD, LeDoux JE (2002) Parallels between cerebellum- and amygdala-dependent conditioning. Nat Rev Neurosci 3(2):122–131. https://doi.org/10.1038/nrn728
Meng Y, Zhang Y, Tregoubov V, Janus C, Cruz L, Jackson M et al (2002) Abnormal spine morphology and enhanced LTP in LIMK-1 knockout mice. Neuron 35(1):121–133
Milad MR, Vidal-Gonzalez I, Quirk GJ (2004) Electrical stimulation of medial prefrontal cortex reduces conditioned fear in a temporally specific manner. Behav Neurosci 118(2):389–394. https://doi.org/10.1037/0735-7044.118.2.389
Miserendino MJ, Sananes CB, Melia KR, Davis M (1990) Blocking of acquisition but not expression of conditioned fear-potentiated startle by NMDA antagonists in the amygdala. Nature 345(6277):716–718. https://doi.org/10.1038/345716a0
Misner DL, Sullivan JM (1999) Mechanism of cannabinoid effects on long-term potentiation and depression in hippocampal CA1 neurons. J Neurosci 19(16):6795–6805
Monory K, Polack M, Remus A, Lutz B, Korte M (2015) Cannabinoid CB1 receptor calibrates excitatory synaptic balance in the mouse hippocampus. J Neurosci 35(9):3842–3850. https://doi.org/10.1523/JNEUROSCI.3167-14.2015
Morrison DJ, Rashid AJ, Yiu AP, Yan C, Frankland PW, Josselyn SA (2016) Parvalbumin interneurons constrain the size of the lateral amygdala engram. Neurobiol Learn Mem 135:91–99. https://doi.org/10.1016/j.nlm.2016.07.007
Mulkey RM, Herron CE, Malenka RC (1993) An essential role for protein phosphatases in hippocampal long-term depression. Science 261(5124):1051–1055

Mulkey RM, Endo S, Shenolikar S, Malenka RC (1994) Involvement of a calcineurin/inhibitor-1 phosphatase cascade in hippocampal long-term depression. Nature 369(6480):486–488. https://doi.org/10.1038/369486a0

Muller J, Corodimas KP, Fridel Z, LeDoux JE (1997) Functional inactivation of the lateral and basal nuclei of the amygdala by muscimol infusion prevents fear conditioning to an explicit conditioned stimulus and to contextual stimuli. Behav Neurosci 111(4):683–691

Muller T, Albrecht D, Gebhardt C (2009) Both NR2A and NR2B subunits of the NMDA receptor are critical for long-term potentiation and long-term depression in the lateral amygdala of horizontal slices of adult mice. Learn Mem 16(6):395–405. https://doi.org/10.1101/lm.1398709

Murakoshi H, Yasuda R (2012) Postsynaptic signaling during plasticity of dendritic spines. Trends Neurosci 35(2):135–143. https://doi.org/10.1016/j.tins.2011.12.002

Myers KM, Davis M (2007) Mechanisms of fear extinction. Mol Psychiatry 12(2):120–150. https://doi.org/10.1038/sj.mp.4001939

Nabavi S, Fox R, Proulx CD, Lin JY, Tsien RY, Malinow R (2014) Engineering a memory with LTD and LTP. Nature 511(7509):348–352. https://doi.org/10.1038/nature13294

Nagerl UV, Kostinger G, Anderson JC, Martin KA, Bonhoeffer T (2007) Protracted synaptogenesis after activity-dependent spinogenesis in hippocampal neurons. J Neurosci 27(30):8149–8156. https://doi.org/10.1523/JNEUROSCI.0511-07.2007

Nedelescu H, Kelso CM, Lazaro-Munoz G, Purpura M, Cain CK, Ledoux JE, Aoki C (2010) Endogenous GluR1-containing AMPA receptors translocate to asymmetric synapses in the lateral amygdala during the early phase of fear memory formation: an electron microscopic immunocytochemical study. J Comp Neurol 518(23):4723–4739. https://doi.org/10.1002/cne.22472

Neumeister A, Normandin MD, Pietrzak RH, Piomelli D, Zheng MQ, Gujarro-Anton A et al (2013) Elevated brain cannabinoid CB1 receptor availability in post-traumatic stress disorder: a positron emission tomography study. Mol Psychiatry 18(9):1034–1040. https://doi.org/10.1038/mp.2013.61

Neumeister A, Seidel J, Ragen BJ, Pietrzak RH (2015) Translational evidence for a role of endocannabinoids in the etiology and treatment of posttraumatic stress disorder. Psychoneuroendocrinology 51:577–584. https://doi.org/10.1016/j.psyneuen.2014.10.012

Ninomiya EM, Martynhak BJ, Zanoveli JM, Correia D, da Cunha C, Andreatini R (2010) Spironolactone and low-dose dexamethasone enhance extinction of contextual fear conditioning. Prog Neuro-Psychopharmacol Biol Psychiatry 34(7):1229–1235. https://doi.org/10.1016/j.pnpbp.2010.06.025

Nowak L, Bregestovski P, Ascher P, Herbet A, Prochiantz A (1984) Magnesium gates glutamate-activated channels in mouse central neurones. Nature 307(5950):462–465

Oberheim NA, Takano T, Han X, He W, Lin JH, Wang F et al (2009) Uniquely hominid features of adult human astrocytes. J Neurosci 29(10):3276–3287. https://doi.org/10.1523/JNEUROSCI.4707-08.2009

Ohno-Shosaku T, Tanimura A, Hashimotodani Y, Kano M (2012) Endocannabinoids and retrograde modulation of synaptic transmission. Neuroscientist 18(2):119–132. https://doi.org/10.1177/1073858410397377

Okamoto K, Nagai T, Miyawaki A, Hayashi Y (2004) Rapid and persistent modulation of actin dynamics regulates postsynaptic reorganization underlying bidirectional plasticity. Nat Neurosci 7(10):1104–1112. https://doi.org/10.1038/nn1311

Ostroff LE, Fiala JC, Allwardt B, Harris KM (2002) Polyribosomes redistribute from dendritic shafts into spines with enlarged synapses during LTP in developing rat hippocampal slices. Neuron 35(3):535–545

Pamplona FA, Prediger RD, Pandolfo P, Takahashi RN (2006) The cannabinoid receptor agonist WIN 55,212-2 facilitates the extinction of contextual fear memory and spatial memory in rats. Psychopharmacology 188(4):641–649. https://doi.org/10.1007/s00213-006-0514-0

Pamplona FA, Bitencourt RM, Takahashi RN (2008) Short- and long-term effects of cannabinoids on the extinction of contextual fear memory in rats. Neurobiol Learn Mem 90(1):290–293. https://doi.org/10.1016/j.nlm.2008.04.003

Panja D, Bramham CR (2014) BDNF mechanisms in late LTP formation: a synthesis and breakdown. Neuropharmacology 76 Pt C:664–676. https://doi.org/10.1016/j.neuropharm.2013.06.024

Pape HC, Pare D (2010) Plastic synaptic networks of the amygdala for the acquisition, expression, and extinction of conditioned fear. Physiol Rev 90(2):419–463. https://doi.org/10.1152/physrev.00037.2009

Pare D, Gaudreau H (1996) Projection cells and interneurons of the lateral and basolateral amygdala: distinct firing patterns and differential relation to theta and delta rhythms in conscious cats. J Neurosci 16(10):3334–3350

Pare D, Quirk GJ, Ledoux JE (2004) New vistas on amygdala networks in conditioned fear. J Neurophysiol 92(1):1–9. https://doi.org/10.1152/jn.00153.2004

Patel S, Roelke CT, Rademacher DJ, Cullinan WE, Hillard CJ (2004) Endocannabinoid signaling negatively modulates stress-induced activation of the hypothalamic-pituitary-adrenal axis. Endocrinology 145(12):5431–5438. https://doi.org/10.1210/en.2004-0638

Patterson SL, Pittenger C, Morozov A, Martin KC, Scanlin H, Drake C, Kandel ER (2001) Some forms of cAMP-mediated long-lasting potentiation are associated with release of BDNF and nuclear translocation of phospho-MAP kinase. Neuron 32(1):123–140

Pavlov IP (1927) Conditioned reflexes: an investigation of the physiological activity of the Cerebral Cortex. Oxford University Press, London

Quirk GJ (2006) Extinction: new excitement for an old phenomenon. Biol Psychiatry 60 (4):317–318. https://doi.org/10.1016/j.biopsych.2006.05.023

Quirk GJ, Armony JL, LeDoux JE (1997) Fear conditioning enhances different temporal components of tone-evoked spike trains in auditory cortex and lateral amygdala. Neuron 19 (3):613–624

Quirk GJ, Russo GK, Barron JL, Lebron K (2000) The role of ventromedial prefrontal cortex in the recovery of extinguished fear. J Neurosci 20(16):6225–6231

Radley JJ, Farb CR, He Y, Janssen WG, Rodrigues SM, Johnson LR et al (2007) Distribution of NMDA and AMPA receptor subunits at thalamo-amygdaloid dendritic spines. Brain Res 1134 (1):87–94. https://doi.org/10.1016/j.brainres.2006.11.045

Rainnie DG, Asprodini EK, Shinnick-Gallagher P (1991) Inhibitory transmission in the basolateral amygdala. J Neurophysiol 66(3):999–1009

Ramikie TS, Patel S (2012) Endocannabinoid signaling in the amygdala: anatomy, synaptic signaling, behavior, and adaptations to stress. Neuroscience 204:38–52. https://doi.org/10.1016/j.neuroscience.2011.08.037

Reibaud M, Obinu MC, Ledent C, Parmentier M, Bohme GA, Imperato A (1999) Enhancement of memory in cannabinoid CB1 receptor knock-out mice. Eur J Pharmacol 379(1):R1–R2

Reijmers LG, Perkins BL, Matsuo N, Mayford M (2007) Localization of a stable neural correlate of associative memory. Science 317(5842):1230–1233. https://doi.org/10.1126/science.1143839

Repa JC, Muller J, Apergis J, Desrochers TM, Zhou Y, LeDoux JE (2001) Two different lateral amygdala cell populations contribute to the initiation and storage of memory. Nat Neurosci 4 (7):724–731. https://doi.org/10.1038/89512

Rex CS, Lin CY, Kramar EA, Chen LY, Gall CM, Lynch G (2007) Brain-derived neurotrophic factor promotes long-term potentiation-related cytoskeletal changes in adult hippocampus. J Neurosci 27(11):3017–3029. https://doi.org/10.1523/JNEUROSCI.4037-06.2007

Rex CS, Gavin CF, Rubio MD, Kramar EA, Chen LY, Jia Y et al (2010) Myosin IIb regulates actin dynamics during synaptic plasticity and memory formation. Neuron 67(4):603–617. https://doi.org/10.1016/j.neuron.2010.07.016

Rogan MT, Staubli UV, LeDoux JE (1997) Fear conditioning induces associative long-term potentiation in the amygdala. Nature 390(6660):604–607. https://doi.org/10.1038/37601

Roitman P, Mechoulam R, Cooper-Kazaz R, Shalev A (2014) Preliminary, open-label, pilot study of add-on oral Delta9-tetrahydrocannabinol in chronic post-traumatic stress disorder. Clin Drug Investig 34(8):587–591. https://doi.org/10.1007/s40261-014-0212-3

Rosenkranz JA, Moore H, Grace AA (2003) The prefrontal cortex regulates lateral amygdala neuronal plasticity and responses to previously conditioned stimuli. J Neurosci 23 (35):11054–11064

Rucker D, Padwal R, Li SK, Curioni C, Lau DCW (2007) Long term pharmacotherapy for obesity and overweight: updated meta-analysis. BMJ 335(7631):1194–1199. https://doi.org/10.1136/bmj.39385.413113.25

Ruehle S, Rey AA, Remmers F, Lutz B (2012) The endocannabinoid system in anxiety, fear memory and habituation. J Psychopharmacol 26(1):23–39. https://doi.org/10.1177/0269881111408958

Ryu J, Liu L, Wong TP, Wu DC, Burette A, Weinberg R et al (2006) A critical role for myosin IIb in dendritic spine morphology and synaptic function. Neuron 49(2):175–182. https://doi.org/10.1016/j.neuron.2005.12.017

Sala C, Piech V, Wilson NR, Passafaro M, Liu G, Sheng M (2001) Regulation of dendritic spine morphology and synaptic function by Shank and Homer. Neuron 31(1):115–130

Sanderson JL, Dell'Acqua ML (2011) AKAP signaling complexes in regulation of excitatory synaptic plasticity. Neuroscientist 17(3):321–336. https://doi.org/10.1177/1073858410384740

Santini E, Ge H, Ren K, Pena de Ortiz S, Quirk GJ (2004) Consolidation of fear extinction requires protein synthesis in the medial prefrontal cortex. J Neurosci 24(25):5704–5710. https://doi.org/10.1523/JNEUROSCI.0786-04.2004

Schafe GE, Nader K, Blair HT, LeDoux JE (2001) Memory consolidation of Pavlovian fear conditioning: a cellular and molecular perspective. Trends Neurosci 24(9):540–546

Sheinin A, Talani G, Davis MI, Lovinger DM (2008) Endocannabinoid- and mGluR5-dependent short-term synaptic depression in an isolated neuron/bouton preparation from the hippocampal CA1 region. J Neurophysiol 100(2):1041–1052. https://doi.org/10.1152/jn.90226.2008

Shin RM, Tully K, Li Y, Cho JH, Higuchi M, Suhara T, Bolshakov VY (2010) Hierarchical order of coexisting pre- and postsynaptic forms of long-term potentiation at synapses in amygdala. Proc Natl Acad Sci USA 107(44):19073–19078. https://doi.org/10.1073/pnas.1009803107

Shonesy BC, Jalan-Sakrikar N, Cavener VS, Colbran RJ (2014) CaMKII: a molecular substrate for synaptic plasticity and memory. Prog Mol Biol Transl Sci 122:61–87. https://doi.org/10.1016/B978-0-12-420170-5.00003-9

Sipe JC, Scott TM, Murray S, Harismendy O, Simon GM, Cravatt BF, Waalen J (2010) Biomarkers of endocannabinoid system activation in severe obesity. PLoS One 5(1):e8792. https://doi.org/10.1371/journal.pone.0008792

Sotres-Bayon F, Cain CK, LeDoux JE (2006) Brain mechanisms of fear extinction: historical perspectives on the contribution of prefrontal cortex. Biol Psychiatry 60(4):329–336. https://doi.org/10.1016/j.biopsych.2005.10.012

Steiner MA, Wotjak CT (2008) Role of the endocannabinoid system in regulation of the hypothalamic-pituitary-adrenocortical axis. Prog Brain Res 170:397–432. https://doi.org/10.1016/S0079-6123(08)00433-0

Stockhorst U, Antov MI (2015) Modulation of fear extinction by stress, stress hormones and estradiol: a review. Front Behav Neurosci 9:359. https://doi.org/10.3389/fnbeh.2015.00359

Sumioka A, Yan D, Tomita S (2010) TARP phosphorylation regulates synaptic AMPA receptors through lipid bilayers. Neuron 66(5):755–767. https://doi.org/10.1016/j.neuron.2010.04.035

Takagi H, Yamamoto K, Shiosaka S, Senba E, Takatsuki K, Inagaki S et al (1981) Morphological study of noradrenaline innervation in the caudal raphe nuclei with special reference to fine structure. J Comp Neurol 203(1):15–22. https://doi.org/10.1002/cne.902030103

Tanaka J, Horiike Y, Matsuzaki M, Miyazaki T, Ellis-Davies GC, Kasai H (2008) Protein synthesis and neurotrophin-dependent structural plasticity of single dendritic spines. Science 319 (5870):1683–1687. https://doi.org/10.1126/science.1152864

Terranova JP, Michaud JC, Le Fur G, Soubrie P (1995) Inhibition of long-term potentiation in rat hippocampal slices by anandamide and WIN55212-2: reversal by SR141716 A, a selective antagonist of CB1 cannabinoid receptors. Naunyn Schmiedeberg's Arch Pharmacol 352 (5):576–579

Terranova JP, Storme JJ, Lafon N, Perio A, Rinaldi-Carmona M, Le Fur G, Soubrie P (1996) Improvement of memory in rodents by the selective CB1 cannabinoid receptor antagonist, SR 141716. Psychopharmacology 126(2):165–172

Tomita S, Chen L, Kawasaki Y, Petralia RS, Wenthold RJ, Nicoll RA, Bredt DS (2003) Functional studies and distribution define a family of transmembrane AMPA receptor regulatory proteins. J Cell Biol 161(4):805–816. https://doi.org/10.1083/jcb.200212116

Tomita S, Adesnik H, Sekiguchi M, Zhang W, Wada K, Howe JR et al (2005) Stargazin modulates AMPA receptor gating and trafficking by distinct domains. Nature 435(7045):1052–1058. https://doi.org/10.1038/nature03624

Tsvetkov E, Carlezon WA, Benes FM, Kandel ER, Bolshakov VY (2002) Fear conditioning occludes LTP-induced presynaptic enhancement of synaptic transmission in the cortical pathway to the lateral amygdala. Neuron 34(2):289–300

Tyler WJ, Pozzo-Miller L (2003) Miniature synaptic transmission and BDNF modulate dendritic spine growth and form in rat CA1 neurones. J Physiol 553(Pt 2):497–509. https://doi.org/10.1113/jphysiol.2003.052639

Varvel SA, Anum E, Niyuhire F, Wise LE, Lichtman AH (2005) Delta(9)-THC-induced cognitive deficits in mice are reversed by the GABA(A) antagonist bicuculline. Psychopharmacology 178 (2–3):317–327. https://doi.org/10.1007/s00213-004-1988-2

Wilensky AE, Schafe GE, LeDoux JE (2000) The amygdala modulates memory consolidation of fear-motivated inhibitory avoidance learning but not classical fear conditioning. J Neurosci 20 (18):7059–7066

Wilker S, Pfeiffer A, Kolassa S, Elbert T, Lingenfelder B, Ovuga E et al (2014) The role of FKBP5 genotype in moderating long-term effectiveness of exposure-based psychotherapy for posttraumatic stress disorder. Transl Psychiatry 4:e403. https://doi.org/10.1038/tp.2014.49

Winder DG, Martin KC, Muzzio IA, Rohrer D, Chruscinski A, Kobilka B, Kandel ER (1999) ERK plays a regulatory role in induction of LTP by theta frequency stimulation and its modulation by beta-adrenergic receptors. Neuron 24(3):715–726

Wong YH, Lee CM, Xie W, Cui B, Poo MM (2015) Activity-dependent BDNF release via endocytic pathways is regulated by synaptotagmin-6 and complexin. Proc Natl Acad Sci USA 112(32):E4475–E4484. https://doi.org/10.1073/pnas.1511830112

Woo NH, Teng HK, Siao CJ, Chiaruttini C, Pang PT, Milner TA et al (2005) Activation of p75NTR by proBDNF facilitates hippocampal long-term depression. Nat Neurosci 8(8):1069–1077. https://doi.org/10.1038/nn1510

Yang SN, Tang YG, Zucker RS (1999) Selective induction of LTP and LTD by postsynaptic [Ca2+] i elevation. J Neurophysiol 81(2):781–787

Yang C, Liu N, Guo X, Qiao C (2006) Cloning of mpd gene from a chlorpyrifos-degrading bacterium and use of this strain in bioremediation of contaminated soil. FEMS Microbiol Lett 265(1):118–125. https://doi.org/10.1111/j.1574-6968.2006.00478.x

Yang YL, Chao PK, Ro LS, Wo YY, Lu KT (2007) Glutamate NMDA receptors within the amygdala participate in the modulatory effect of glucocorticoids on extinction of conditioned fear in rats. Neuropsychopharmacology 32(5):1042–1051. https://doi.org/10.1038/sj.npp.1301215

Yang J, Harte-Hargrove LC, Siao CJ, Marinic T, Clarke R, Ma Q et al (2014) proBDNF negatively regulates neuronal remodeling, synaptic transmission, and synaptic plasticity in hippocampus. Cell Rep 7(3):796–806. https://doi.org/10.1016/j.celrep.2014.03.040

Yi B, Nichiporuk I, Nicolas M, Schneider S, Feuerecker M, Vassilieva G et al (2016) Reductions in circulating endocannabinoid 2-arachidonoylglycerol levels in healthy human subjects exposed to chronic stressors. Prog Neuro-Psychopharmacol Biol Psychiatry 67:92–97. https://doi.org/10.1016/j.pnpbp.2016.01.004

Yin Y, Edelman GM, Vanderklish PW (2002) The brain-derived neurotrophic factor enhances synthesis of Arc in synaptoneurosomes. Proc Natl Acad Sci USA 99(4):2368–2373. https://doi.org/10.1073/pnas.042693699

Ying SW, Futter M, Rosenblum K, Webber MJ, Hunt SP, Bliss TV, Bramham CR (2002) Brain-derived neurotrophic factor induces long-term potentiation in intact adult hippocampus: requirement for ERK activation coupled to CREB and upregulation of Arc synthesis. J Neurosci 22 (5):1532–1540

Yoshimura A, Fujii R, Watanabe Y, Okabe S, Fukui K, Takumi T (2006) Myosin-Va facilitates the accumulation of mRNA/protein complex in dendritic spines. Curr Biol 16(23):2345–2351. https://doi.org/10.1016/j.cub.2006.10.024

Young KA, Thompson PM, Cruz DA, Williamson DE, Selemon LD (2015) BA11 FKBP5 expression levels correlate with dendritic spine density in postmortem PTSD and controls. Neurobiol Stress 2:67–72. https://doi.org/10.1016/j.ynstr.2015.07.002

Yu SY, Wu DC, Zhan RZ (2010) GluN2B subunits of the NMDA receptor contribute to the AMPA receptor internalization during long-term depression in the lateral amygdala of juvenile rats. Neuroscience 171(4):1102–1108. https://doi.org/10.1016/j.neuroscience.2010.09.038

Zhao L, Yeh ML, Levine ES (2015) Role for endogenous BDNF in endocannabinoid-mediated long-term depression at neocortical inhibitory synapses. eNeuro 2(2). https://doi.org/10.1523/ENEURO.0029-14.2015

Zito K, Scheuss V, Knott G, Hill T, Svoboda K (2009) Rapid functional maturation of nascent dendritic spines. Neuron 61(2):247–258. https://doi.org/10.1016/j.neuron.2008.10.054

Functioning of the Core Neural Network in Fear and Extinction

6

6.1 Fear

Evidence suggests that mechanisms that underlie anxiety have much in common with Pavlovian fear conditioning (Pitman et al. 1999; LeDoux 2000; Bouton et al. 2001; Sullivan et al. 2003; Maren and Quirk 2004; Shin et al. 2006; Bremner et al. 2008; Graham and Milad 2011) and is conserved across species (Phelps and LeDoux 2005; LeDoux 2014). The Pavlovian fear response to a conditioned stimulus (CS; say a tone) in conjunction with a unconditioned stimulus (US; say a shock) is learnt in dorsal lateral amygdala (LAd) and retained for long periods up to years (Quirk et al. 1995; Rogan et al. 1997; Pape and Pare 2010) following only a few US-CS pairings (McAllister et al. 1986; Gale et al. 2004). This learning involves the phenomenon of long-term potentiation (LTP) (McKernan and Shinnick-Gallagher 1997; Tsvetkov et al. 2002; Cho et al. 2013). On the other hand, extinction occurs with repeated presentations of the CS in the absence of the US, which leads to gradual decay in the fear response (Pavlov 1927; Rescorla 2002; Maren and Quirk 2004; Myers and Davis 2007). This is dependent on the establishment of an extinction memory rather than decay of the fear memory (Herry et al. 2008). There is a considerable literature implicating *N*-methyl-D-aspartate (NMDA) receptors, together with LTP, in the formation of these memories (see, e.g. Schafe et al. 2001; Lin et al. 2003a, b; Quirk et al. 2010; Dalton et al. 2012) and pharmacological treatments sought to potentiate LTP in extinction (see, e.g. Lin et al. 2003a, b; Ressler et al. 2004; Davis et al. 2006). In this work a review is first presented on the mechanisms in the amygdala involved in establishing and stabilizing this extinction memory (Sects. 6.1–6.3). Existing models of these processes are next considered (Sect. 6.4) before a new model is presented (Sect. 6.5) that incorporates the most recent observations in a coherent framework.

Fear conditioning is dependent on neural networks in the lateral nucleus (LA) and the basal nucleus (BA) of the amygdala (Maren 2001), with fear learning occurring in LA (Pare et al. 2004; Sigurdsson et al. 2007) and fear extinction inhibited there or

M. Bennett, J. Lagopoulos, *Stress, Trauma and Synaptic Plasticity*,
https://doi.org/10.1007/978-3-319-91116-8_6

in BA (Myers and Davis 2002; Pare et al. 2004; Sotres-Bayon et al. 2004; Sigurdsson et al. 2007). There are two main types of neurons in LA and BA, pyramidal-like projection neurons and local circuit gamma-aminobutyric acid (GABAergic) interneurons (McDonald 1984). The BLA (i.e. LA and BA) is like the cortex (Ehrlich et al. 2009) in as much as it possesses an abundance of excitatory as compared with inhibitory neurons (Carlsen et al. 1985; Smith and Pare 1994). On the other hand, the central nucleus of the amygdala (Ce) is more like the striatum in as much as most of the neurons are inhibitory (GABAergic) (McDonald 1984) and of medium spiny shape like those in the striatum (Martina et al. 1999; Schiess et al. 1999; Markram et al. 2004; Ascoli et al. 2008). The present model (Sect. 6.5) attempts to show how these different neuron types in the different nuclei of the amygdala orchestrate and maintain the extinction memories.

6.1.1 Fear Activation

6.1.1.1 Principal Pathways

Three pathways leading from LAd are activated following CS exciting LAd neurons after CS/US conditioning and that lead to enhanced activity of the medial central nucleus (CeM) output from the amygdala (Pare and Duvarci 2012; Duvarci and Pare 2014). The principal one of these is a direct glutamatergic input to BA neurons, called fear neurons (BAF), which project to form glutamatergic synapses on the amygdala output CeM neurons (Fig. 6.1a). The second most important one of these involves the projection of LAd neurons to dorsal-intercalated inhibitory neurons (ITCd or ICMMD) that in turn inhibit ventral-intercalated inhibitory neurons (ITCv or ICMMV), thereby disinhibiting ITCv on CeM neurons (Fig. 6.1a). Finally, LAd neurons project to inhibitory neurons in the lateral central nucleus (CeL), namely, CeL on neurons characterized by PKCδ− (lacking protein kinase cells) which in turn inhibit CeL off neurons characterized by PKCδ+ (cells possessing protein kinase and lack somatostatin expression) (Ciocchi et al. 2010; Haubensak et al. 2010; Li et al. 2013); these normally inhibit the output of the CEM cells but in this case they are disinhibited (Fig. 6.1a).

6.1.2 Fear Memory

6.1.2.1 The LAd Module

The LAd contains neurons that are critical for the laying down of Pavlovian fear memories acquired in a classical US/CS paradigm (LeDoux et al. 1990; Quirk et al. 1995; Repa et al. 2001). CS inputs and UC inputs converge on neurons at the input to the LAd (LeDoux 2000; Walker and Davis 2000), a site known to contain neurons that are capable of sustained forms of synaptic potentiation, or plasticity (LeDoux 2000; Malkani and Rosen 2000; Blair et al. 2001). As a consequence of such potentiation mechanisms, overlap in activity initiated in the CS and US pathways can lead to the CS on its own providing large responses at the input

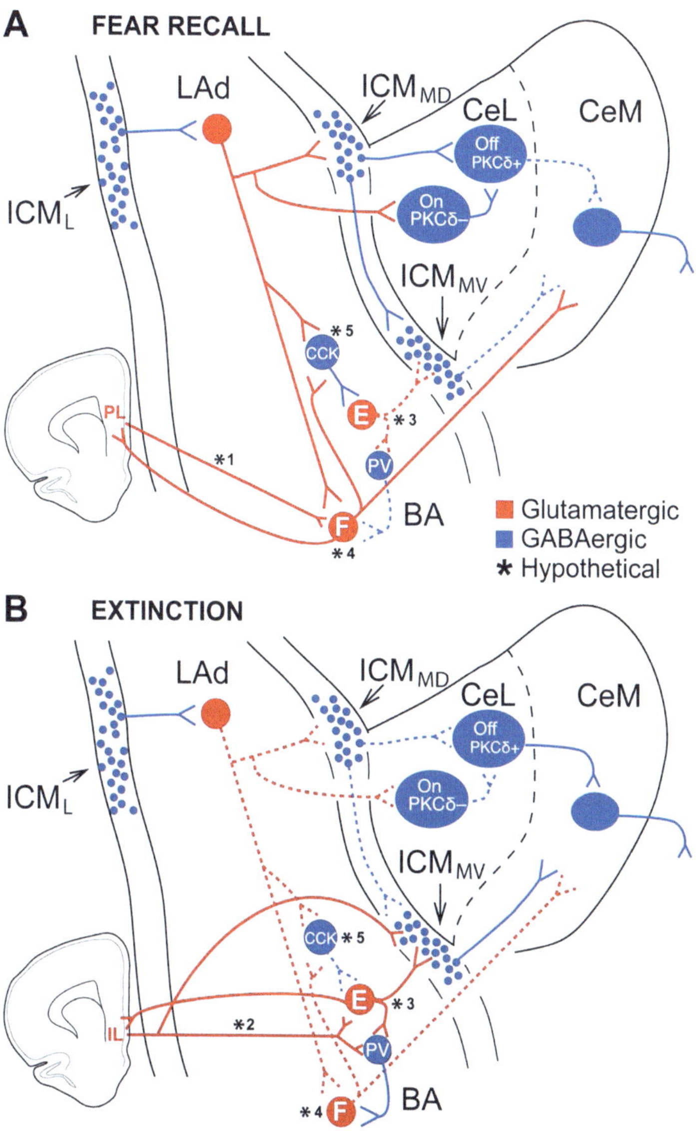

Fig. 6.1 Neural networks subserving the expression of conditioned fear (**a**) and of its extinction (**b**). 'The networks include known pathways and hypothetical links (marked by asterisk) that collectively account for most of the available evidence reviewed in Chaps. 2–4. Solid and dashed lines represent connections that are more or less active, respectively. (**a**) The increased CS responsiveness of CeM output neurons after conditioning likely results from two parallel mechanisms: excitation by glutamatergic BA neurons plus disinhibition from CeL and ICM_{mv} inputs. CeM excitation: CS-induced LA activation causes a BA neuron subtype ('fear neurons', F) to fire and excite CeM cells, whereas another type of BA neuron ('extinction cells', E) is inhibited, possibly by CCK^+ interneurons. Although LA neurons respond transiently to the CS, BA fear neurons, through excitatory interactions with each other and/or with prelimbic (PL) (lower left), would prolong the transient tone signal emanating from LA into persistent CS responses. CeM disinhibition: the excitation of LA cells also leads to the recruitment of ICM_{MD} neurons and of a subgroup of CeL

to LAd (Quirk et al. 1995; Blair et al. 2001). About 25% of the principal neurons in the LAd increase their activity in such a way as to imply that they constitute an engram for a fearful stimulus such as a shock (Quirk et al. 1995; Blair et al. 2001; Repa et al. 2001; Rumpel et al. 2005). Indeed those in the ventral part of the LAd (LAdv) show long-lasting increases in firing that are sustained even in the face of extinction trials (Repa et al. 2001).

Strong evidence for an engram of randomly distributed LA neurons has come from Tanogawa's laboratory. For example, optogenetic excitement of BLA neurons, which were first excited in a context where foot shocks were delivered, determined that the mice avoided the context (Repa et al. 2001). This is almost certainly due to these neurons being recruited into the engram by the development of stronger synaptic connections, as shown by enhanced field potentials recorded in LA to the CS (Quirk et al. 1995; McKernan and Shinnick-Gallagher 1997; Rogan et al. 1997; Goosens et al. 2003; Tye et al. 2008). Recruitment of neurons into engrams is such that engrams laid down by distinct fear conditioning events that are close in time come to share some neurons whereas those distant in time do not (Rashid et al. 2016).

Fig. 6.1 (continued) cells, likely PKCδ− (CeL-On) cells. ICM_{MD} neurons would then inhibit ICM_{MV} cells, disinhibiting CeM neurons. In addition, ICM_{MD} cells would inhibit subgroups of CeL neurons, possibly PKCδ+ (CeL-Off) cells. The recruitment of PKCδ− (CeL-On) cells by LAd neurons would cause a further inhibition of PKCδ+ neurons and disinhibition of CeM cells. (**b**) The decreased CS responsiveness of CeM output neurons after extinction likely depends on two parallel mechanisms: disfacilitation of CeM cells and increased feedforward inhibition of CeM neurons. CeM disfacilitation: the rapid extinction of LAd responses to the CS results in a diminished recruitment of BA fear neurons, disfacilitation of CeM neurons, and, possibly, of CCKδ+ interneurons. As a result, BA extinction cells are disinhibited. Reciprocal excitatory interactions between IL (lower left) and BA might also contribute to enhance the excitability of extinction cells. The disinhibition of extinction cells causes increased excitation of a different set of BA interneurons, possibly PV^+ interneurons, controlling fear cells. CeM inhibition: the reduced CS responsiveness of LAd neurons would cause a disfacilitation of ICM_{MD} neurons and consequent disinhibition of ICM_{MV} neurons. This effect would coincide with an increased excitation of ICM_{MV} cells by inputs from BA extinction cells, thus resulting in an increased feedforward inhibition of CeM cells. The disfacilitation of ICM_{MD} neurons would also cause a disinhibition of subsets of CeL cells, possibly corresponding to PKCδ+ neurons. This effect would be reinforced by the reduced activation of PKC^- cells secondary to reduced LAd inputs. Hypothetical connections (marked by asterisks): (*1 and *2) While it was shown that BA fear and extinction cells differentially project to PL and IL, respectively, whether return mPFC projections are similarly segregated is unknown. (*3 and *4) Currently, there is no data available on the connections of fear and extinction neurons with other amygdala neurons. The differential connections shown are hypotheses based on the available literature. (*5) Little data is available on the connectivity of CCK cells with different types of BA neurons. Trouche et al. (2013) reported that they contact fear (not shown here) and extinction-resistant neurons. CCK synapses to extinction-resistant (but nor fear) neurons showed an upregulation of CB1 receptors after extinction training. The input and output connections of CCK^+ interneurons shown in the figure are all hypothetical. It is possible that other subtypes of interneurons are differentially connected to fear and extinction neurons.' [from Duvarci and Pare (2014); legend to their Fig. 1]

The recruitment of engram neurons is dependent on LTP and the excitability of the neurons, which is in turn modulated by the release of neuromodulators, the efficacy of the synapses and the activity of the local GABAergic neurons mentioned above. These engram neurons are reactivated on remembering (retrieval) of the memory (Redondo et al. 2014; Yiu et al. 2014). The GABAergic interneurons contain either parvalbumin (PV), somatostatin (SOM) or both vasoactive intestinal peptide (VIP) and cholecystokinin (CCK). PV neurons respond to a CS (auditory) and thereby disinhibit SOM neurons mediating inhibition of dendrites of the primary excitatory neurons, so enhancing the CS response of these. On the other hand, a US (shock) excites VIP/CCK inhibitory neurons to inhibit PV and SOM interneurons, thereby enhancing the US response (Wolff et al. 2014). The size of the engram of distributed neurons laid down in LA is determined by these GABAergic neurons, in particular the PV-containing interneurons (Morrison et al. 2016).

6.1.2.2 The BAF Module

A possible contradiction between the conceptual scheme of the engram, consisting of a random distribution of cells, and that of identifiable fear neurons, namely, BAFs, has been noted in a recent review (Bocchio et al. 2017): "The random nature of the Engram neuron distribution is at odds with data showing that certain neuron populations are assigned to specific behavioural outputs. For example, BLA neurons projecting to the prelimbic mPFC (medial prefrontal cortex) promote fear expression (BAF, fear neurons) whereas the axons projecting to the infralimbic region (IL) promote fear extinction (extinction neurons (BAE) (Senn et al. 2014)". However the two schemes can be reconciled, as they are in the present work, if the engram neurons are located at the input to LA, namely, in LAd, with these distinguished from the BAFs located in BA and receive projections from the engram neurons.

The BAF neurons receive input from local GABAergic inhibitory neurons, as do all principal excitatory neurons, and those that contain PV are of particular interest as they receive input from extinction neurons in BA, namely, BAE neurons, as mentioned below. Thus the BAF module possesses BAF neurons with inputs from LAd as well as from inhibitory PV cells that in turn receive inputs from LAd as well as from BAE.

6.2 Extinction

6.2.1 Extinction Activation

6.2.1.1 Principal Pathways

One pathway leading from LAd is activated following CS exciting LAd neurons after conditioning and which leads to decreased activity of CeM output from the amygdala (Pare and Duvarci 2012; Duvarci and Pare 2014). This involves a direct glutamatergic input to BA neurons, called (BAE), which project to form glutamatergic synapses on intercalated inhibitory ITCv neurons that then inhibit the amygdala output CeM neurons (Fig. 6.1b). The BAE neurons also change the response of BAF cells (Herry et al. 2008; Sierra-Mercado et al. 2011; Senn et al. 2014) consequent to changes in the activity of the GABAergic interneurons containing PV that control BAF cells (Pare and Duvarci 2012). These changes in activity contribute to a decrease in the CS fear response (Bouton and King 1983; Quirk 2006).

The extinction memory associated with BAE neurons (see Sect. 6.2.2.1) decreases the activity of the BAF fear neurons (Bouton et al. 2006; Maren 2011). Extinction is then dependent on depression of activity in BAF neurons as well as enhanced activity in ITCv inhibitory neurons. In addition there is some evidence that the LTP responsible for setting up a fear engram in LAd (Mahanty and Sah 1998; Bauer and LeDoux 2004) is depontentiated in extinction (Kim et al. 2007; Hong et al. 2009), decreasing the drive on BAF neurons.

6.2.2 Extinction Memory

6.2.2.1 The BAE Module

The infralimbic (IL) part of the medial prefrontal cortex (mPFC) projection to some BLA neurons increases activity during extinction training that does not develop in its absence (Morgan et al. 1993; Quirk et al. 2000, 2003; Maren and Quirk 2004; Milad et al. 2004; Santini et al. 2004; Holmes et al. 2012). Thus lesions to this part of the cortex inhibit extinction memory development (Maren and Quirk 2004; Burgos-Robles et al. 2009). These observations indicate the existence of discrete extinction neurons in the BLA, namely, BAE neurons, and this is supported by the observation that extinction memory is accompanied by an increase in excitability of about 17% of the principal neurons in the BLA, which are probably BAE neurons (Bouton et al. 2006; Herry et al. 2008; Amano et al. 2010; Maren 2011).

An extinction memory in the BLA, dependent on BAE neurons, involves plastic changes, including the activation of NMDA receptors (Falls et al. 1992; Huang et al. 1994; Baker and Azorlosa 1996; Santini et al. 2001; Bauer et al. 2002; Lin et al. 2003a, b; Zimmerman and Maren 2010) as would be expected if it is LTP dependent. Hence modelling the extinction process mediated by BAE neurons at the synaptic level requires consideration of calcium influx through NMDA channels triggering a cascade of intracellular changes leading to short- and/or long-term forms of LTP (Shouval et al. 2002a, b; Li et al. 2009; Ball et al. 2012). Long-LTP is dependent on protein synthesis (Huang et al. 1994; Nguyen and Kandel 1996; Huang and Kandel 1998). Another plastic change involves excited BAE neurons releasing endocannabinoids onto receptors (CB1R) located on local GABAergic (CCK containing) interneurons, so blocking their activity and the inhibition of BAE neurons (Lafenetre et al. 2007); for a review see (Bennett et al. 2017).

These various plastic changes involving BAE neurons, together with the dependence of extinction on input from IL (Adhikari et al. 2015), suggest that BAE cells participate in the plastic circuitry as follows: BAE neurons are weakly excited by input from fear engram neurons in LAd, and GABA (CCK containing) local inhibitory interneurons are strongly excited by the LAd input, all in the absence of an input from IL; subsequently, during extinction training, the excited IL input to BAE neurons concomitantly with the CS input mediated by CS excited engram cells in LAd leads through LTP to the strengthening of the LAd input to BAE principal neurons; this is accompanied by the enhanced release of endocannabinoids from the excited BAE neurons and so their disinhibition. BAE neurons are now engaged each time a CS is presented at the input to LAd, providing an extinction memory. It is to be noted that the extinction memory outlined here is separate from the fear engram and does not erase the fear engram; indeed it depends on it, in order to be established (Pavlov 1927; Bouton and King 1983; Rescorla 2002).

A key question here is whether a gradual failure of the plasticity mechanisms involved in the activity of IL projections to what we have termed the BAE and ITC modules contributes in important way to the spontaneous return of fear responses over time (Maren 2015).

6.2.2.2 The ITC Module

A number of studies suggest that a cause of inhibition of CeM in extinction experiments is direct excitation of ITCv GABAergic neurons (Amano et al. 2010; Busti et al. 2011) by principal neurons in the IL (Pavlov 1927; Bouton and King 1983; McDonald et al. 1996; Rescorla 2002) during extinction training (Cho et al. 2013; Quirk et al. 2003; Pare and Duvarci 2012). The inhibition of CeM cells is mediated by projections of ITCv cells onto the dendrites of CeM neurons (Royer et al. 1999; Delaney and Sah 2001). It has recently been shown that this effect of IL inputs onto ITCv cells is not direct, but rather through the IL input to ITCv increasing the efficacy of BAF terminals on ITCv neurons through a mechanism likely to involve LTP (Do-Monte et al. 2015). It now seems very likely from combined behavioural and in vitro studies that the principal extinction memory, due to an input from IL, involves BAE neurons (see Sect. 6.2.2.1) and it is the projection from BAE neurons to ITCv that is critical in driving inhibition of the CeM amygdala output in extinction (Amano et al. 2010; reviewed in Pare and Duvarci 2012; Bocchio et al. 2017).

6.3 Synaptic Plasticity

The mechanisms of synaptic plasticity at a synapse have been explored using phenomenological (Pfister and Gerstner 2006; Clopath and Gerstner 2010; El Boustani et al. 2012), functional (Pfister and Gerstner 2006; Sprekeler et al. 2007; Clopath and Gerstner 2010; Pool and Mato 2011; El Boustani et al. 2012) and biophysical (Lisman 1989; Artola and Singer 1993; Shouval et al. 2002a, b) approaches. These have been used to explore models concerning the changes in the LAd that might be involved in establishing a fear engram there (see, for instance, Li et al. 2009; Ball et al. 2012). Details of the biochemical pathways that support these plasticity mechanisms have been recently reviewed for both relatively short-lasting changes in synaptic efficacy arising from plasticity (Bennett et al. 2017) as well as for very long changes lasting years (Hsieh et al. 2017). Given the emphasis on LTP plasticity in establishing the extinction memory involving BAE (see Sect. 6.2.2.1), which has been highlighted as a possible focus for investigations on how to prevent the decline of extinction, it is important to consider plasticity mechanisms, as given below.

6.3.1 Conditions for Plasticity

The distribution of synaptic efficacies over synapses in a population is skewed, with relatively high numbers of low-efficacy synapses (van Rossum et al. 2000; Loewenstein et al. 2011; Leen and Friel 2012; Zheng et al. 2013), a condition that should hold for energy conservation, among other considerations (Krieg and Triesch 2014). Such a distribution of low efficacy synapses is also important in order for synaptic plasticity mechanisms to encode a fear engram (Olshausen and Field 1996; Hromadka et al. 2008) resulting from associative fear learning (e.g. Rogan et al. 1997; Nabavi et al. 2014). However there are a variety of plasticity mechanism at different synapses, such as those dependent on the timing of presynaptic and postsynaptic action potentials (Artola et al. 1990; Artola and Singer 1993; Gerstner et al. 1996; Markram et al. 1997), on the frequency of these action potentials (Bliss and Lømo 1973; Sjostrom et al. 2001), as well as on previous synaptic activity (Bienenstock et al. 1982; Wang and Wagner 1999) and the extent of synaptic efficacy before the plasticity is engendered as noted above (Ngezahayo et al. 2000). At this time it is not clear which of these are required for plasticity at the different synapses in the amygdala.

6.3.2 Neuromodulators

Neuromodulators released from subcortical inputs to the BLA assist in the induction of LTP associated with fear learning. Thus both NMDA receptor-mediated LTP and neuromodulators are required for associative fear learning under behavioural conditions that approach the normal (Johansen et al. 2014). These neuromodulators include noradrenaline (Hu et al. 2007; Johansen et al. 2014) and dopamine (Bissiere et al. 2003; Chen et al. 2003; Hu et al. 2007; Tully et al. 2007; Fadok et al. 2009; Johnson et al. 2015; Jiang et al. 2016).

6.3.3 Theta and Gamma Oscillations

A CS after fear learning increases the synchrony and power of theta oscillations between BLA and mPFC (Seidenbecher et al. 2003; Lesting et al. 2011a, b; Likhtik et al. 2014), especially during freezing (Karalis et al. 2016). Although gamma power is low in these circumstances, it is phase locked with the theta oscillations (Stujenske et al. 2014). Synchronization of theta oscillations between BLA and mPFC during fear memory recall is reflected in the synchronization of their local field potentials (Seidenbecher et al. 2003; Narayanan et al. 2007; Lesting et al. 2011a, b), probably indicating specific inputs from mPFC principal neurons to the BLA (Bienvenu et al. 2012, Manko et al. 2012, Lesting et al. 2013). Extinction stimuli are accompanied by an increase in the directionality of theta synchronicity between mPFC and BLA (Bocchio et al. 2017) although a decrease in theta synchrony occurs between BLA and hippocampal CA1 regions (Pape and Pare 2010; Lesting et al. 2011a, b). In general these changes in synchronicity of theta oscillations are consistent with the known inputs from mPFC to the BLA in fear learning and extinction; however an important question remains, namely, are the plasticity changes mediated by LTP induced by such synchronicity or does the synchronicity arise as a consequence of such plastic changes (Bocchio et al. 2017).

6.3.4 Excitability

Neurons with high excitability are favoured for recruitment into engrams during formation of fear memories (Kim et al. 2013). Such neurons can be identified as possessing relatively high concentrations of CREB (cAMP response element-binding protein) (Han et al. 2007, 2009). Activation of CREB leads to a decrease in the slow after-hyperpolarization current in these neurons (Viosca et al. 2009; Benito and Barco 2010), increasing their excitability and so enhancing their recruitment into the fear engram (Zhou et al. 2009). Removing this set of neurons with relatively high levels of excitability by targeting CREB gene deletion then blocks the formation of a fear engram (Han et al. 2009).

6.4 Models of the Amygdala in Fear and Extinction

6.4.1 Fear Engram

Modelling of the fear engram in LAd has been carried out by Kim, Pare and colleagues using a biophysical computational model of large numbers of principal excitatory neurons, each with one of three different classes of excitability characterized by different spike frequency adaptations as well as inhibitory interneurons, the connections between these capable of undergoing changes in Hebbian plasticity (Kim et al. 2013, 2016; Feng et al. 2016). This model showed that di-synaptic inhibition (that is between inhibitory neurons and principal neurons) is a dominant factor in LAd determining which principal neurons with high intrinsic excitability are grouped together through their collateral excitatory connections to form an engram (Feng et al. 2016). They show the essential role of plasticity at the synapses both from the thalamic input to LAd and between the principal neurons in LAd (Kim et al. 2013).

6.4.2 Context

The above considerations do not incorporate mechanisms that account for contextual clues involved in fear and extinction learning. Such clues are incorporated in the model of Carrere and Alexandre (2015). After US/CS fear learning in the LAd, following sensory inputs from the thalamus and cortex concerning the US (foot shock) and the CS (tone), BAF and CeL on cells are directly excited leading to enhanced activity of CeM as in the present model. But in addition, BAF neurons receive inputs from the hippocampus relaying information about the context of the conditioning. A projection of BAF neurons to CeL on cells or of BAE neurons to CeL off cells and mutual inhibition of CeL on and CeL off cells can be incorporated. This model of Carrere and Alexandre (2015) replicated a number of animal experiments concerning Pavlovian conditioning in different contexts.

An early model that does not emphasize the networks of the amygdala but does the hippocampal inputs carrying contextual information, as well as the plasticity of synapses mediating US/CS and context-specific inputs, showed that two populations of BA excitatory neurons could be recruited during conditioning and extinction (Vlachos et al. 2011). These presumably can be identified as BAF and BAE neurons, which the later literature has emphasized. The importance of neuromodulators, such as noradrenaline, is also emphasized in this work.

Some models place emphasis on the neurons in the BLA (BAF neurons) that directly excite the CeM output neurons of the amygdala, whereas ventral mPFC (infralimbic (IL) neurons) projections to ITC inhibitory neurons that in turn directly inhibit CeM are emphasized. Context is then accounted for by hippocampal neuron projections to both ventral mPFC and BLA (BAF neurons) (Moustafa et al. 2013). In this case account is not made of the more specific circuitry of the amygdala, such as that involving BAE and CeL (on or off) neurons. More particularly, consideration of the IL projections to BAE as perhaps the most important pathway mediating extinction (not IL to ITC) is not canvassed.

6.4.3 Endocannabinoids

The release of endocannabinoids in the BLA is required for extinction, and this involves their release from excited BAE neurons onto the terminals of GABAergic cholecystokinin containing interneurons to block the release of the inhibitory transmitter (for a review, see Bennett et al. 2017). This mechanism has been incorporated into a model of connectivity between different amygdala nuclei (Duvarci and Pare 2014; Anastasio 2013). Using a computational approach, the model was used to indicate that the endocannabinoid-mediated depression of the GABAergic synapses could lead to changes in synaptic efficacy that might support extinction.

6.4.4 Hebbian Plasticity

Wilson–Cowan representations of different pyramidal neuron types in the BLA (Wilson and Cowan 1972; Destexhe and Sejnowski 2009), characterized by their action potential firing patterns, together with Hebbian calcium mechanisms for LTP according to Cooper and his colleagues (Shouval et al. 2002a, b), have been used to provide a general network of the amygdala during fear learning. However this work does not take into account the different neuron types classified on the basis of their known function and connectivity (Ball et al. 2012; Li et al. 2009).

6.5 A New Model of the Amygdala in Extinction

The amygdala networks arrived at on consideration of the literature are summarized in the following model.

6.5.1 Incorporating Local Inhibitory Mechanisms for LAd, BAF and BAE Neurons in the Model

The principal excitatory neurons in LAd are the primary sites of inputs from US and CS, and it is through the mechanism of LTP that the US drives the CS synapses on these neurons to an enhanced level of efficacy. This classical Hebbian mechanism is modulated however by a number of different classes of GABAergic interneurons in LAd (Wolff et al. 2014). During a US (say a foot shock), both PV-containing and SOM-containing GABAergic interneurons are inhibited by CCK/VIP-containing GABAergic interneurons, thus inhibiting both SOM and PV inhibition of principal excitatory neurons, so enhancing the foot shock response. During a CS (say auditory cue), there is excitation of PV-containing GABAergic interneurons that in turn disinhibit SOM-containing GABAergic interneurons which normally inhibit inputs to dendrites, so enhancing the major CS (auditory) input to these dendrites, resulting in auditory cue-foot shock associations.

BAF neurons, besides receiving inputs from LAd, also receive inputs from local PV containing GABAergic interneurons, which in turn are innervated by BAE neurons. On the other hand, BAE neurons, besides receiving inputs from LAd and IL neurons, also receive inputs from CCK-containing GABAergic interneurons, which in turn are innervated by LAd neurons and BAE neurons (Duvarci and Pare 2014). It is argued that endocannabinoids released from principal BAE neurons during extinction act to disinhibit themselves through block of GABA release from CCK-containing interneurons (see Lafenetre et al. 2007).

6.5.2 Control of the CeM Output Neurons in the Model

Observations suggest that there are four pathways that determine an increased CeM output in fear conditioning. LAd neurons, following CS, excite BAF neurons that directly increase CeM activity as well as, via PV inhibitory interneurons, silencing BAE neurons. A second pathway involves LAd neurons providing direct stimulation of PKCδ− (CeL on) cells, which in turn inhibit PKCδ+ (CeL off), thus leading to disinhibition of CeM output neurons. A third pathway involves direct stimulation of ITCd neurons that in turn inhibit ITCv neurons leading to further disinhibition of CeM. In addition, the excited ITCd neurons also inhibit PKCδ+ (CeL off) cells, enhancing disinhibition of the CeM output neurons further.

Other observations indicate there are several pathways that determine a diminished CeM output in fear extinction. One of these involves enhanced activity of BAE neurons that directly excite ITCv neurons producing inhibition of CeM activity. Another is due to inhibition of BAF neurons through PV inhibitory interneurons as a consequence of their input from the excited BAE neurons. Further, the ITCv neurons receive a direct excitatory drive from excited IL neurons in the mPFC, leading to inhibition of CeM neurons. Finally, and most importantly, the drive from BAE neurons to ITCv neurons is greatly enhanced by disinhibition of the former by local GABA (CCK) interneurons consequent on the release of endocannabinoids released from the excited BAE neurons due to their also receiving input from IL neurons in the mPFC.

6.6 Considerations of the Model

The present review set out with the objective of determining the most likely site(s) at which extinction memories are laid down and what the mechanisms are which support these memories. According to our present knowledge, the critical sites are in BLA, most likely involving the IL-induced LTP of LAd synapses on BAE neurons located in BLA, with the possibility of forming an engram for extinction in the BLA. This site is favoured over the alternative, namely, at the IL input to ITCv neurons, for on present evidence IL enhanced activity is required for extinction training but not for extinction retrieval (Do-Monte et al. 2015). Once the extinction engram is established through the action of IL in conjunction with LAd synapses, IL is no longer required in the process of retrieving an extinction memory. Inhibitory neurons do not form engrams, so the only mechanism open for IL's action on ITCv neurons is to potentiate synaptic transmission from BAE neurons to ITCv neurons. According to this scenario, the most likely site at which failure to retrieve extinction occurs, such as in the failure of treatment for patients with post-traumatic stress disorder, is at the synapses of LAd neurons on BAE neurons, following their potentiation during extinction training by IL synapses of IL neurons on BAE neurons. This, it is argued, should be the site requiring detailed study in the context of the failure of extinction memory.

References

Adhikari A, Lerner TN, Finkelstein J, Pak S, Jennings JH, Davidson TJ et al (2015) Basomedial amygdala mediates top-down control of anxiety and fear. Nature 527(7577):179–185. https://doi.org/10.1038/nature15698

Amano T, Unal CT, Pare D (2010) Synaptic correlates of fear extinction in the amygdala. Nat Neurosci 13(4):489–494. https://doi.org/10.1038/nn.2499

Anastasio TJ (2013) Computational search for hypotheses concerning the endocannabinoid contribution to the extinction of fear conditioning. Front Comput Neurosci 7:74. https://doi.org/10.3389/fncom.2013.00074

Artola A, Singer W (1993) Long-term depression of excitatory synaptic transmission and its relationship to long-term potentiation. Trends Neurosci 16(11):480–487

Artola A, Brocher S, Singer W (1990) Different voltage-dependent thresholds for inducing long-term depression and long-term potentiation in slices of rat visual cortex. Nature 347(6288):69–72. https://doi.org/10.1038/347069a0

Ascoli GA, Alonso-Nanclares L, Anderson SA, Barrionuevo G, Benavides-Piccione R, Burkhalter A et al (2008) Petilla terminology: nomenclature of features of GABAergic interneurons of the cerebral cortex. Nat Rev Neurosci 9(7):557–568. https://doi.org/10.1038/nrn2402

Baker JD, Azorlosa JL (1996) The NMDA antagonist MK-801 blocks the extinction of Pavlovian fear conditioning. Behav Neurosci 110(3):618–620

Ball JM, Hummos AM, Nair SS (2012) Role of sensory input distribution and intrinsic connectivity in lateral amygdala during auditory fear conditioning: a computational study. Neuroscience 224:249–267. https://doi.org/10.1016/j.neuroscience.2012.08.030

Bauer EP, LeDoux JE (2004) Heterosynaptic long-term potentiation of inhibitory interneurons in the lateral amygdala. J Neurosci 24(43):9507–9512. https://doi.org/10.1523/jneurosci.3567-04.2004

Bauer EP, Schafe GE, LeDoux JE (2002) NMDA receptors and L-type voltage-gated calcium channels contribute to long-term potentiation and different components of fear memory formation in the lateral amygdala. J Neurosci 22(12):5239–5249

Benito E, Barco A (2010) CREB's control of intrinsic and synaptic plasticity: implications for CREB-dependent memory models. Trends Neurosci 33(5):230–240. https://doi.org/10.1016/j.tins.2010.02.001

Bennett MR, Arnold J, Hatton SN, Lagopoulos J (2017) Regulation of fear extinction by long-term depression: the roles of endocannabinoids and brain derived neurotrophic factor. Behav Brain Res 319:148–164. https://doi.org/10.1016/j.bbr.2016.11.029

Bienenstock E, Cooper L, Munro P (1982) Theory for the development of neuron selectivity: orientation specificity and binocular interaction in visual cortex. J Neurosci 2(1):32–48

Bienvenu TC, Busti D, Magill PJ, Ferraguti F, Capogna M (2012) Cell-type-specific recruitment of amygdala interneurons to hippocampal theta rhythm and noxious stimuli in vivo. Neuron 74(6):1059–1074. https://doi.org/10.1016/j.neuron.2012.04.022

Bissiere S, Humeau Y, Luthi A (2003) Dopamine gates LTP induction in lateral amygdala by suppressing feedforward inhibition. Nat Neurosci 6(6):587–592. https://doi.org/10.1038/nn1058

Blair HT, Schafe GE, Bauer EP, Rodrigues SM, LeDoux JE (2001) Synaptic plasticity in the lateral amygdala: a cellular hypothesis of fear conditioning. Learn Mem 8(5):229–242. https://doi.org/10.1101/lm.30901

Bliss TVP, Lømo T (1973) Long-lasting potentiation of synaptic transmission in the dentate area of the anaesthetized rabbit following stimulation of the perforant path. J Physiol 232(2):331–356. https://doi.org/10.1113/jphysiol.1973.sp010273

Bocchio M, Nabavi S, Capogna M (2017) Synaptic plasticity, engrams, and network oscillations in amygdala circuits for storage and retrieval of emotional memories. Neuron 94(4):731–743. https://doi.org/10.1016/j.neuron.2017.03.022

Bouton ME, King DA (1983) Contextual control of the extinction of conditioned fear: tests for the associative value of the context. J Exp Psychol Anim Behav Process 9(3):248–265

Bouton ME, Mineka S, Barlow DH (2001) A modern learning theory perspective on the etiology of panic disorder. Psychol Rev 108(1):4–32

Bouton ME, Westbrook RF, Corcoran KA, Maren S (2006) Contextual and temporal modulation of extinction: behavioral and biological mechanisms. Biol Psychiatry 60(4):352–360. https://doi.org/10.1016/j.biopsych.2005.12.015

Bremner JD, Elzinga B, Schmahl C, Vermetten E (2008) Structural and functional plasticity of the human brain in posttraumatic stress disorder. Prog Brain Res 167:171–186. https://doi.org/10.1016/s0079-6123(07)67012-5

Burgos-Robles A, Vidal-Gonzalez I, Quirk GJ (2009) Sustained conditioned responses in prelimbic prefrontal neurons are correlated with fear expression and extinction failure. J Neurosci 29 (26):8474–8482. https://doi.org/10.1523/jneurosci.0378-09.2009

Busti D, Geracitano R, Whittle N, Dalezios Y, Manko M, Kaufmann W et al (2011) Different fear states engage distinct networks within the intercalated cell clusters of the amygdala. J Neurosci 31(13):5131–5144. https://doi.org/10.1523/jneurosci.6100-10.2011

Carlsen J, Zaborszky L, Heimer L (1985) Cholinergic projections from the basal forebrain to the basolateral amygdaloid complex: a combined retrograde fluorescent and immunohistochemical study. J Comp Neurol 234(2):155–167. https://doi.org/10.1002/cne.902340203

Carrere M, Alexandre F (2015) A pavlovian model of the amygdala and its influence within the medial temporal lobe. Front Syst Neurosci 9:41. https://doi.org/10.3389/fnsys.2015.00041

Chen A, Hough CJ, Li H (2003) Serotonin type II receptor activation facilitates synaptic plasticity via N-methyl-D-aspartate-mediated mechanism in the rat basolateral amygdala. Neuroscience 119(1):53–63

Cho JH, Deisseroth K, Bolshakov VY (2013) Synaptic encoding of fear extinction in mPFC-amygdala circuits. Neuron 80(6):1491–1507. https://doi.org/10.1016/j.neuron.2013.09.025

Ciocchi S, Herry C, Grenier F, Wolff SB, Letzkus JJ, Vlachos I et al (2010) Encoding of conditioned fear in central amygdala inhibitory circuits. Nature 468(7321):277–282. https://doi.org/10.1038/nature09559

Clopath C, Gerstner W (2010) Voltage and spike timing interact in STDP—a unified model. Front Synaptic Neurosci 2:25. https://doi.org/10.3389/fnsyn.2010.00025

Dalton GL, Wu DC, Wang YT, Floresco SB, Phillips AG (2012) NMDA GluN2A and GluN2B receptors play separate roles in the induction of LTP and LTD in the amygdala and in the acquisition and extinction of conditioned fear. Neuropharmacology 62(2):797–806. https://doi.org/10.1016/j.neuropharm.2011.09.001

Davis M, Myers KM, Chhatwal J, Ressler KJ (2006) Pharmacological treatments that facilitate extinction of fear: relevance to psychotherapy. NeuroRx 3(1):82–96. https://doi.org/10.1016/j.nurx.2005.12.008

Delaney AJ, Sah P (2001) Pathway-specific targeting of GABA(A) receptor subtypes to somatic and dendritic synapses in the central amygdala. J Neurophysiol 86(2):717–723

Destexhe A, Sejnowski TJ (2009) The Wilson-Cowan model, 36 years later. Biol Cybern 101 (1):1–2. https://doi.org/10.1007/s00422-009-0328-3

Do-Monte FH, Manzano-Nieves G, Quinones-Laracuente K, Ramos-Medina L, Quirk GJ (2015) Revisiting the role of infralimbic cortex in fear extinction with optogenetics. J Neurosci 35 (8):3607–3615. https://doi.org/10.1523/jneurosci.3137-14.2015

Duvarci S, Pare D (2014) Amygdala microcircuits controlling learned fear. Neuron 82(5):966–980. https://doi.org/10.1016/j.neuron.2014.04.042

Ehrlich I, Humeau Y, Grenier F, Ciocchi S, Herry C, Luthi A (2009) Amygdala inhibitory circuits and the control of fear memory. Neuron 62(6):757–771. https://doi.org/10.1016/j.neuron.2009.05.026

El Boustani S, Yger P, Fregnac Y, Destexhe A (2012) Stable learning in stochastic network states. J Neurosci 32(1):194–214. https://doi.org/10.1523/jneurosci.2496-11.2012

Fadok JP, Dickerson TM, Palmiter RD (2009) Dopamine is necessary for cue-dependent fear conditioning. J Neurosci 29(36):11089–11097. https://doi.org/10.1523/JNEUROSCI.1616-09.2009
Falls WA, Miserendino MJ, Davis M (1992) Extinction of fear-potentiated startle: blockade by infusion of an NMDA antagonist into the amygdala. J Neurosci 12(3):854–863
Feng F, Samarth P, Pare D, Nair SS (2016) Mechanisms underlying the formation of the amygdalar fear memory trace: a computational perspective. Neuroscience 322:370–376. https://doi.org/10.1016/j.neuroscience.2016.02.059
Gale GD, Anagnostaras SG, Godsil BP, Mitchell S, Nozawa T, Sage JR et al (2004) Role of the basolateral amygdala in the storage of fear memories across the adult lifetime of rats. J Neurosci 24(15):3810–3815. https://doi.org/10.1523/jneurosci.4100-03.2004
Gerstner W, Kempter R, van Hemmen JL, Wagner H (1996) A neuronal learning rule for sub-millisecond temporal coding. Nature 383(6595):76–81. https://doi.org/10.1038/383076a0
Goosens KA, Hobin JA, Maren S (2003) Auditory-evoked spike firing in the lateral amygdala and Pavlovian fear conditioning: mnemonic code or fear bias? Neuron 40(5):1013–1022
Graham BM, Milad MR (2011) The study of fear extinction: implications for anxiety disorders. Am J Psychiatry 168(12):1255–1265. https://doi.org/10.1176/appi.ajp.2011.11040557
Han JH, Kushner SA, Yiu AP, Cole CJ, Matynia A, Brown RA et al (2007) Neuronal competition and selection during memory formation. Science 316(5823):457–460. https://doi.org/10.1126/science.1139438
Han JH, Kushner SA, Yiu AP, Hsiang HL, Buch T, Waisman A et al (2009) Selective erasure of a fear memory. Science 323(5920):1492–1496. https://doi.org/10.1126/science.1164139
Haubensak W, Kunwar PS, Cai H, Ciocchi S, Wall NR, Ponnusamy R et al (2010) Genetic dissection of an amygdala microcircuit that gates conditioned fear. Nature 468(7321):270–276. https://doi.org/10.1038/nature09553
Herry C, Ciocchi S, Senn V, Demmou L, Muller C, Luthi A (2008) Switching on and off fear by distinct neuronal circuits. Nature 454(7204):600–606. https://doi.org/10.1038/nature07166
Holmes A, Fitzgerald PJ, MacPherson KP, DeBrouse L, Colacicco G, Flynn SM et al (2012) Chronic alcohol remodels prefrontal neurons and disrupts NMDAR-mediated fear extinction encoding. Nat Neurosci 15(10):1359–1361. https://doi.org/10.1038/nn.3204
Hong I, Song B, Lee S, Kim J, Kim J, Choi S (2009) Extinction of cued fear memory involves a distinct form of depotentiation at cortical input synapses onto the lateral amygdala. Eur J Neurosci 30(11):2089–2099. https://doi.org/10.1111/j.1460-9568.2009.07004.x
Hromadka T, Deweese MR, Zador AM (2008) Sparse representation of sounds in the unanesthetized auditory cortex. PLoS Biol 6(1):e16. https://doi.org/10.1371/journal.pbio.0060016
Hsieh C, Tsokas P, Serrano P, Hernandez AI, Tian D, Cottrell JE et al (2017) Persistent increased PKMzeta in long-term and remote spatial memory. Neurobiol Learn Mem 138:135–144. https://doi.org/10.1016/j.nlm.2016.07.008
Hu H, Real E, Takamiya K, Kang MG, Ledoux J, Huganir RL, Malinow R (2007) Emotion enhances learning via norepinephrine regulation of AMPA-receptor trafficking. Cell 131(1):160–173. https://doi.org/10.1016/j.cell.2007.09.017
Huang YY, Kandel ER (1998) Postsynaptic induction and PKA-dependent expression of LTP in the lateral amygdala. Neuron 21(1):169–178
Huang YY, Li XC, Kandel ER (1994) cAMP contributes to mossy fiber LTP by initiating both a covalently mediated early phase and macromolecular synthesis-dependent late phase. Cell 79(1):69–79
Jiang L, Kundu S, Lederman JD, Lopez-Hernandez GY, Ballinger EC, Wang S et al (2016) Cholinergic signaling controls conditioned fear behaviors and enhances plasticity of cortical-amygdala circuits. Neuron 90(5):1057–1070. https://doi.org/10.1016/j.neuron.2016.04.028
Johansen JP, Diaz-Mataix L, Hamanaka H, Ozawa T, Ycu E, Koivumaa J et al (2014) Hebbian and neuromodulatory mechanisms interact to trigger associative memory formation. Proc Natl Acad Sci USA 111(51):E5584–E5592. https://doi.org/10.1073/pnas.1421304111

Johnson PL, Molosh A, Fitz SD, Arendt D, Deehan GA, Federici LM et al (2015) Pharmacological depletion of serotonin in the basolateral amygdala complex reduces anxiety and disrupts fear conditioning. Pharmacol Biochem Behav 138:174–179. https://doi.org/10.1016/j.pbb.2015.09.021

Karalis N, Dejean C, Chaudun F, Khoder S, Rozeske RR, Wurtz H et al (2016) 4-Hz oscillations synchronize prefrontal-amygdala circuits during fear behavior. Nat Neurosci 19(4):605–612. https://doi.org/10.1038/nn.4251

Kim J, Lee S, Park K, Hong I, Song B, Son G et al (2007) Amygdala depotentiation and fear extinction. Proc Natl Acad Sci USA 104(52):20955–20960. https://doi.org/10.1073/pnas.0710548105

Kim D, Pare D, Nair SS (2013) Mechanisms contributing to the induction and storage of Pavlovian fear memories in the lateral amygdala. Learn Mem 20(8):421–430. https://doi.org/10.1101/lm.030262.113

Kim D, Samarth P, Feng F, Pare D, Nair SS (2016) Synaptic competition in the lateral amygdala and the stimulus specificity of conditioned fear: a biophysical modeling study. Brain Struct Funct 221(4):2163–2182. https://doi.org/10.1007/s00429-015-1037-4

Krieg D, Triesch J (2014) A unifying theory of synaptic long-term plasticity based on a sparse distribution of synaptic strength. Front Synaptic Neurosci 6:3. https://doi.org/10.3389/fnsyn.2014.00003

Lafenetre P, Chaouloff F, Marsicano G (2007) The endocannabinoid system in the processing of anxiety and fear and how CB1 receptors may modulate fear extinction. Pharmacol Res 56(5):367–381. https://doi.org/10.1016/j.phrs.2007.09.006

LeDoux JE (2000) Emotion circuits in the brain. Annu Rev Neurosci 23:155–184. https://doi.org/10.1146/annurev.neuro.23.1.155

LeDoux JE (2014) Coming to terms with fear. Proc Natl Acad Sci USA 111(8):2871–2878. https://doi.org/10.1073/pnas.1400335111

LeDoux JE, Cicchetti P, Xagoraris A, Romanski LM (1990) The lateral amygdaloid nucleus: sensory interface of the amygdala in fear conditioning. J Neurosci 10(4):1062–1069

Leen TK, Friel R (2012) Stochastic perturbation methods for spike-timing-dependent plasticity. Neural Comput 24(5):1109–1146. https://doi.org/10.1162/NECO_a_00267

Lesting J, Geiger M, Narayanan RT, Pape HC, Seidenbecher T (2011a) Impaired extinction of fear and maintained amygdala-hippocampal theta synchrony in a mouse model of temporal lobe epilepsy. Epilepsia 52(2):337–346. https://doi.org/10.1111/j.1528-1167.2010.02758.x

Lesting J, Narayanan RT, Kluge C, Sangha S, Seidenbecher T, Pape HC (2011b) Patterns of coupled theta activity in amygdala-hippocampal-prefrontal cortical circuits during fear extinction. PLoS One 6(6):e21714. https://doi.org/10.1371/journal.pone.0021714

Lesting J, Daldrup T, Narayanan V, Himpe C, Seidenbecher T, Pape H-C (2013) Directional theta coherence in prefrontal cortical to amygdalo-hippocampal pathways signals fear extinction. PLoS One 8(10):e77707

Li G, Nair SS, Quirk GJ (2009) A biologically realistic network model of acquisition and extinction of conditioned fear associations in lateral amygdala neurons. J Neurophysiol 101(3):1629–1646. https://doi.org/10.1152/jn.90765.2008

Li H, Penzo MA, Taniguchi H, Kopec CD, Huang ZJ, Li B (2013) Experience-dependent modification of a central amygdala fear circuit. Nat Neurosci 16(3):332–339. https://doi.org/10.1038/nn.3322

Likhtik E, Stujenske JM, Topiwala MA, Harris AZ, Gordon JA (2014) Prefrontal entrainment of amygdala activity signals safety in learned fear and innate anxiety. Nat Neurosci 17(1):106–113. https://doi.org/10.1038/nn.3582

Lin CH, Lee CC, Gean PW (2003a) Involvement of a calcineurin cascade in amygdala depotentiation and quenching of fear memory. Mol Pharmacol 63(1):44–52

Lin CH, Yeh SH, Leu TH, Chang WC, Wang ST, Gean PW (2003b) Identification of calcineurin as a key signal in the extinction of fear memory. J Neurosci 23(5):1574–1579

Lisman J (1989) A mechanism for the Hebb and the anti-Hebb processes underlying learning and memory. Proc Natl Acad Sci USA 86(23):9574–9578

Loewenstein Y, Kuras A, Rumpel S (2011) Multiplicative dynamics underlie the emergence of the log-normal distribution of spine sizes in the neocortex in vivo. J Neurosci 31(26):9481–9488. https://doi.org/10.1523/jneurosci.6130-10.2011
Mahanty NK, Sah P (1998) Calcium-permeable AMPA receptors mediate long-term potentiation in interneurons in the amygdala. Nature 394(6694):683–687. https://doi.org/10.1038/29312
Malkani S, Rosen JB (2000) Specific induction of early growth response gene 1 in the lateral nucleus of the amygdala following contextual fear conditioning in rats. Neuroscience 97 (4):693–702
Manko M, Bienvenu TC, Dalezios Y, Capogna M (2012) Neurogliaform cells of amygdala: a source of slow phasic inhibition in the basolateral complex. J Physiol 590(22):5611–5627. https://doi.org/10.1113/jphysiol.2012.236745
Maren S (2001) Neurobiology of Pavlovian fear conditioning. Annu Rev Neurosci 24:897–931. https://doi.org/10.1146/annurev.neuro.24.1.897
Maren S (2011) Seeking a spotless mind: extinction, deconsolidation, and erasure of fear memory. Neuron 70(5):830–845. https://doi.org/10.1016/j.neuron.2011.04.023
Maren S (2015) Out with the old and in with the new: synaptic mechanisms of extinction in the amygdala. Brain Res 1621:231–238. https://doi.org/10.1016/j.brainres.2014.10.010
Maren S, Quirk GJ (2004) Neuronal signalling of fear memory. Nat Rev Neurosci 5(11):844–852. https://doi.org/10.1038/nrn1535
Markram H, Lubke J, Frotscher M, Sakmann B (1997) Regulation of synaptic efficacy by coincidence of postsynaptic APs and EPSPs. Science 275(5297):213–215
Markram H, Toledo-Rodriguez M, Wang Y, Gupta A, Silberberg G, Wu C (2004) Interneurons of the neocortical inhibitory system. Nat Rev Neurosci 5(10):793–807. https://doi.org/10.1038/nrn1519
Martina M, Royer S, Pare D (1999) Physiological properties of central medial and central lateral amygdala neurons. J Neurophysiol 82(4):1843–1854
McAllister WR, McAllister DE, Scoles MT, Hampton SR (1986) Persistence of fear-reducing behavior: relevance for the conditioning theory of neurosis. J Abnorm Psychol 95(4):365–372
McDonald AJ (1984) Neuronal organization of the lateral and basolateral amygdaloid nuclei in the rat. J Comp Neurol 222(4):589–606. https://doi.org/10.1002/cne.902220410
McDonald AJ, Mascagni F, Guo L (1996) Projections of the medial and lateral prefrontal cortices to the amygdala: a Phaseolus vulgaris leucoagglutinin study in the rat. Neuroscience 71(1):55–75
McKernan MG, Shinnick-Gallagher P (1997) Fear conditioning induces a lasting potentiation of synaptic currents in vitro. Nature 390(6660):607–611. https://doi.org/10.1038/37605
Milad MR, Vidal-Gonzalez I, Quirk GJ (2004) Electrical stimulation of medial prefrontal cortex reduces conditioned fear in a temporally specific manner. Behav Neurosci 118(2):389–394. https://doi.org/10.1037/0735-7044.118.2.389
Morgan MA, Romanski LM, LeDoux JE (1993) Extinction of emotional learning: contribution of medial prefrontal cortex. Neurosci Lett 163(1):109–113
Morrison DJ, Rashid AJ, Yiu AP, Yan C, Frankland PW, Josselyn SA (2016) Parvalbumin interneurons constrain the size of the lateral amygdala engram. Neurobiol Learn Mem 135:91–99. https://doi.org/10.1016/j.nlm.2016.07.007
Moustafa AA, Gilbertson MW, Orr SP, Herzallah MM, Servatius RJ, Myers CE (2013) A model of amygdala-hippocampal-prefrontal interaction in fear conditioning and extinction in animals. Brain Cogn 81(1):29–43. https://doi.org/10.1016/j.bandc.2012.10.005
Myers KM, Davis M (2002) Behavioral and neural analysis of extinction. Neuron 36(4):567–584
Myers KM, Davis M (2007) Mechanisms of fear extinction. Mol Psychiatry 12(2):120–150. https://doi.org/10.1038/sj.mp.4001939
Nabavi S, Fox R, Proulx CD, Lin JY, Tsien RY, Malinow R (2014) Engineering a memory with LTD and LTP. Nature 511(7509):348–352. https://doi.org/10.1038/nature13294
Narayanan RT, Seidenbecher T, Kluge C, Bergado J, Stork O, Pape HC (2007) Dissociated theta phase synchronization in amygdalo-hippocampal circuits during various stages of fear memory. Eur J Neurosci 25(6):1823–1831

Ngezahayo A, Schachner M, Artola A (2000) Synaptic activity modulates the induction of bidirectional synaptic changes in adult mouse hippocampus. J Neurosci 20(7):2451–2458
Nguyen PV, Kandel ER (1996) A macromolecular synthesis-dependent late phase of long-term potentiation requiring cAMP in the medial perforant pathway of rat hippocampal slices. J Neurosci 16(10):3189–3198
Olshausen BA, Field DJ (1996) Emergence of simple-cell receptive field properties by learning a sparse code for natural images. Nature 381(6583):607–609. https://doi.org/10.1038/381607a0
Pape HC, Pare D (2010) Plastic synaptic networks of the amygdala for the acquisition, expression, and extinction of conditioned fear. Physiol Rev 90(2):419–463. https://doi.org/10.1152/physrev.00037.2009
Pare D, Duvarci S (2012) Amygdala microcircuits mediating fear expression and extinction. Curr Opin Neurobiol 22(4):717–723. https://doi.org/10.1016/j.conb.2012.02.014
Pare D, Quirk GJ, Ledoux JE (2004) New vistas on amygdala networks in conditioned fear. J Neurophysiol 92(1):1–9. https://doi.org/10.1152/jn.00153.2004
Pavlov IP (1927) Conditioned reflexes: an investigation of the physiological activity of the Cerebral Cortex. Oxford University Press, London
Pfister JP, Gerstner W (2006) Triplets of spikes in a model of spike timing-dependent plasticity. J Neurosci 26(38):9673–9682. https://doi.org/10.1523/jneurosci.1425-06.2006
Phelps EA, LeDoux JE (2005) Contributions of the amygdala to emotion processing: from animal models to human behavior. Neuron 48(2):175–187. https://doi.org/10.1016/j.neuron.2005.09.025
Pitman RK, Orr SP, Shalev AY, Metzger LJ, Mellman TA (1999) Psychophysiological alterations in post-traumatic stress disorder. Semin Clin Neuropsychiatry 4(4):234–241. https://doi.org/10.153/SCNP00400234
Pool RR, Mato G (2011) Spike-timing-dependent plasticity and reliability optimization: the role of neuron dynamics. Neural Comput 23(7):1768–1789. https://doi.org/10.1162/NECO_a_00140
Quirk GJ (2006) Extinction: new excitement for an old phenomenon. Biol Psychiatry 60(4):317–318. https://doi.org/10.1016/j.biopsych.2006.05.023
Quirk GJ, Repa C, LeDoux JE (1995) Fear conditioning enhances short-latency auditory responses of lateral amygdala neurons: parallel recordings in the freely behaving rat. Neuron 15(5):1029–1039
Quirk GJ, Russo GK, Barron JL, Lebron K (2000) The role of ventromedial prefrontal cortex in the recovery of extinguished fear. J Neurosci 20(16):6225–6231
Quirk GJ, Likhtik E, Pelletier JG, Pare D (2003) Stimulation of medial prefrontal cortex decreases the responsiveness of central amygdala output neurons. J Neurosci 23(25):8800–8807
Quirk GJ, Pare D, Richardson R, Herry C, Monfils MH, Schiller D, Vicentic A (2010) Erasing fear memories with extinction training. J Neurosci 30(45):14993–14997. https://doi.org/10.1523/jneurosci.4268-10.2010
Rashid AJ, Yan C, Mercaldo V, Hsiang HL, Park S, Cole CJ et al (2016) Competition between engrams influences fear memory formation and recall. Science 353(6297):383–387. https://doi.org/10.1126/science.aaf0594
Redondo RL, Kim J, Arons AL, Ramirez S, Liu X, Tonegawa S (2014) Bidirectional switch of the valence associated with a hippocampal contextual memory engram. Nature 513(7518):426–430. https://doi.org/10.1038/nature13725
Repa JC, Muller J, Apergis J, Desrochers TM, Zhou Y, LeDoux JE (2001) Two different lateral amygdala cell populations contribute to the initiation and storage of memory. Nat Neurosci 4(7):724–731. https://doi.org/10.1038/89512
Rescorla RA (2002) Comparison of the rates of associative change during acquisition and extinction. J Exp Psychol Anim Behav Process 28(4):406–415
Ressler KJ, Rothbaum BO, Tannenbaum L, Anderson P, Graap K, Zimand E et al (2004) Cognitive enhancers as adjuncts to psychotherapy: use of D-cycloserine in phobic individuals to facilitate extinction of fear. Arch Gen Psychiatry 61(11):1136–1144. https://doi.org/10.1001/archpsyc.61.11.1136

Rogan MT, Staubli UV, LeDoux JE (1997) Fear conditioning induces associative long-term potentiation in the amygdala. Nature 390(6660):604–607. https://doi.org/10.1038/37601

Royer S, Martina M, Pare D (1999) An inhibitory interface gates impulse traffic between the input and output stations of the amygdala. J Neurosci 19(23):10575–10583

Rumpel S, LeDoux J, Zador A, Malinow R (2005) Postsynaptic receptor trafficking underlying a form of associative learning. Science 308(5718):83–88. https://doi.org/10.1126/science.1103944

Santini E, Muller RU, Quirk GJ (2001) Consolidation of extinction learning involves transfer from NMDA-independent to NMDA-dependent memory. J Neurosci 21(22):9009–9017

Santini E, Ge H, Ren K, Pena de Ortiz S, Quirk GJ (2004) Consolidation of fear extinction requires protein synthesis in the medial prefrontal cortex. J Neurosci 24(25):5704–5710. https://doi.org/10.1523/JNEUROSCI.0786-04.2004

Schafe GE, Nader K, Blair HT, LeDoux JE (2001) Memory consolidation of Pavlovian fear conditioning: a cellular and molecular perspective. Trends Neurosci 24(9):540–546

Schiess MC, Callahan PM, Zheng H (1999) Characterization of the electrophysiological and morphological properties of rat central amygdala neurons in vitro. J Neurosci Res 58(5):663–673

Seidenbecher T, Laxmi TR, Stork O, Pape HC (2003) Amygdalar and hippocampal theta rhythm synchronization during fear memory retrieval. Science 301(5634):846–850. https://doi.org/10.1126/science.1085818

Senn V, Wolff SB, Herry C, Grenier F, Ehrlich I, Grundemann J et al (2014) Long-range connectivity defines behavioral specificity of amygdala neurons. Neuron 81(2):428–437. https://doi.org/10.1016/j.neuron.2013.11.006

Shin LM, Rauch SL, Pitman RK (2006) Amygdala, medial prefrontal cortex, and hippocampal function in PTSD. Ann N Y Acad Sci 1071:67–79. https://doi.org/10.1196/annals.1364.007

Shouval HZ, Bear MF, Cooper LN (2002a) A unified model of NMDA receptor-dependent bidirectional synaptic plasticity. Proc Natl Acad Sci USA 99(16):10831–10836. https://doi.org/10.1073/pnas.152343099

Shouval HZ, Castellani GC, Blais BS, Yeung LC, Cooper LN (2002b) Converging evidence for a simplified biophysical model of synaptic plasticity. Biol Cybern 87(5–6):383–391. https://doi.org/10.1007/s00422-002-0362-x

Sierra-Mercado D, Padilla-Coreano N, Quirk GJ (2011) Dissociable roles of prelimbic and infralimbic cortices, ventral hippocampus, and basolateral amygdala in the expression and extinction of conditioned fear. Neuropsychopharmacology 36(2):529–538. https://doi.org/10.1038/npp.2010.184

Sigurdsson T, Doyere V, Cain CK, LeDoux JE (2007) Long-term potentiation in the amygdala: a cellular mechanism of fear learning and memory. Neuropharmacology 52(1):215–227. https://doi.org/10.1016/j.neuropharm.2006.06.022

Sjostrom PJ, Turrigiano GG, Nelson SB (2001) Rate, timing, and cooperativity jointly determine cortical synaptic plasticity. Neuron 32(6):1149–1164

Smith Y, Pare D (1994) Intra-amygdaloid projections of the lateral nucleus in the cat: PHA-L anterograde labeling combined with postembedding GABA and glutamate immunocytochemistry. J Comp Neurol 342(2):232–248. https://doi.org/10.1002/cne.903420207

Sotres-Bayon F, Bush DE, LeDoux JE (2004) Emotional perseveration: an update on prefrontal-amygdala interactions in fear extinction. Learn Mem 11(5):525–535. https://doi.org/10.1101/lm.79504

Sprekeler H, Michaelis C, Wiskott L (2007) Slowness: an objective for spike-timing-dependent plasticity? PLoS Comput Biol 3(6):e112. https://doi.org/10.1371/journal.pcbi.0030112

Stujenske JM, Likhtik E, Topiwala MA, Gordon JA (2014) Fear and safety engage competing patterns of theta-gamma coupling in the basolateral amygdala. Neuron 83(4):919–933. https://doi.org/10.1016/j.neuron.2014.07.026

Sullivan GM, Apergis J, Gorman JM, LeDoux JE (2003) Rodent doxapram model of panic: behavioral effects and c-Fos immunoreactivity in the amygdala. Biol Psychiatry 53 (10):863–870

Trouche S, Sasaki JM, Tu T, Reijmers LG (2013) Fear extinction causes target-specific remodeling of perisomatic inhibitory synapses. Neuron 80(4):1054–1065. https://doi.org/10.1016/j.neuron.2013.07.047

Tsvetkov E, Carlezon WA, Benes FM, Kandel ER, Bolshakov VY (2002) Fear conditioning occludes LTP-induced presynaptic enhancement of synaptic transmission in the cortical pathway to the lateral amygdala. Neuron 34(2):289–300

Tully K, Li Y, Tsvetkov E, Bolshakov VY (2007) Norepinephrine enables the induction of associative long-term potentiation at thalamo-amygdala synapses. Proc Natl Acad Sci USA 104(35):14146–14150. https://doi.org/10.1073/pnas.0704621104

Tye KM, Stuber GD, de Ridder B, Bonci A, Janak PH (2008) Rapid strengthening of thalamo-amygdala synapses mediates cue-reward learning. Nature 453(7199):1253–1257. https://doi.org/10.1038/nature06963

van Rossum MC, Bi GQ, Turrigiano GG (2000) Stable Hebbian learning from spike timing-dependent plasticity. J Neurosci 20(23):8812–8821

Viosca J, Lopez de Armentia M, Jancic D, Barco A (2009) Enhanced CREB-dependent gene expression increases the excitability of neurons in the basal amygdala and primes the consolidation of contextual and cued fear memory. Learn Mem 16(3):193–197. https://doi.org/10.1101/lm.1254209

Vlachos I, Herry C, Luthi A, Aertsen A, Kumar A (2011) Context-dependent encoding of fear and extinction memories in a large-scale network model of the basal amygdala. PLoS Comput Biol 7 (3):e1001104. https://doi.org/10.1371/journal.pcbi.1001104

Walker DL, Davis M (2000) Involvement of NMDA receptors within the amygdala in short- versus long-term memory for fear conditioning as assessed with fear-potentiated startle. Behav Neurosci 114(6):1019–1033

Wang H, Wagner JJ (1999) Priming-induced shift in synaptic plasticity in the rat hippocampus. J Neurophysiol 82(4):2024–2028

Wilson HR, Cowan JD (1972) Excitatory and inhibitory interactions in localized populations of model neurons. Biophys J 12(1):1–24. https://doi.org/10.1016/s0006-3495(72)86068-5

Wolff SB, Grundemann J, Tovote P, Krabbe S, Jacobson GA, Muller C et al (2014) Amygdala interneuron subtypes control fear learning through disinhibition. Nature 509(7501):453–458. https://doi.org/10.1038/nature13258

Yiu AP, Mercaldo V, Yan C, Richards B, Rashid AJ, Hsiang HL et al (2014) Neurons are recruited to a memory trace based on relative neuronal excitability immediately before training. Neuron 83(3):722–735. https://doi.org/10.1016/j.neuron.2014.07.017

Zheng P, Dimitrakakis C, Triesch J (2013) Network self-organization explains the statistics and dynamics of synaptic connection strengths in cortex. PLoS Comput Biol 9(1):e1002848. https://doi.org/10.1371/journal.pcbi.1002848

Zhou Y, Won J, Karlsson MG, Zhou M, Rogerson T, Balaji J et al (2009) CREB regulates excitability and the allocation of memory to subsets of neurons in the amygdala. Nat Neurosci 12(11):1438–1443. https://doi.org/10.1038/nn.2405

Zimmerman JM, Maren S (2010) NMDA receptor antagonism in the basolateral but not central amygdala blocks the extinction of Pavlovian fear conditioning in rats. Eur J Neurosci 31 (9):1664–1670. https://doi.org/10.1111/j.1460-9568.2010.07223.x

Modulation of the Core Synaptic Network in Extinction: The Role of Brain-Derived Neurotrophic Factor

7

7.1 BDNF Function at Synapses in the Core Neural Network for Extinction

7.1.1 BDNF-Mediated Synaptic Interactions Between the BLA, IL and HPC in Extinction

Microinfusion of BDNF or its neutralizing antibody into the basolateral amygdala (BLA), infralimbic cortex (IL) or hippocampus (HPC) points to synapses in the IL as requiring this neurotrophin for the relatively long-duration long-term potentiation (lLTP) supporting fear extinction. During fear extinction the increase of BDNF in the IL follows the increase of BDNF in ventral HPC (Chhatwal et al. 2006; Peters et al. 2010). The excitability of IL neurons is enhanced by microinfusion of BDNF into the ventral HPC following fear conditioning, (Rosas-Vidal et al. 2014) suggesting that BDNF has been transported from the latter to the former (Burgos-Robles et al. 2007; Chang et al. 2010; Knapska and Maren 2009; Peters et al. 2010; Rosas-Vidal et al. 2014). This is supported by the observation that co-microinfusion of BDNF into the hippocampus together with BDNF antibody into IL blocks the BDNF-generated extinction (Peters et al. 2010). Rats that are opaque to extinction training show poor BDNF projections from the middle HPC to the IL (Peters et al. 2010).

Further arguments placing the ventral HPC and BLA as primary sites of BDNF synthesis in fear extinction and not medial prefrontal cortex (Heldt et al. 2007; Peters et al. 2010; Sakata et al. 2013) are that BDNF knockdowns in dorsal HPC or the BLA block or interrupt fear extinction (Chhatwal et al. 2006; Heldt et al. 2007, 2014) and increases in BDNF following fear extinction are only found in BLA and ventral HPC, but not IL or prelimbic cortex (Rosas-Vidal et al. 2014). BDNF increase in the BLA seems to lead to the increase in the IL, perhaps through a TrkB signalling mechanism from BLA to IL. The extinction due to microinjection of BDNF into the BLA can be blocked by microinfusion of BDNF neutralizing antibody into IL,

M. Bennett, J. Lagopoulos, *Stress, Trauma and Synaptic Plasticity*,
https://doi.org/10.1007/978-3-319-91116-8_7

suggesting that IL is the main area for the action of BDNF in extinction. Increases of BDNF in the IL then becomes the principal site of BDNF action in extinction. Taken together, these results indicate that BDNF is primarily synthesized in the HPC and BLA, from whence it is then transported to synapses in the IL for regulating lLTP, for example, in extinction.

7.1.2 BDNF Autoregulatory Loops

BDNF can be secreted from either pre- or postsynaptic elements (dendritic spines) at the synapse (Amaral and Pozzo-Miller 2007; Hartmann et al. 2001; Kohara et al. 2001; Kojima et al. 2001; Lessmann et al. 2003), to which it can be specifically targeted (Lessmann and Brigadski 2009). Whether BDNF is secreted in a regulatory or constitutive manner, the secretion is enhanced by a variety of neurotrophins, including BDNF itself, suggesting the existence of a BDNF autoregulatory loop (Kruttgen et al. 1998). There is a BDNF autoregulatory feedback loop in the HPC that engages the enhancer-binding protein (βC/EBP) (Bambah-Mukku et al. 2014) as well as activation of cyclic AMP response element-binding protein (CREB) that leads to expression of BDNF transcription. Very interestingly, at 12 h after the beginning of a memorizing session, CREB and βC/EBP binding are detected at the BDNF exon IV promoter, but this is substantially decreasing by 48 h, with concomitant increase in methyl-CpG-binding protein 2 (MeCp2) attachment (Bambah-Mukku et al. 2014), indicating that the latter has displaced the former.

7.1.3 BDNF Support for Long Periods of Long-Term Potentiation (lLTP)

Calcium entry through *N*-methyl-D-aspartate (NMDA) receptors on synaptic release of glutamate immediately activates a calcium calmodulin kinase ($CaMK_{II}$)-dependent pathway that leads to the recruitment of alpha-amino-3-hydroxy-5-methyl-4-isoxazolepropionic (AMPA) receptor subunits through phosphorylation of membrane-bound transmembrane AMPA receptor regulatory proteins (TARP). It also leads to the phosphorylation of AMPA subunits, increasing their open times under glutamate, synergistically with catecholamine action on G-protein-coupled (GPC) receptors. The effect of these changes is to lead to relatively short periods of LTP (sLTP).

A slower pathway, taking some 60 min, is the mitogen-activated protein kinase (MAPK) pathway (Ras, Rak, Erk1/2), especially brought into play by large calcium currents through NMDA receptors leading to phosphorylation of cyclic AMP (cAMP) response element-binding protein CREB, that is, pCREB (Katche et al. 2013; Taubenfeld et al. 2001a, b). The increase in pCREB and c/EBP beta binding to the BDNF exon IV promoter at 12 h, is followed by phsopshorylation of Ser133 (pCREB) with increased transcription of BDNF exon IV and translation of BDNF protein for more than 24 h. After about 48 h, the binding of pCREB and βc/EBP at the BDNF promoter has decreased to such an extent that a baseline production of BDNF is established once more (Bambah-Mukku et al. 2014; Bekinschtein et al. 2008). The critical question is what determines this time course of action of BDNF exon IV activation (Cardinaux et al. 2000)? It is critical, because ultimately this is the determining factor for BDNF-guided dendritic spine growth, and the relatively long periods of LTP (lLTP) underlying extinction, principally engineered, according to the current hypothesis, by lLTP at synapses in the IL and to some extent in the BLA. To provide a satisfactory answer to this question requires a detailed consideration of the kinetics of the BDNF exon IV promoter with the MeCP2–histone deacetylase (HDAC)–mSin3a complex.

7.2 The Role of the MeCP2 Complex in Extinction

Calcium entry into neurons and subsequent synthesis of BDNF is correlated with a decrease in the extent of CpG methylation in the promoter region for BDNF as a consequence of dissociation of the MeCP2–histone deacetylase (HDAC)–mSin3a complex (Hsu et al. 2008; Martinowich et al. 2003). In the case of BDNF, MeCP2 seems to act as a repressor (Chen et al. 2003), with the response role assigned to the BDNF IV gene in neurons (Fig. 1 in S4 Rousseaud et al. 2015). However there is some evidence that MeCP2 can act as an activator and further that it has a genome-wide function (Gonzales et al. 2012).

7.2.1 MeCP2 Phosphorylation and Neural Activity

The phosphorylation and dephosphorylation of different sites on MeCP2 determine the extent of its binding affinity to promoters (Li and Chang 2014). Activity-dependent phosphorylation of MeCP2 can occur at residues S86, S274, TE08, S424, and S421 (Li and Chang 2014; Tao et al. 2009). Neural activity releases MeCP2 from the BDNF promoter allowing CREB to replace it (Chen et al. 2003; Zhou et al. 2006). Interestingly, it also leads to phosphorylation of MeCP2 at sites across the whole genome (Cohen et al. 2011).

7.2.2 MeCP2^{S421} Activity

The modification or deletion of the phosphorylation of S421 on the regulation of MeCP2 function results in interference with BDNF transcription and synaptic spine growth (Cohen et al. 2011). This points to a role for S421 phosphorylation in MeCP2 dissociation from DNA in active neurons. On the other hand, S421 phosphorylation occurs on increased neural activity across the genome, not just at a few specific genes like that for BDNF, with little difference in the expression of messenger RNAs (Cohen et al. 2011; Hutchinson et al. 2012; Huttlin et al. 2010; Tao et al. 2009; Zhou et al. 2006). Some conclude that activity-dependent phosphorylation of MeCP2 at

S421 is insufficient to lead to dissociation of MeCP2 from DNA (Cohen et al. 2011). How these observations are to be reconciled with the observed large increase in MeCP2^{S421} phosphorylation in the hippocampus after appropriate training is not clear (Li et al. 2011).

7.2.3 MeCP2$^{S421/S424}$ Activity

These observations on S421 are in contrast to those concerning the combination of activity-regulated residues S421 and S424. In this case, transcription of at least two genes are increased, including BDNF, and two are decreased with increases in neural activity (Huttlin et al. 2010; Li et al. 2011; Tao et al. 2009). Specific knock-in of MeCP2$^{S421A/S424A}$ enhances BDNF transcription and so stronger lLTP in the HPC at synapses in CA1 and CA3 (Li et al. 2011; Tao et al. 2009).

7.2.4 MeCP2^{S80} Activity

S80 is the most abundantly phosphorylated residue of MeCP2 in the resting state neuron (Tao et al. 2009), and is found at a number of gene promoters, with increasing neuronal activity decreasing phosphorylation that is causing dephosphorylation. This leads to increased transcription at 149 genes across the genome and a decrease of 56 others (Bracaglia et al. 2009; Gonzales et al. 2012; Huttlin et al. 2010; Tao et al. 2009; Young et al. 2005; Zhou et al. 2006). S80 mutation into alanine (S80A) antagonizes the association of MeCP2 and cofactors such as Sin3A at a number of gene promoters (Tao et al. 2009). This suggests that the normal phosphorylation of MeCP2^{S80} in the resting state is important for regulating the activity-dependent functioning of these genes.

7.2.5 MeCP2 Residues at S80, S421 and S424

It seems likely that phosphorylation of residues S421 and S424 occurs in concert during neural activity and that phosphorylation of residue S80 occurs during the resting state. This suggests that a push–pull relationship exists between these phosphorylated residues in controlling MeCP2 function on DNA, so modulating the attachment of MeCP2 at specific genes, including BDNF.

7.2.6 The MeCP2 Hypothesis for Activity-Dependent Transcription of BDNF

Repression of BDNF transcription is through the binding of MeCP2 demethylated CpGs, which recruit a complex of histone deacetylase (HDAC), Sin3A and CBP to the BDNF promoter. The calcium increase in stimulated neurons leads to calcium calmodulin kinase phosphorylation of MeCP2 at specific residues (S421/S424) and dephosphorylation at others (S80), so antagonizing MeCP2 attachment to DNA, leading to DNA demethylation (Tao et al. 2009; Zhou et al. 2006). This calcium increase also leads to phosphorylation of S133 on CREB, giving pCREB, so facilitating its binding to acetyltransferase CBP, and thus activating the BDNF promoter in the relaxed chromatin conformation. However the phosphorylation of these different residues only lasts for several hours after the neuronal activity has ceased, so that eventually the MeCP2 complex occupies the BDNF promoter once more, with the consequences that the BDNF returns to about baseline conditions after its action in promoting dendritic spine growth is complete.

7.3 Hypotheses Concerning Mechanisms for Extending Periods of Extinction

7.3.1 Spine Loss and the Need to Block Glucocorticoids

Spine loss in the hippocampus leads to the loss of contextual memory. Spine loss in the anterior cingulate cortex leads to disinhibition (through ITC and BAE) of the amygdala output. Spine loss in the infralimbic (IL) cortex leads to failure of extinction through BAE.

The mechanism of spine loss is through the action of glucocorticoids, in particular cortisol, acting to block the RAS/RAF/MEK1/2 pathway, the principal path for activating BDNF synthesis. Glucocorticoids also decrease production of microRNAs that lead to inhibition of p250GAP 3UTR and hence enhanced Rac1 and LIMK1 inhibition of cofilin. So a main effort should be directed to preventing the action of cortisol and hence spine loss.

7.3.2 Maintaining lLTP Through Enhanced Basal Levels of BDNF

It is argued above that the extinction memory for non-contextual fear resides at IL and perhaps BAE synapses. The aim here is to sustain lLTP over very long times, which requires maintaining enlarged dendritic spines. This involves increasing the autoregulatory loop for BDNF production. Maintaining the displacement of MeCP2 by CREB holds out the prospect of maintaining enhanced basal levels of BDNF production. This will in turn maintain F-actin polymerization against degradation (e.g. via cofilin). The supply of G-actin and calcium calmodulin kinase 2 ($CaMK_{II}$) from the dendrites to the spine head needs to be sufficient to support the enlarged spine.

Although, from the arguments above, potentiation of the action of endocannabinoids and so LTD in the BLA seems a promising pathway for elevating the activity of extinction neurons, this is unlikely to be the case. The principal reason for this is that endocannabinoids are only active during the day of the extinction test but not during the subsequent days when ILTD is prominent (Plendl and Wotjak 2010), suggesting that one should seek to extend the latter rather than the effects of the former.

7.3.3 Enhancing Basal Levels of BDNF Through Phosphorylation of MeCP2 Residues

The present hypothesis is that in the resting state, the MeCP2 is phosphorylation at residue S80 and so blocks p-CREB from the BDNF promoter. When neuronal activity is initiated, S80 becomes dephosphorylating and residues S421/S424 phosphorylated, blocking the repressive role of MeCP2 and allowing CREB (C/EBP, CBP) to attach to the promoter and initiate BDNF transcription. This occurs in a time window of about 12 h. However, by 48 h of low neural activity, MeCP2 is again phosphorylated at S80 and dephosphorylated at S421/S424, leading to the displacement of CREB (C/EBP, CBP) once more by the repressing MeCP2.

An apparent problem with this hypothesis is that neuronal activity leads to phosphorylation of MeCP2 across the genome, implicating transcription initiation for many genes (Cohen and Greenberg 2010). Of course, this does not preclude the activity regulation of BDNF through MeCP2, described above, but it does suggest that many other gene products are regulated by activity of MeCP2.

References

Amaral MD, Pozzo-Miller L (2007) TRPC3 channels are necessary for brain-derived neurotrophic factor to activate a nonselective cationic current and to induce dendritic spine formation. J Neurosci 27(19):5179–5189. https://doi.org/10.1523/JNEUROSCI.5499-06.2007

Bambah-Mukku D, Travaglia A, Chen DY, Pollonini G, Alberini CM (2014) A positive autoregulatory BDNF feedback loop via C/EBPβ mediates hippocampal memory consolidation. J Neurosci 34(37):12547–12559. https://doi.org/10.1523/JNEUROSCI.0324-14.2014

Bekinschtein P, Cammarota M, Katche C, Slipczuk L, Rossato JI, Goldin A et al (2008) BDNF is essential to promote persistence of long-term memory storage. Proc Natl Acad Sci 105 (7):2711–2716. https://doi.org/10.1073/pnas.0711863105

Bracaglia G, Conca B, Bergo A, Rusconi L, Zhou Z, Greenberg ME et al (2009) Methyl-CpG-binding protein 2 is phosphorylated by homeodomain-interacting protein kinase 2 and contributes to apoptosis. EMBO Rep 10(12):1327–1333. https://doi.org/10.1038/embor.2009.217

Burgos-Robles A, Vidal-Gonzalez I, Santini E, Quirk GJ (2007) Consolidation of fear extinction requires NMDA receptor-dependent bursting in the ventromedial prefrontal cortex. Neuron 53 (6):871–880. https://doi.org/10.1016/j.neuron.2007.02.021

Cardinaux JR, Notis JC, Zhang Q, Vo N, Craig JC, Fass DM et al (2000) Recruitment of CREB binding protein is sufficient for CREB-mediated gene activation. Mol Cell Biol 20 (5):1546–1552

Chang CH, Berke JD, Maren S (2010) Single-unit activity in the medial prefrontal cortex during immediate and delayed extinction of fear in rats. PLoS One 5(8):e11971. https://doi.org/10.1371/journal.pone.0011971

Chen WG, Chang Q, Lin Y, Meissner A, West AE, Griffith EC et al (2003) Derepression of BDNF transcription involves calcium-dependent phosphorylation of MeCP2. Science 302 (5646):885–889. https://doi.org/10.1126/science.1086446

Chhatwal JP, Stanek-Rattiner L, Davis M, Ressler KJ (2006) Amygdala BDNF signaling is required for consolidation but not encoding of extinction. Nat Neurosci 9(7):870–872. https://doi.org/10.1038/nn1718

Cohen S, Greenberg ME (2010) A bird's-eye view of MeCP2 binding. Mol Cell 37(4):451–452. https://doi.org/10.1016/j.molcel.2010.02.006

Cohen S, Gabel HW, Hemberg M, Hutchinson AN, Sadacca LA, Ebert DH et al (2011) Genome-wide activity-dependent MeCP2 phosphorylation regulates nervous system development and function. Neuron 72(1):72–85. https://doi.org/10.1016/j.neuron.2011.08.022

Gonzales ML, Adams S, Dunaway KW, LaSalle JM (2012) Phosphorylation of distinct sites in MeCP2 modifies cofactor associations and the dynamics of transcriptional regulation. Mol Cell Biol 32(14):2894–2903. https://doi.org/10.1128/mcb.06728-11

Hartmann M, Heumann R, Lessmann V (2001) Synaptic secretion of BDNF after high-frequency stimulation of glutamatergic synapses. EMBO J 20(21):5887–5897. https://doi.org/10.1093/emboj/20.21.5887

Heldt SA, Stanek L, Chhatwal JP, Ressler KJ (2007) Hippocampus-specific deletion of BDNF in adult mice impairs spatial memory and extinction of aversive memories. Mol Psychiatry 12 (7):656–670. https://doi.org/10.1038/sj.mp.4001957

Heldt SA, Zimmermann K, Parker K, Gaval M, Ressler KJ (2014) Bdnf deletion or TrkB impairment in amygdala inhibits both appetitive and aversive learning. J Neurosci 34 (7):2444–2450. https://doi.org/10.1523/JNEUROSCI.4085-12.2014

Hsu C-C, Lu C-W, Huang B-M, Wu M-H, Tsai S-J (2008) Cyclic adenosine 3′,5′-monophosphate response element-binding protein and CCAAT/enhancer-binding protein mediate prostaglandin E(2)-induced steroidogenic acute regulatory protein expression in endometriotic stromal cells. Am J Pathol 173(2):433–441. https://doi.org/10.2353/ajpath.2008.080199

Hutchinson AN, Deng JV, Cohen S, West AE (2012) Phosphorylation of MeCP2 at Ser421 contributes to chronic antidepressant action. J Neurosci 32(41):14355–14363. https://doi.org/10.1523/jneurosci.2156-12.2012

Huttlin EL, Jedrychowski MP, Elias JE, Goswami T, Rad R, Beausoleil SA et al (2010) A tissue-specific atlas of mouse protein phosphorylation and expression. Cell 143(7):1174–1189. https://doi.org/10.1016/j.cell.2010.12.001

Katche C, Cammarota M, Medina JH (2013) Molecular signatures and mechanisms of long-lasting memory consolidation and storage. Neurobiol Learn Mem 106:40–47. https://doi.org/10.1016/j.nlm.2013.06.018

Knapska E, Maren S (2009) Reciprocal patterns of c-Fos expression in the medial prefrontal cortex and amygdala after extinction and renewal of conditioned fear. Learn Mem 16(8):486–493. https://doi.org/10.1101/lm.1463909

Kohara K, Kitamura A, Morishima M, Tsumoto T (2001) Activity-dependent transfer of brain-derived neurotrophic factor to postsynaptic neurons. Science 291(5512):2419–2423. https://doi.org/10.1126/science.1057415

Kojima M, Takei N, Numakawa T, Ishikawa Y, Suzuki S, Matsumoto T et al (2001) Biological characterization and optical imaging of brain-derived neurotrophic factor-green fluorescent protein suggest an activity-dependent local release of brain-derived neurotrophic factor in neurites of cultured hippocampal neurons. J Neurosci Res 64(1):1–10. https://doi.org/10.1002/jnr.1080

Kruttgen A, Moller JC, Heymach JV Jr, Shooter EM (1998) Neurotrophins induce release of neurotrophins by the regulated secretory pathway. Proc Natl Acad Sci USA 95(16):9614–9619

Lessmann V, Brigadski T (2009) Mechanisms, locations, and kinetics of synaptic BDNF secretion: an update. Neurosci Res 65(1):11–22. https://doi.org/10.1016/j.neures.2009.06.004

Lessmann V, Gottmann K, Malcangio M (2003) Neurotrophin secretion: current facts and future prospects. Prog Neurobiol 69(5):341–374

Li H, Chang Q (2014) Regulation and function of stimulus-induced phosphorylation of MeCP2. Front Biol (Beijing) 9(5):367–375. https://doi.org/10.1007/s11515-014-1330-2

Li H, Zhong X, Chau KF, Williams EC, Chang Q (2011) Loss of activity-induced phosphorylation of MeCP2 enhances synaptogenesis, LTP and spatial memory. Nat Neurosci 14(8):1001–1008. https://doi.org/10.1038/nn.2866

Martinowich K, Hattori D, Wu H, Fouse S, He F, Hu Y et al (2003) DNA methylation-related chromatin remodeling in activity-dependent BDNF gene regulation. Science 302 (5646):890–893. https://doi.org/10.1126/science.1090842

Peters J, Dieppa-Perea LM, Melendez LM, Quirk GJ (2010) Induction of fear extinction with hippocampal-infralimbic BDNF. Science 328(5983):1288–1290. https://doi.org/10.1126/science.1186909

Plendl W, Wotjak CT (2010) Dissociation of within- and between-session extinction of conditioned fear. J Neurosci 30(14):4990–4998. https://doi.org/10.1523/JNEUROSCI.6038-09.2010

Rosas-Vidal LE, Do-Monte FH, Sotres-Bayon F, Quirk GJ (2014) Hippocampal—prefrontal BDNF and memory for fear extinction. Neuropsychopharmacology 39(9):2161–2169. https://doi.org/10.1038/npp.2014.64

Rousseaud A, Delepine C, Nectoux J, Billuart P, Bienvenu T (2015) Differential expression and regulation of brain-derived neurotrophic factor (BDNF) mRNA isoforms in brain cells from Mecp2(308/y) mouse model. J Mol Neurosci 56(4):758–767. https://doi.org/10.1007/s12031-014-0487-0

Sakata K, Martinowich K, Woo NH, Schloesser RJ, Jimenez DV, Ji Y et al (2013) Role of activity-dependent BDNF expression in hippocampal-prefrontal cortical regulation of behavioral perseverance. Proc Natl Acad Sci USA 110(37):15103–15108. https://doi.org/10.1073/pnas.1222872110

Tao J, Hu K, Chang Q, Wu H, Sherman NE, Martinowich K et al (2009) Phosphorylation of MeCP2 at Serine 80 regulates its chromatin association and neurological function. Proc Natl Acad Sci USA 106(12):4882–4887. https://doi.org/10.1073/pnas.0811648106

Taubenfeld SM, Milekic MH, Monti B, Alberini CM (2001a) The consolidation of new but not reactivated memory requires hippocampal C/EBPbeta. Nat Neurosci 4(8):813–818. https://doi.org/10.1038/90520

Taubenfeld SM, Wiig KA, Monti B, Dolan B, Pollonini G, Alberini CM (2001b) Fornix-dependent induction of hippocampal CCAAT enhancer-binding protein [beta] and [delta] Co-localizes with phosphorylated cAMP response element-binding protein and accompanies long-term memory consolidation. J Neurosci 21(1):84–91

Young JI, Hong EP, Castle JC, Crespo-Barreto J, Bowman AB, Rose MF et al (2005) Regulation of RNA splicing by the methylation-dependent transcriptional repressor methyl-CpG binding protein 2. Proc Natl Acad Sci USA 102(49):17551–17558. https://doi.org/10.1073/pnas.0507856102

Zhou Z, Hong EJ, Cohen S, Zhao WN, Ho HY, Schmidt L et al (2006) Brain-specific phosphorylation of MeCP2 regulates activity-dependent Bdnf transcription, dendritic growth, and spine maturation. Neuron 52(2):255–269. https://doi.org/10.1016/j.neuron.2006.09.037

Appendix A: F-Actin Determination of Dendritic Spine Integrity[1]

A.1 Introduction

The induction of synaptic plasticity, such as accompanies long-term potentiation, alters the size of synaptic spines on dendrites, these being the principal postsynaptic structures on excitatory pyramidal neurons (Hosokawa et al. 1995; Buchs and Muller 1996). *N*-methyl-D-aspartate (NMDA) receptors, when activated by glutamate during synaptic transmission, lead to changes in the size of spines (Ultanir et al. 2007). However, no quantitative model exists of how the synaptic plasticity manifest in spine changes is linked to NMDA receptor activation, although there is a growing experimental literature on the subject (see, for example, Okamoto et al. 2004, 2007). In particular, NMDA receptor control of F-actin treadmilling in the spines has not been treated in detail, although the mechanism of treadmilling in lamellipodia is well characterized and forms the basis of the present inquiry.

In lamellipodia, treadmilling of F-actin occurs in a dynamic pool (Higgs and Pollard 2001). Lamellipodia extend as a consequence of the formation and elongation of actin filaments, namely F-actin, that push against the plasmalemma. F-actin grows by incorporating monomeric actin subunits (G-actin) with bound ATP at the so-called barbed (plus, +) end of the F-actin. Older actin subunits are removed from the pointed (minus, −) end of F-actin in a treadmilling process (Fig. A.1). The older subunits are identified as a consequence of a process in which ATP-actin slowly hydrolyzes to ADP-actin. An F-actin severing protein, ADF/cofilin, has high affinity for ADP-actin, so promoting disassembly of the older pointed (−) ends of the actin filaments (for a review, see Van Troys et al. 2008). Profilin catalyses the exchange of ADP for ATP on these ADP-actin monomers. In this way, ATP-actin monomers are made available for addition to the barbed (+) ends of F-actin, so completing the treadmilling process. Filament growth is blocked by capping their barbed (+) ends,

[1]Reprinted from: Bull Math Biol. Bennett MR, Farnell L, Gibson WG. A model of neuregulin control of NMDA receptors on synaptic spines. Vol. 74(3):717–735. Copyright (2012). With permission from Springer.

Bull Math Biol. Bennett MR, Farnell L, Gibson WG. A model of NMDA receptor control of F-actin treadmilling in synaptic spines and their growth. 73(9):2109–2131. Copyright (2011). With permission from Springer.

M. Bennett, J. Lagopoulos, *Stress, Trauma and Synaptic Plasticity*,
https://doi.org/10.1007/978-3-319-91116-8

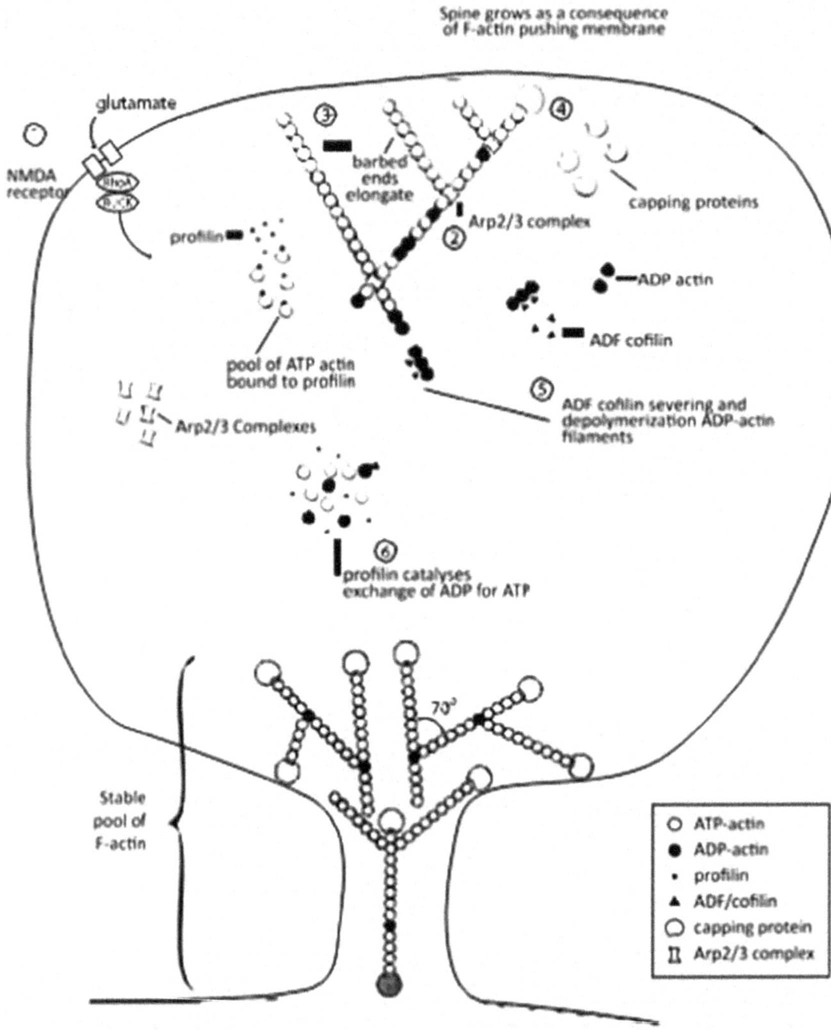

Fig. A.1 The actin nucleation model of growth and regression of synaptic spines. There are two pools of F-actin: a stable pool that is retained at the base of the spine head and a dynamic pool, where growth or regression can occur. Processes occurring in the dynamic pool are as follows: *1*, glutamate binds to NMDA receptors and activates a signalling pathway that leads via RhoA and ROCK to the release of profilin which then binds ADP-actin to give ATP-actin-profilin. *2*, activated Arp2/3 complexes form new actin filaments by attachment to an existing filament to form a new side branch. *3*, filaments grow by the attachment of profilin-ATP-actin to their barbed ends, thus pushing the membrane forward and leading to spine growth. *4*, capping proteins can bind to the barbed ends, thus terminating growth. *5*, actin-depolymerizing factor ADF/cofilin severs and depolymerizes the ADP-actin in the older regions of the filaments. *6*, profilin promotes dissociation of ADP and binding of ATP to the dissociated subunits. (Diagram and description after Fig. 1 in Pollard and Borisy 2003)

for instance with gelsolin. The filament branching network is enabled by actin-related protein (Arp2/3), which binds to the barbed end of F-actin and acts as a template for the assembly of new filaments, eventually becoming the minus-end cap as filaments disappear during the treadmilling process (Fig. A.1; Carlisle and Kennedy 2005). This network is stabilized by F-actin binding proteins (Cingolani and Goda 2008), principal amongst which are alpha-actinin as well as drebrin, spinophilin and Abp1 (Nakagawa et al. 2004; Haeckel et al. 2008).

Existing models of the treadmilling process developed for lamellipodia can be applied to the case of synaptic spines, structures of only about 1 μm^3 in volume. Such work is necessary before the main aim of this section can be achieved, which is to model how activation of NMDA receptors controls actin dynamics in the spine head.

Two principal pathways allow the NMDA receptor to modulate F-actin treadmilling in synaptic spines. In one of these paths, activation of this receptor by glutamate leads to the largest calcium influx that any receptor can generate. This calcium activates calcium–calmodulin kinase (CaMKII), which in turn phosphorylates and activates calmodulin-dependent kinase kinase (CaMKK–CaMK1). These kinases form a multimolecular complex with the guanine nucleotide exchange factor β PIX (Cingolani and Goda 2008; Saneyoshi et al. 2008). Phosphorylation of this results in activation of Rac1 and Pak1. Pak1-mediated phosphorylation then activates LIM kinase (LIMK) which inactivates actin-depolymerizing protein ADP/cofilin, so promoting stabilization of F-actin and giving spine enlargement (Edwards et al. 1999). The other principal means of NMDA receptor control of F-actin treadmilling involves the receptor regulation of Rho A (Fig. A.1; Nakayama and Luo 2000). Rho A recruits and activates its specific kinase ROCK which in turn complexes with profilin IIa so modifying actin stability through a RhoA/ROCK/profilin IIa pathway (Fig. A.1; Da Silva et al. 2003). It is known that RhoA associates with ionotropic glutamate receptors (iGluRs) and metabotropic glutamate receptors (mGluRs) at the plasma membrane of synaptic spines with activation of the iGluRs, such as the NMDA receptor, leading to detachment of RhoA from these receptors and their recruitment to the mGluRs (Schubert et al. 2006). This triggers local reduction of RhoA activity at NMDA receptors resulting in inactivation of RhoA-specific kinase ROCK there and so disruption of the ROCK-profilin IIa, release of profilin and so rapid changes in the underlying actin cytoskeleton and remodelling of spine shape (Schubert et al. 2006).

In the second part of this section, a model is presented of the effect of NMDA receptors in promoting profilin and inhibiting cofilin and thence on the growth of synaptic spines. These results are then both compared with Monte Carlo simulations and with the time course of changes in action observed experimentally in synaptic spines following stimulation of NMDA receptors. This emphasis on the NMDA receptor control of spine-head dynamics as a consequence of synaptic transmission provides a quantitative account of mechanisms that control synaptic spine dynamics.

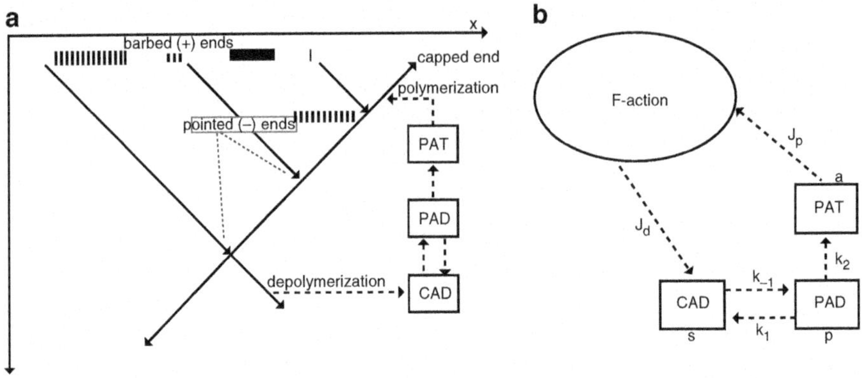

Fig. A.2 (**a**) Schematic diagram showing the geometry and interactions in the model. The model is two-dimensional, with the y-axis downwards as shown and the membrane at $y = 0$. The solid lines represent the F-actin filaments, with the barbed (+) ends near the membrane and the pointed (−) ends in the interior (cf. the actin filaments in Fig. A.1). The broken lines show the recycling of actin: depolymerization gives cofilin-ADP-actin (CAD), leading (reversibly) to profilin-ADP-actin (PAD) and then to profilin-ATP-actin which polymerizes the filaments. (Cf. Mogilner and Edelstein-Keshet 2002, Fig. 7) (**b**) Exchange reactions for the G-actin complexes of various forms shown in (**a**). J_p and J_d are the polymerization and depolymerization rates, respectively; k_1, k_{-1}, k_2 are rate constants for the other reactions; and a, p and s are the concentrations of PAT, PAD and CAD, respectively. (Cf. Mogilner and Edelstein-Keshet 2002, Fig. 3) (from Bennett et al. 2011)

A.2 Treadmilling Model

Assume that the spine head contains two pools of F-actin: a stable pool near the spine neck which remains fixed in size and a dynamic pool above this where growth can occur (Fig. A.1; see also Fig. 8 in Honkura et al. 2008). In the dynamic pool, actin filaments (F-actin) are highly branched with their barbed (plus, +) ends oriented towards to the membrane and their pointed (minus, −) ends oriented away. Following Mogilner and Edelstein-Keshet (2002), the model of the dynamic compartment is three-dimensional, with the y-axis vertical (positive direction downwards) and the origin ($y = 0$) at the spine membrane (Fig. A.2a); the analytic theory, however, involves only one dimension (y), and quantities can be looked upon as averaged over the x- and z-directions. The three-dimensional nature of the model does show up in the units used for various quantities; the Monte Carlo simulation method, described below, is two-dimensional. The origin is fixed on the membrane and may therefore move outwards with velocity V, so a moving coordinate system is being used. However, in the steady state, the membrane does not move, and V is to be interpreted as the treadmilling speed (Mogilner and Edelstein-Keshet 2002). The barbed ends can be terminated by a capping protein that prevents further attachment of monomers. It is assumed that uncapped barbed ends are close to the membrane. The pointed ends usually terminate at an Arp2/3 complex where a filament forms a branch at 70° to another filament, and the Arp2/3 complex acts as a cap (Figs A.1 and A.2a). Uncapped pointed ends rapidly disappear due to depolymerization.

A.2.1 Polymerization and Depolymerization

Monomers of ATP-actin bound to profilin (PAT) attach to uncapped barbed ends (polymerization). Also, new branches are constantly being initiated as activated Arp2/3 complexes attached to uncapped barbed ends. As the monomers move down the filament, they are hydrolysed to ADP-actin. The minus ends can become uncapped, and at such an end ADF/cofilin accelerates depolymerization, and the actin monomers separate from the filament and are recycled (Figs. A.1 and A.2).

A.2.2 Density of Minus Ends

Analytic theory of the processes can now be given. Let $m_c \equiv m_c\,(y,\,t)$ be the density of capped minus ends, and $m_u \equiv m_u\,(y,\,t)$ be the density of uncapped minus ends. Then [cf. Mogilner and Edelstein-Keshet 2002, Eqs. (A.7) and (A.8)]:

$$\frac{\partial m_c}{\partial t} + V\frac{\partial m_c}{\partial y} = -\frac{m_c}{\tau_1}, \tag{A.1}$$

$$\frac{\partial m_u}{\partial t} + V\frac{\partial m_u}{\partial y} = \frac{m_c}{\tau_1} - \frac{m_u}{\tau_2}, \tag{A.2}$$

where $1/\tau_1$ is the uncapping rate, and $1/\tau_2$ is the depolymerization rate.

The boundary conditions are

$$m_u = 0 \text{ when } y = 0, \tag{A.3}$$

$$m_c = \frac{n}{V} \text{ when } y = 0, \tag{A.4}$$

where n, the branching rate, is also the rate at which new barbed ends are being created. The $m_u = 0$ condition follows because there are no uncapped minus ends at the membrane. $m_c = n/V$ follows from the observation that a new capped minus end is created at $y = 0$ whenever a branching takes place, since this occurs whenever an Arp2/3 complex is inserted. Thus, in time Δt, $n\Delta t$ new ends are formed in a volume $V\Delta t$ leading to a density of $\frac{n\Delta t}{V\Delta t} = n/V$.

A.2.3 Steady-State Solutions

The above equations describe a dynamic process. However, at any instant of time we can regard the system as being in a quasi-steady state, in that average properties of the system remain constant. This is because changes brought about by external agencies, such as NMDA receptors, are on a much slower timescale than polymer recycling. In particular, the total number of G-actin monomers attaching to uncapped barbed ends must equal the number leaving uncapped pointed ends. Setting time derivatives to zero in Eqs. (A.1) and (A.2) allows them to be solved, using the boundary conditions (A.3) and (A.4), to give

$$m_c(y) = {}^{n}_{V}e^{-y/(V\tau_1)V}, \tag{A.5}$$

$$m_u(y) = \frac{n\tau_2}{V_{(\tau 1-\tau 2)}} e^{-y/(V\tau_1)} - e^{-y/(V\tau_2)}, \tag{A.6}$$

Thus, m_c is an exponentially decreasing function of distance, whereas m_u starts from 0 at $y = 0$, rises to a maximum at $y = [V_{\tau 1 \tau 2}/(\tau_1 - \tau_2)]$ In (τ_1/τ_2) and then decreases. By straightforward integration, using Eqs. (A.5) and (A.6), we can find the total number of minus ends:

$$\widehat{m}_c = \int_0^\infty m_c(y)dy = n\tau_1 \tag{A.7}$$

$$\widehat{m}_u = \int_0^\infty m_u(y)dy = n\tau_2 \tag{A.8}$$

In the steady state, the membrane is not moving out, and V is to be interpreted as the "treadmilling" speed, that is, the speed at which actin monomers move down a filament. Thus,

$$V = g\delta \tag{A.9}$$

where g is the rate at which G-actin monomers are added at the barbed end of an actin filament, and δ is the monomer length.

The average length of an actin filament is

$$= \frac{g}{y}\delta \tag{A.10}$$

where γ is the barbed-end capping rate.

The inner boundary of the dynamic pool is not well defined (the histogram of end positions has an exponential tail). In order to have a well-defined measure, we introduce the *mean transit length*, defined to be the average value of y, weighted by the density of uncapped minus ends:

$$\bar{y} = {}^{\infty}_{0} y m_u(y)dy^{0}{}^{0\infty}_{0} m_u(y)dy^{0}, \tag{A.11}$$

and using Eq. (A.6) for $m_u(y)$, this gives

$$\bar{y} = V(\tau_1 + \tau_2), \tag{A.12}$$

The total number of F-actin monomers in the spine is

$$\text{Total actin} = (\widehat{m}_c + \widehat{m}_u)\frac{g}{y} = \frac{ng}{y}(\tau_1 + \tau_2), \tag{A.13}$$

where Eqs. (A.7) and (A.8) have been used to get the second form.

A.2.4 Monte Carlo Simulation

The above theory produces averages, which are indeed the appropriate quantities to compare with experimental results. However, in order to gain insight into the microscopic processes that occur in synaptic spines, and also to verify that the theoretical formulas give correct averages, an MC simulation needs to be implemented, and this is now described.

The simulation is performed in two dimensions, with the y-axis vertically downwards and the x-axis horizontal, as in Fig. A.2. The filaments move away from the membrane in discrete steps, governed by a time increment $1/g$ and spatial increments $\delta x = 0.0012$ μm and $\delta y \equiv \delta = 0.0022$ μm, due to polymerization at the barbed (+) ends; this process is treated deterministically, with the polymerization rate being g monomers/s. Branching also occurs at the membrane, and this and other processes are treated probabilistically using:

$$\text{Plus } (+) \text{ end capping probability} = \gamma/g, \tag{A.14}$$

$$\text{Minus } (-)\text{end uncapping probability} = \frac{1}{\tau_1 g}, \tag{A.15}$$

$$\text{Minus } (-)\text{end depolymerization probability} = \frac{1}{\tau_2 g}, \tag{A.16}$$

$$\text{Branching probability for each filament} = n/\left(g\,N_p\right). \tag{A.17}$$

In the last expression, N_p is the number of filaments available for branching at each time step (those with uncapped + ends); the remaining parameters are as defined earlier.

The simulation starts with a number of monomers at the membrane; there is then a preliminary phase in which the number of filaments builds to approximately a steady state, and then a quasi-steady phase during which statistical information is gathered.

Table A.1 Model parameter values

Symbol	Definition	Value	Notes
$\delta(\mu m)$	Length of actin monomer	0.0022	ME-K
$g_0(s^{-1})$	Unstimulated polymerization rate at barbed (+) end	4	See text
$\gamma(s^{-1})$	Barbed (+) end capping rate	1	ME-K
$\tau_1(s)$	Time constant for uncapping at pointed (−) end	30	ME-K
$\tau_2(s)$	Time constant for depolymerization at pointed (−) end	1	ME-K
$n(\mu m^{-2}s^{-1})$	Branching rate	500	see text
$k_2(s^{-1})$	PAD → PAT reaction rate	20	ME-K
$k_+^{PD}(\mu M^{-1}s^{-1})$	PAD association rate	15	ME-K
$k_-^{PD}(s^{-1})$	PAD dissociation rate	10	ME-K
$k_+^{C}(\mu M^{-1}s^{-1})$	CAD association rate	150	ME-K
$k_-^{C}(s^{-1})$	CAD dissociation rate	20	ME-K
$G(\mu M)$	Concentration of total G-actin	250	ME-K
$j(s^{-1})$	Proportionality constant [Eq. (A.26)]	20	fitted
$\eta(\mu M^{-1}\mu m^{-2})$	Conversion factor [Eq. (A.30)]	100	ME-K

Many values are available in the literature, as indicated; others have been chosen to fit experimental data. ME-K is Mogilner and Edelstein-Keshet (2002)

A.2.5 Parameter Values

Parameter values are listed in Table A.1. Many of the values are taken directly from the lamellipodia model of Mogilner and Edelstein-Keshet (2002). Others that are not well determined experimentally have been given plausible values. The polymerization rate g in the absence of stimulation has been set to $g_0 = 4\ s^{-1}$ leading to a treadmilling speed $V = g\delta$ of 0.53 μm min^{-1}, which can be compared to the observed experimental range of 0–0.8 μm min^{-1} (see Fig. 3C in Honkura et al. 2008). The choice $\tau 1 = 30$ s for the depolymerization time constant is compatible with Fig. 3E of Honkura et al. (2008), which shows a comparable rate of decay. The mean transit length $\bar{y} = V\ (\tau 1 + \tau 2)$ of 0.27 μm is comparable to the spine dimensions. Note also that Eqs. (A.12) and (A.13) give a total actin directly proportional to $\bar{y}$, which agrees with Honkura et al. (2008) Fig. 3D, provided that we assume that the spine head has an approximately constant cross section so that the volume is proportional to height. The branching rate n is assigned a value of 100 $\mu m^{-1}\ s^{-1}$ in Mogilner and Edelstein-Keshet (2002); this is for a thin region of width ~0.2 μm, so to allow for the thicker spine head, this has been scaled to 500 $\mu m^{-2}\ s^{-1}$.

A.2.6 Results for Treadmilling Model

In order to illustrate the MC method, a simulation can be run with unrealistic parameters to show more clearly the behaviour of individual filaments (Fig. A.3). The simulation starts with a single monomer, and as a result of branching and

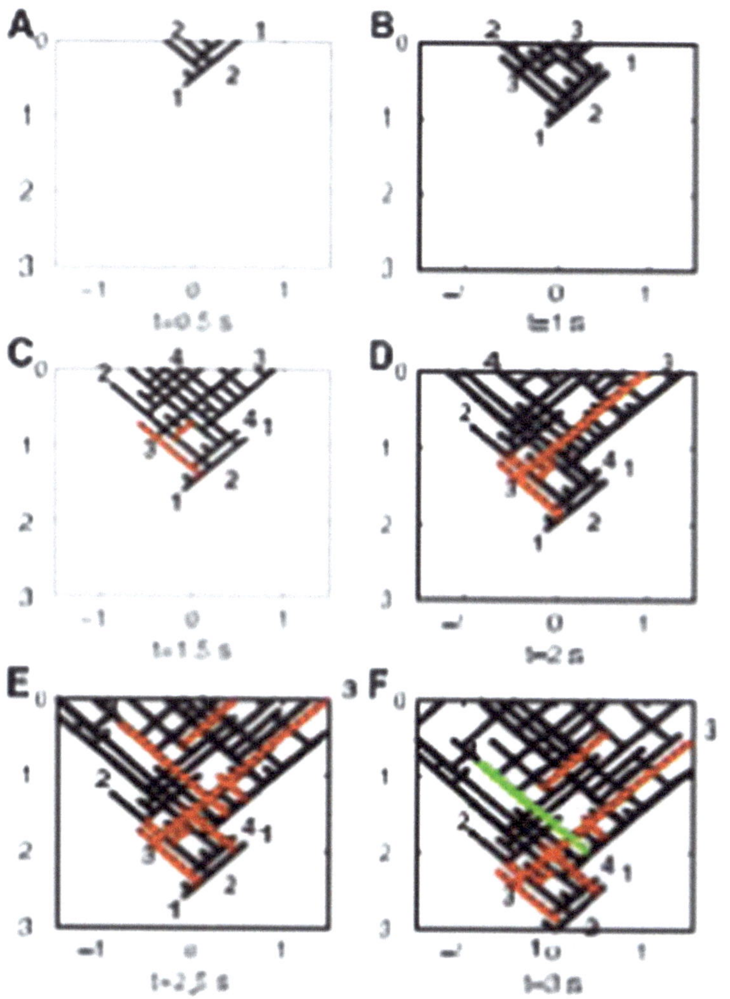

Fig. A.3 Growth, branching and polymerization of actin filaments (F-actin) in a synaptic spine according to an MC simulation, following initiation of treadmilling (see Figs. A.1 and A.2). The lines indicate F-actin with capped minus ends (black), uncapped minus ends (red) and the "ghosts" of depolymerized filaments with uncapped minus ends (green); these latter filaments are ones that have recently depolymerized and are no longer present, but are included for illustrative purposes. Depolymerization is rapid (time constant $\tau 2$) compared to other processes and so is assumed to occur in one time step. The filaments are treated as independent entities, so the fact that a red filament can abut a black one does not imply a physical join. Also, because the panels show a two-dimensional representation of a three-dimensional process, some filaments are overlaid and appear to change colour at some point along their length. The panels show the configuration at the times indicated after initiation at a single point. The parameters used are as given in Table A.1, except for the following changes: $\tau 1 = 10$ s, $\tau 2 = 2$ s, $n = 20\ \mu m^{-2}\ s^{-1}$, $g = 10\ s^{-1}$, $\delta x = \delta y = 0.1\ \mu m$; the ordinate gives distance (μm). These modified values are not realistic but have been chosen to most clearly show the various filaments. Four filaments have been labelled: *1*, the initial monomer, becomes capped after 7 steps and remains that way (black); *2*, initiated by branching at the 3rd step and becomes uncapped (red) just before frame (**c**); *3*, initiated by branching at the 7th step and is

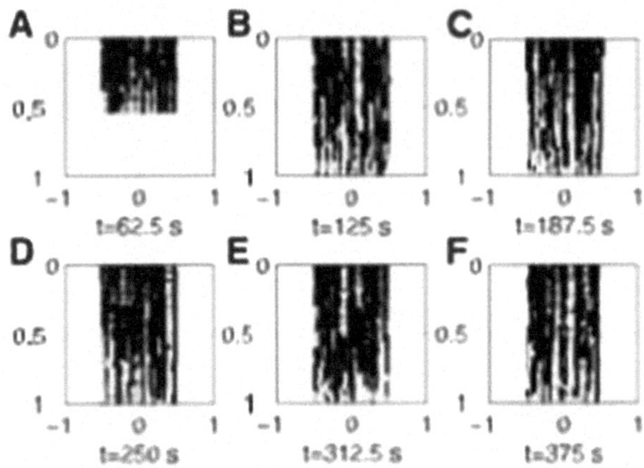

Fig. A.4 Growth, branching and polymerization of actin filaments (F-actin) in a synaptic spine according to an MC simulation, following initiation of treadmilling and growth of the filament network (see Figs. A.1 and A.2). The lines indicate F-actin. The ordinate is the distance y (μm), measured downwards, from the surface membrane (at $y = 0$). The parameter values used are in Table A.1. The simulation is initiated at $t = 0$ with monomers placed at 10 points on the membrane, between −0.5 and 0.5 μm, and the spatial distribution of F-actin in the synaptic spine is shown at the six different times indicated (from Bennett et al. 2011)

polymerization, both of which occur only at the membrane, the network of filaments moves downwards as shown at selected times in Fig. A.3a–f. Uncapping at a minus end is indicated by the filament changing from black to red; uncapped filaments can then be depolymerized as indicated by a change to green. Selected filaments have been labelled 1–4 to facilitate identification.

Figure A.4 shows a simulation with more realistic parameters over a longer period of time, with the polymerization rate g fixed at its unstimulated value g_0. The branching and polymerization processes are now initiated at a number of points on the membrane. The width is restricted to 1 μm and is implemented with periodic boundary conditions (wraparound). Figure A.5 graphs data obtained from the MC calculations shown in Fig. A.4, with columns A, B and C in Fig. A.5 corresponding to Fig. A.4a, c and f, respectively. Values calculated using the theoretical equations are also indicated. The capped end positions (row b) show only a small fluctuation about the theoretical value (black line), whereas the uncapped end positions (row c) show considerably more fluctuation, reflecting the much smaller number of such filaments. Cumulative total for filaments with capped and uncapped ends is shown in Fig. A.6a, b, respectively; now the fluctuation about the theoretical values (black lines) is much less. Figure A.6c–e shows that the mean filament length, mean transit

---

Fig. A.3 (continued) uncapped (red) by frame (**d**); *4*, initiated by branching at the 10th step, is uncapped (red) by frame (**e**) and depolymerized (green) by frame (**f**) (from Bennett et al. 2011)

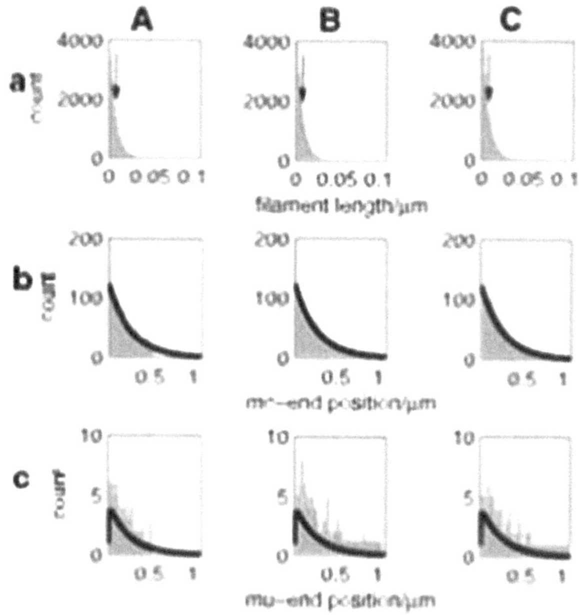

Fig. A.5 Comparison between changes in filaments in synaptic spines according to the MC simulations and those predicted by the theory. Columns A, B and C show changes in F-actin distributions in the spines for the cases given in Fig. A.4 a, c and e, respectively, that is, at 62.5, 187.5 and 312.5 s after initiation. In each case, the black arrow or line gives the analytical prediction according to the following equations in the theory: (a) distribution of filament lengths (given in numbers of actin monomers) at the final time step, with the theoretical average, as calculated using Eq. (A.10), indicated by the vertical arrow at 0.0088 µm; the averages from the MC simulation are 0.0086, 0.0086 and 0.0087 µm for A, B and C, respectively; (b) histogram of the number of capped ends at position y, the line being given by Eq. (A.5); (c) histogram of the number of uncapped ends at position y, the line being given by Eq. (A.6) (from Bennett et al. 2011)

length and total number of filaments, respectively, all approach the theoretical steady-state values (horizontal lines) with increasing time.

A.3 Recycling of F-actin

The recycling scheme is (Fig. A.2)

$$\text{(F-actin)} \xrightarrow{J_\mathrm{d}} \text{CAD} \underset{k_{-1}}{\overset{k_1}{\rightleftarrows}} \text{PAD} \xrightarrow{k_2} \text{PAT} \xrightarrow{J_\mathrm{p}} \text{(F-actin)}, \tag{A.18}$$

where CAD = cofilin-ADP-actin, PAD = profilin-ADP-actin and PAT = profilin-ATP-actin (Fig. A.2). J_d is the depolymerization rate, and J_p is the polymerization rate.

Let $s = [\text{CAD}]$, $p = [\text{PAD}]$, $a = [\text{PAT}]$. At equilibrium $J_\mathrm{p} = J_\mathrm{d} = J$ and

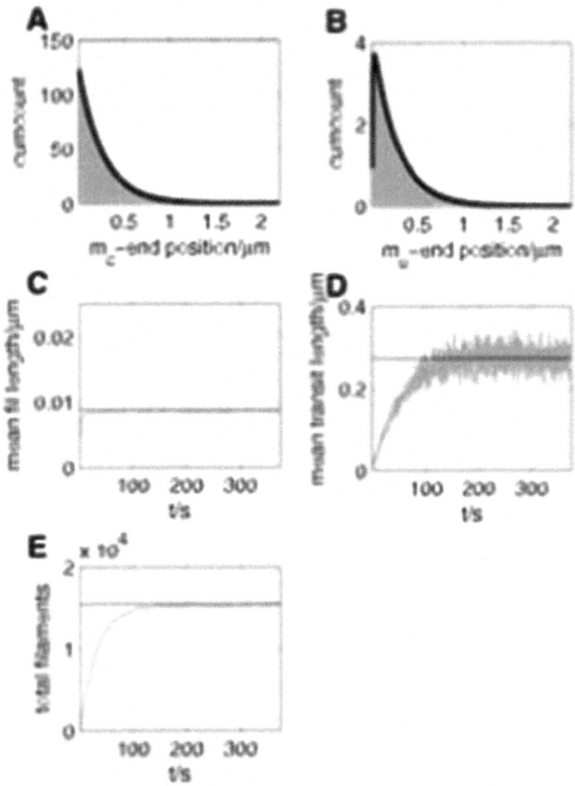

Fig. A.6 Comparison between the numbers of different filament types (capped or uncapped) in synaptic spines according to the MC simulation and the predictions of the theory given by the lines (**a**) and (**b**), respectively, show histograms of the number of capped and uncapped minus ends as a function of their end coordinate, y; (**c**) gives the mean filament length as a function of time, analytic value $g\delta/\gamma$ [Eq. (A.10)], as indicated by the line; (**d**) shows the mean transit length vs. time, and the line is $V(\tau_1 + \tau_2)$ [Eq. A.12)]; (**e**) shows the total number of F-actin filaments versus time, and the line is $n(\tau_1 + \tau_2)$ [cf. Eq. (A.13)] (from Bennett et al. 2011)

$$p = \frac{J}{k_2}, \tag{A.19}$$

$$s = \frac{J}{k_1} 1 + \frac{k_{-1}}{k_2}, \tag{A.20}$$

$$a = G - s - p, \tag{A.21}$$

where G is the concentration of total G-actin.

PAD and CAD dissociate according to

$$\mathrm{PAD} \underset{k_+^{\mathrm{PD}}}{\overset{k_-^{\mathrm{PD}}}{\rightleftarrows}} \mathrm{P} + \mathrm{AD} \tag{A.22}$$

$$\mathrm{CAD} \underset{k_+^{\mathrm{C}}}{\overset{k_-^{\mathrm{C}}}{\rightleftarrows}} \mathrm{C} + \mathrm{AD}, \tag{A.23}$$

It follows that (cf. Mogilner and Edelstein-Keshet 2002, Appendix):

$$k_1 = \frac{k_+^{\mathrm{PD}}[P]k_-^{\mathrm{C}}}{k_+^{\mathrm{C}}[S] + k_+^{\mathrm{PD}}[P]}, \tag{A.24}$$

$$k_{-1} = \frac{k_+^{\mathrm{C}}[S]k_-^{\mathrm{PD}}}{k_+^{\mathrm{C}}[S] + k_+^{\mathrm{PD}}[P]}, \tag{A.25}$$

where $[P]$ = total free profilin concentration and $[S]$ = total free cofilin concentration. These equations enable us to calculate the variation in k_1 and k_{-1} as $[P]$ and $[S]$ change over time.

An increase in profilin leads to an increase in polymerization, so we set

$$J = ja, \tag{A.26}$$

where a = [PAT] is the profilin-ATP-actin concentration, and j is a constant. Then Eqs. (A.19)–(A.21) can be written

$$a = \frac{G}{1 + \frac{j}{k_2} + \frac{j}{k_1}\left(1 + \frac{k_{-1}}{k_2}\right)}, \tag{A.27}$$

$$p = \frac{ja}{k_2}, \tag{A.28}$$

$$s = \frac{ja}{k_1} 1 + \frac{k_{-1}}{k_2}, \tag{A.29}$$

As the concentrations $[P]$ and/or $[S]$ vary with time, Eqs. (A.24) and (A.25) give the time dependence of k_1 and k_{-1}, and hence Eqs. (A.27)–(A.29) give a, p and s as functions of t. The polymerization rate is

$$g = J\,\eta/B = J\,\eta\gamma/n, \tag{A.30}$$

where $B = n/\gamma$ is the density of uncapped barbed ends, and $\eta \approx 600D$, where D is the depth of the profilin compartment, is a conversion factor (Mogilner and Edelstein-Keshet 2002). Here, it is assumed that profilin is confined to this compartment, but cofilin is distributed throughout the extensible compartment, and both it and G-actin can be replenished via the spine neck. Figure A.7 shows the dependence of the polymerization rate g on the concentrations of profilin and cofilin. It is seen that quite large concentrations of both are required to give a substantial polymerization rate.

A concentration of 200 μM in a spine head corresponds to about 1200 molecules; this is larger than concentrations reported in lamellipodia (Pollard and Borisy 2003) and in the ME-K model (Mogilner and Edelstein-Keshet 2002). However, this difference is largely due to the presence of thymosin β4 in the lamellipodia model which is omitted in our spine model due to the absence of experimental evidence.

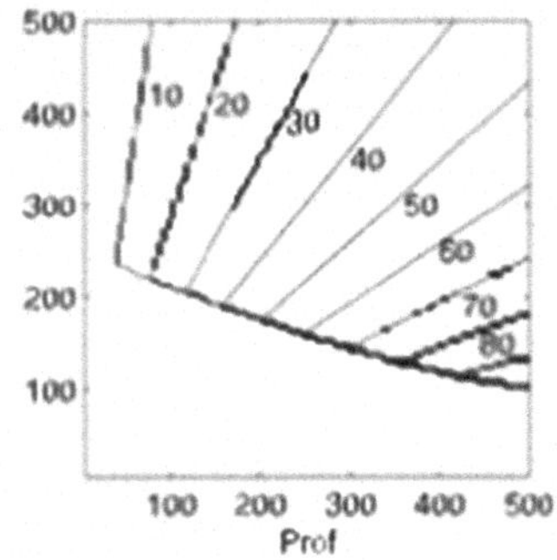

Fig. A.7 The variation in polymerization rate g, given by the values labelling the lines of constant g (s_{-1}), as a function of the concentrations (μM) of cofilin (vertical axis) and profilin (horizontal axis). The permissible ranges of concentrations are determined by the total concentration of G-actin. The curves are calculated by putting the concentrations of profilin and cofilin into for $k1$ and k_{-1}

A.3.1 Inclusion of NMDA

NMDA receptors on the spine are activated by a ligand (NMDA or glutamate) according to

$$\mathrm{R} + \mathrm{L} \underset{k_-}{\overset{k_+}{\rightleftarrows}} \mathrm{RL}, \tag{A.31}$$

where R = NMDA receptor and L = ligand concentration. At equilibrium, the fraction of bound receptors is

$$\sigma = \frac{[\mathrm{RL}]}{\mathrm{R_{tot}}} = \frac{[\mathrm{L}]}{[\mathrm{L}] + K}, \tag{A.32}$$

where $R_{tot} = [\mathrm{R}] + [\mathrm{RL}]$ is the total number of NMDA receptors, and $K = k_-/k_+$ is the dissociation constant for ligand binding to NMDA receptors. The activated NMDA receptors work via RhoA and ROCK to increase profilin in the spine, and this leads to increased F-actin (Fig. A.1).

A.3.2 NMDA and Profilin

RhoA exists in an active (A) and an inactive (I) form:

$$\mathrm{I} \underset{k_i}{\overset{k_a}{\rightleftarrows}} \mathrm{A}, \tag{A.33}$$

so at equilibrium the fraction of active RhoA is $\rho = [A]/\text{RhoA}_{\text{tot}} = 1/(1 + K_\rho)$, where $\text{RhoA}_{\text{tot}} = [A] + [I]$ is the total RhoA, and $K_\rho = k_i/k_a$. If it is assumed that RhoA inactivates quickly, then this expression can be used for ρ. Bound NMDA receptors inactivate RhoA, and this is incorporated into the model by simply setting $K_\rho = u\sigma$ where u is a constant. Thus,

$$\rho = \frac{1}{1 + u\sigma}, \tag{A.34}$$

ROCK and profiling bind reversibly to form a complex:

$$\text{Rock} + P \underset{k_d}{\overset{k_c}{\rightleftarrows}} \text{Comp}, \tag{A.35}$$

Giving

$$\frac{d[P]}{dt} = k_d[\text{Comp}] - k_c[\text{ROCK}][P] = k_d P_{\text{tot}} - \left(\text{k}_d + \bar{k}_c\right)[P], \tag{A.36}$$

where $P_{\text{tot}} = [P] + [\text{Comp}]$ is the total profilin. We have assumed that there is excess ROCK, so we have replaced [ROCK] by ROCK_{tot}, the total ROCK, and $\bar{k}_c = k_c\text{ROCK}_{\text{tot}}$

It remains to relate the rate constants k_c and k_d to ρ. RhoA promotes the binding of ROCK and profilin, and this is accounted for in two different ways. In the first, the simple assumption is made that k_d is constant and $\bar{k}_c$ is proportional to ρ:

$$\bar{k}_c = \nu\rho, \tag{A.37}$$

where ν is a constant. However, this turns out to give a rather small variation in the F/G ratio (see Fig. A.8f), so a second model is implemented in which $\bar{k}_c$ is again taken to be proportional to ρ as in Eq. (A.37), but k_d is allowed to vary inversely with ρ:

$$k_d = \frac{\omega}{\rho}, \tag{A.38}$$

where ω is a constant.

Equation (A.36) can be written as

$$\frac{d\varphi}{dt} = q_1 - q_2\varphi, \tag{A.39}$$

where $\varphi = [P]/P_{\text{tot}}$ is the fraction of free profilin, $q_1 = k_d$ and $q_2 = k_d + \bar{k}_c$. This has solution (for constant q_1 and q_2)

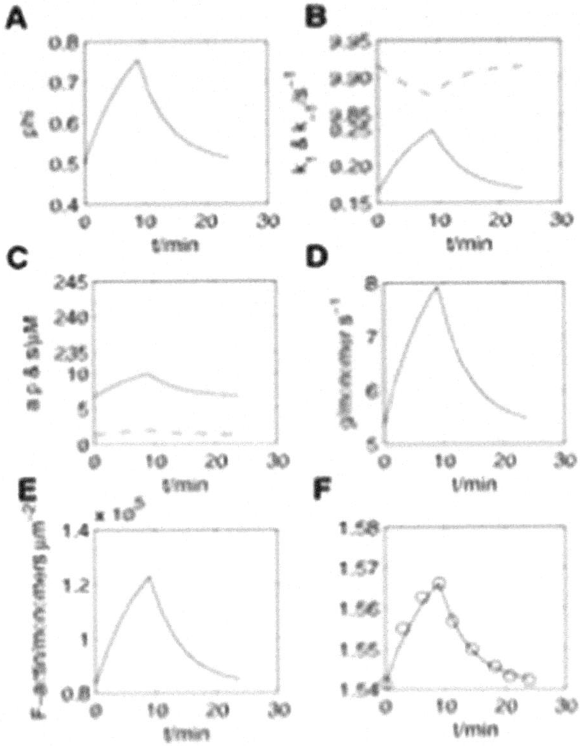

Fig. A.8 Comparison between the experimentally determined changes in F-actin in a synaptic spine following glutamate stimulation and those determined by the NMDA receptor-treadmilling model, using the first scheme where k_d is taken to be constant ($k_d = 0.1\ \mathrm{s}^{-1}$) and $\bar{k}_c$ is proportional to ρ [Eq. (A.37) with $\nu = 0.1\ \mathrm{s}^{-1}$]. The total free cofilin is fixed at $[S] = 300\ \mu\mathrm{M}$ and the total profilin at 50 μM; $u = 10$ [Eq. (A.34)]. Other parameters are as in Table A.1. (**a**) Changes in profilin in the spine following glutamate stimulation of NMDA receptors for a period of 9 min, according to Eq. (A.40); in these calculations, σ [Eq. (A.32)] is taken to be 0.9 during the rising phase ($t < 9$ min) and 0 during the falling phase ($t > 9$ min). (**b**) Changes, following glutamate stimulation, in the rate constants k_1 (solid line) and k_{-1} (broken line) [Eqs. (A.24) and (A.25)], for the forward and reverse reactions [Eq. (A.18)] of CAD (cofilin-ADP-actin) to PAD (profilin-ADP-actin). (**c**) Changes, following glutamate stimulation, in PAT (profilin-ATP-actin) (a) (solid line), PAD (p) (dashed line) and CAD (s) (dotted line), according to Eqs. (A.27)–(A.29). In panels (**b**) and (**c**), a broken scale has been used on the vertical axes for clarity. (**d**) Changes, following glutamate stimulation, in the polymerization rate g, given by Eq. (A.30). (**e**) Changes, following glutamate stimulation, in the total F-actin in the spine, according to Eq. (A.13) with g inserted from panel (**e**) above. (**f**) The ratio of F-actin to G-actin, according to Eq. (A.41); the open circles are the experimental observations of Okamoto et al. (2004). Here, the amplitude of the experimental data has been multiplied by a constant scaling factor to match it to the theoretical values, as the experimental data does not measure the ratio directly (from Bennett et al. 2011)

$$\varphi = \frac{q_1}{q_2} + \varphi 0 - \frac{q_1}{q_2} e^{-q_2 t}, \tag{A.40}$$

where $\varphi 0 = \varphi(t = 0)$.

We wish to make contact with the experimental results of Okamoto et al. (2004); in particular their Fig. 2f, which gives the time course of a fluorescence measurement (FRET) that represents the ratio of F-actin to G-actin. Thus, we need a theoretical expression for this ratio. Profilin is restricted to a compartment of depth D near the membrane, and the density of profilin-ADP-actin in this compartment is $\eta J/k_2$ [cf. Eq. (A.19)]. Cofilin is present throughout the spine, and this has depth $g\delta$ $(\tau_1+\tau_2)$(Eq. 12) giving $(\eta/D)\ g\delta(\tau_1+\tau_2))(J/k_1)(1 + k_{-1}/k_2)$ for the cofilin-ADP-actin density [cf. Eq. (A.20)]. Adding these densities and using Eq. (A.13) for the total actin gives the ratio of G-actin to F-actin as

$$\begin{aligned} &\frac{\eta J/k_2 + \eta/Dg\delta(\tau_1+\tau_2)(J/k_1)(1+k_{-1}/k_2)}{(ng/y)(\tau_1+\tau_2)}, \\ &= \frac{1}{k_2(\tau_1+\tau_2)} + \frac{g\delta}{Dk_1} 1 + \frac{k_{-1}}{k_2}, \end{aligned} \tag{A.41}$$

where Eq. (A.30) has been used to obtain the second form.

A.3.3 Steps in the Calculation

The activation of the NMDA receptors determines σ; ρ then follows from Eq. (A.34), $\bar{k}_c$, k_d and hence q_1, q_2 from Eqs. (A.37) and (A.38) and φ and hence $[P]$ from Eq. (A.40). This allows the calculation of k_1 and k_{-1} from Eqs. (A.24) and (A.25), a, p and s from Eqs. (A.27)–(A.29), J and g from Eqs. (A.26) and (A.30) and finally the ratio of F-actin to G-actin from Eq. (A.41).

A.3.4 NMDA and Cofilin

Activated NMDA receptors also have the effect of increasing the cofilin concentration in the synaptic spine via a pathway involving a LIM kinase (Delorme et al. 2007; see Fig. 3 in Bennett 2009). There are so many steps in this pathway that detailed modelling is not feasible. However, by omitting the initial steps leading from activation of the NMDA receptors and starting by assigning plausible values to q_1 and q_2, a calculation can be performed with the profilin concentration $[P]$ kept constant and the cofilin concentration $[S]$ allowed to vary. Note that this model still assumes that polymerization is rate-determining [via Eq. (A.26)]. A suitable model in which the rate-determining step is the control of depolymerization by cofilin is not available at this time.

A.3.5 Results for Recycling Model

The effect of glutamate acting on NMDA receptors was investigated using the model. Fig. A.8 shows results for the case where k_d is kept constant and k_c is proportional to ρ [Eq. (A.37)]. Fig. A.8a shows the function φ as given by Eq. (A.40), with q_1 and q_2 calculated using the fraction of bound NMDA receptors, σ, to be 0.9 during the rising phase and zero afterwards. Assuming that cofilin remains constant and taking the profilin concentration to be proportional to φ, as described above, then allows calculation of intermediate variables (Fig. A.8b–d) and finally the total actin with a time course in agreement with the experimental data of Okamoto et al. (2004) (Fig. A.8e). With this model, the range of variation of the polymerization rate g (Fig. A.8d) is small, leading to a correspondingly small variation in the ratio of F- to G-actin (Fig. A.8f; here, the experimental data has been multiplied by a constant scaling factor to match it to the theoretical values, since the experimental data does not measure this ratio directly). Figure A.9 shows the results of a parallel calculation, but now using the model where k_d varies inversely with ρ [Eq. (A.38)]. Again, there is agreement with the experimental results (Okamoto et al. 2004) with a bigger range of g values (cf. Figs. A.8d and A.9d) and a corresponding larger variation in the ratio of F- to G-actin (cf. Figs. A.8f and A.9f). Note that profilin and cofilin concentrations are higher than in the lamellipodia model of Mogilner and Edelstein-Keshet (2002), this being a consequence of not including thymosin β4 in the spine-head model.

Fig. A.10 shows a parallel calculation under the assumption that the profilin concentration remains fixed and the F-actin recycling is driven by changes in the cofilin concentration. There is no direct link to NMDA receptor activation; rather, q_1 and q_2 have been assigned values that lead to reasonable agreement with experimental values (Fig. A.10f), but not as good as the cases where profilin was the driving force (Figs. A.8f and A.9f).

An MC simulation of the same protocol as used in Fig. A.9 is shown in Fig. A.11. Fig. A.12 shows data obtained from the above MC simulation. Again, the mean filament length (Fig. A.12a), mean transit length (Fig. A.12b) and total amount of actin in the spine head (Fig. A.12c) all increase with the 9-min application of glutamate and then decrease back to equilibrium after about another 10 min. In particular, the mean filament length increases from its unstimulated length of 0.0086 μm (Fig. A.6c) to a maximum of about 0.022 μm (Fig. A.12a).

A.4 Discussion

A model of F-actin treadmilling at the apex of lamellipodia has been developed for application to synaptic spines. This has been shown to adequately predict MC simulations of treadmilling in the spines. It is notable that although a spine head is much smaller than a lamellipodium, the model does scale successfully, and parameter values can be taken over unchanged (Table A.1). On the assumption that NMDA

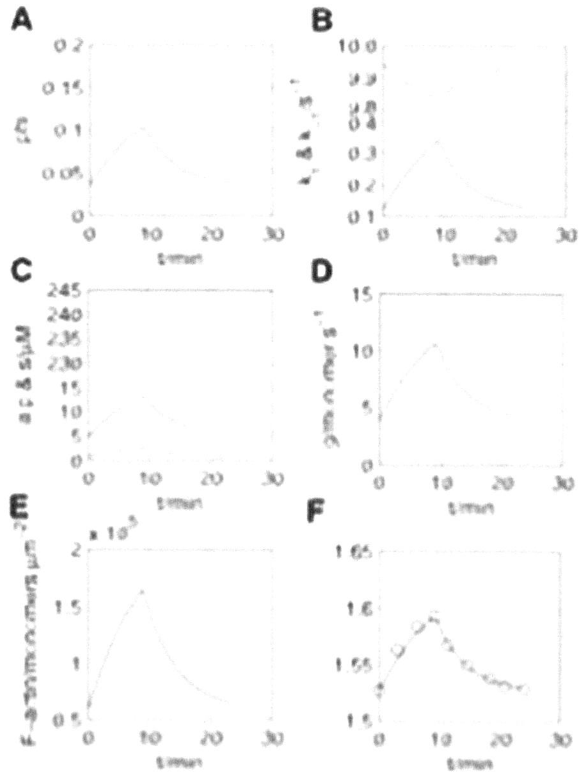

Fig. A.9 As for Fig. A.8, only now the alternative scheme in which k_d is taken to be inversely proportional to ρ is used. As for Fig. A.8, only now the alternative scheme in which k_d is taken to be inversely proportional to ρ is used Eq. (A.38). Parameters are $u = 3$, $\nu = 0.193\ \mathrm{s}^{-1}$, $\omega = 0.007\ \mathrm{s}^{-1}$, σ is 0.4 during the rising phase and the total profiling concentration is 500 μM. The remaining parameters are as for Fig. A.8 (from Bennett et al. 2011)

receptor excitation increases profilin in the spines, the theory is shown to give a quantitative account of observed changes in F-actin dynamics. The results suggest that experimental observations and theoretical descriptions of F-actin changes in lamellipodia might be applicable to the growth of synaptic spines.

The synaptic spine has a spine head of about 0.7-μm diameter and a spine neck of about 0.2-μm length and less than 0.1-μm diameter. At the base of the spine head is a stable pool of F-actin with a lifetime of about 17 min, with stability maintained through cross-binding of F-actin by synaptopodin (Asanuma et al. 2005; Okubo-Suzuki et al. 2008; Honkura et al. 2008). This stable pool supports a dynamic pool of F-actin which has a short lifetime of about 40 s due to constant flow of G-actin from the membrane end of the F-actin towards the cytoplasmic end (Honkura et al. 2008). The mechanical pressure exerted on the spine head by the extracellular matrix, primarily through proteoglycans, resists the expansive force due to the dynamic pool of F-actin such that an equilibrium is reached in spine volume. The spine volume stability is buttressed by a pool of F-actin cross-linked by calcium–calmodulin kinase II (CaMKIIa) found throughout the spine head (Okamoto et al. 2007; Rostaing et al. 2006). If it is enhanced by a stimulus, such as that arising from excitation of *N*-methyl-D-aspartate (NMDA) receptors by glutamate, then the expansive force exerted by the dynamic F-actin pool may become greater than that due to the surrounding proteoglycans, leading to an increase in volume

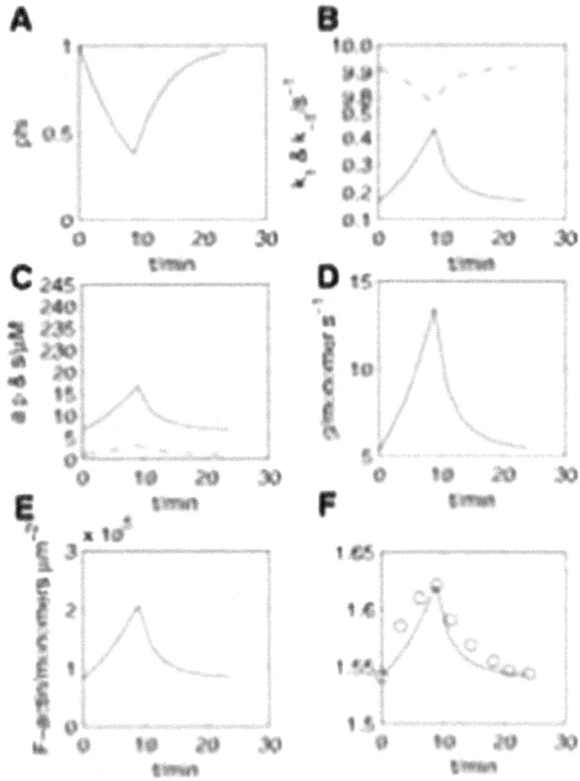

Fig. A.10 As for Fig. A.8, only now profilin is kept constant ($[P] = 50$ μM) and the cofilin concentration $[S]$ allowed to vary (with the total cofilin concentration set at 600 μM). There is no detailed link to NMDA receptor activation, and the calculation starts by assigning values to q_1 and q_2 [see Eq. (A.40)], namely $q_1 = 0.00037\ s^{-1}$, $q_2 = 0.11\ s^{-1}$ during NMDA receptor activation and $q_1 = 0.2\ s^{-1}$, $q_2 = 0.2\ s^{-1}$ afterwards (from Bennett et al. 2011)

of the spine head. This enlarged head is then maintained by rapid growth of the CaMKIIa cross-linked F-actin pool, through recruitment of both G-actin and CaMKIIa from the dendrite through the spine neck (Shen et al. 1998; Sharma et al. 2006; Ahmed et al. 2006). Thus, the amount of bound CaMKIIa in spines is proportional to spine size (Asrican et al. 2007).

In synaptic spines, cofilin is concentrated in the spine shell, particularly within the postsynaptic density, consistent with the dynamic F-actin pool occurring in the periphery of the spine (Racz and Weinberg 2006, 2008). Calcium–calmodulin kinase II (CMKII) is found extensively in the spine head (Rostaing et al. 2006), consistent with its role in maintaining the stability of the F-actin enlargement pool, although it is still at higher concentration near the postsynaptic density at the apex of the spine. The vasodilator-stimulated phosphoprotein (VASP) is attached to the cytoplasmic face of the plasmalemma and acts to protect the barbed (+) ends of F-actin from capping (Applewhite et al. 2007; Svitkina et al. 2003), consistent with the apex being a site of favourable F-actin polymerization (Rostaing et al. 2006).

The sequence of events associated with the growth of lamellipodia, based on the dendritic-nucleation array-treadmilling model (Higgs and Pollard 2001; Pollard and Borisy 2003), has been quantitatively described by Mogilner and Edelstein-Keshet (the ME-K model; Mogilner and Edelstein-Keshet 2002; Mogilner 2006). In this

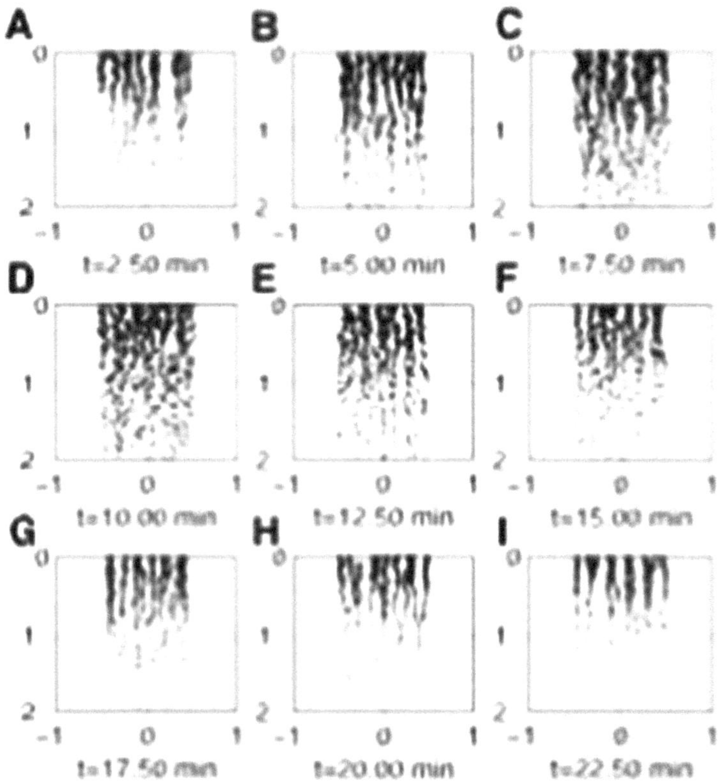

Fig. A.11 Growth, branching and polymerization of actin filaments (F-actin) in a synaptic spine according to an MC simulation, following initiation of treadmilling by a transient change in glutamate acting on NMDA receptors on the spine. The same protocol is used as in Figs. A.8, A.9 and A.10; namely the binding rate σ is taken to be 0.9 during glutamate application ($t < 9$ min) and 0 thereafter. The lines indicate F-actin filaments. The ordinate is the distance y (μm) from the surface membrane (at $y = 0$). Parameter values are as for Fig. A.9. The spatial distribution of F-actin in the synaptic spine is shown at nine different times, as indicated, after the beginning of glutamate action (from Bennett et al. 2011)

model, ADF/cofilin accelerates depolymerization of actin filaments at their older, ADP-actin positions. ADF/cofilin and profilin form complexes with G-actin, in which profilin catalyses exchange of ADP for ATP on the actin monomers. Profilin-ATP-G- actin intercalates into the gap between a filament and the plasma membrane, and assembles into the barbed (+) ends of the filament, so pushing the plasmalemma forward (Figs. A.1 and A.2). The actin dynamics described by the ME-K model have been developed into a Monte Carlo (MC) model of the same dynamics but in two dimensions (Schaus et al. 2007). More recently, a stochastic two-dimensional simulation, which incorporates appropriate chemical reaction kinetics, G-actin diffusion and filament transport, has been presented which shows how self-organization within a lamellipodia can give rise to two distinct F-actin

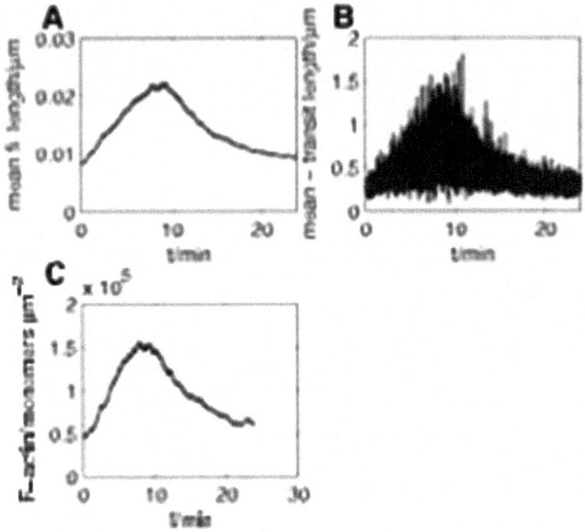

Fig. A.12 MC predictions of the changes in mean F-actin length (**a**), mean transit length (**b**) and total actin (**c**) during and after stimulation of NMDA receptors with glutamate for the period of 0–9 min; data used is from Fig. A.11. The changes in F-actin in the spine following glutamate stimulation are shown in (**c**), where the dotted line is the trace from Fig. A.9e (from Bennett et al. 2011)

networks which, in the case of the spine head, might be equivalent to the stable and dynamic F-actin pools (Huber et al. 2008). Following observations on lamellipodia, it might be proposed that the mechanism of spine enlargement involves increasing F-actin polymerization at its barbed (+) ends at the apex of the spine head, where the F-actin is probably protected from capping by VASP on the plasmalemma. Two principal ways of achieving this are, first, to increase profilin which enhances spine growth through accelerating ADP-profilin to ATP-profilin, so making the G-actin available for attachment at the barbed (+) ends of F-actin (Fig. A.1), and, second, to inhibit ADF/cofilin and so prevent depolymerization of F-actin whilst allowing proliferation to proceed (Fig. A.1). The principal means of obtaining spine shortening is to activate myosin in the spine, so increasing the retrograde movement of F-actin from the apex of the spine head (Fifkova and Morales 1989; Lin et al. 1996; Svitkina et al. 1997; Schmidt et al. 2002; Cai et al. 2006). Spine stabilization is affected by local translation of activity-regulated cytoskeletal-associated protein (Arc) which consolidates F-actin by phosphorylating and so terminating cofilin (Bramham and Wells, 2007, Fig. 3), a process triggered by BDNF action on spine TrkB receptors (Carvalho et al. 2008).

The main receptors at the plasmalemma which activate signalling cascades that lead to changes in F-actin treadmilling and spine shape are neurotransmitter receptors as well as voltage-dependent calcium-ion channels on the one hand and surface membrane receptors which activate intracellular signalling cascades that control activity of Rho small GTPases such as Rho A on the other (Fig. 1; Ethell and Pasquale 2005). The main transmitter-activated calcium-permeable channels are the NMDA receptors, which are modulated by ErbB4 receptors, and the main receptors controlling the RhoGTPases are glutamate receptors (Schubert et al.

2006), ephrin (Schubert and Dotti 2007), brain-derived neurotrophic factor (BDNF; Schubert and Dotti 2007) and plexin (Rohm et al. 2000).

The treadmilling theory of F-actin dynamics at the apex of lamellipodia, involving the catalysis of profilin, has been adapted here for the first time to growth of synaptic-spine heads. It is used to show how enhanced profilin activity following NMDA receptor stimulation can change the ratio of filamentous to globular actin in the spine head and hence changes in dimensions of the spine head. These calculations give a good fit to experimental observations of changes in spine-head actin following application of glutamate to the spine head and are well fitted by Monte Carlo simulations. These results support the concept that treadmilling of F-actin in the spine head is the key process in synaptic-spine growth.

Appendix B: Regulation of NMDA Receptors[2]

B.1 NMDA Receptor Control of Dendritic Spine Formation and Regression

The main mechanisms of spine enlargement involve increasing F-actin polymerization at the apex of the spine head (Fig. B.1). The two principal ways of achieving this are to either increase Arp2/3 and so F-actin proliferation and branching or inhibit adenosine diphosphate (ADF)/cofilin and so prevent depolymerization of F-actin whilst allowing proliferation to proceed. Increases in profilin will also assist spine growth through accelerating adenosine diphosphate (ADP)-profilin to adenosine-5′-triphosphate (ATP)-profilin, so making the G-actin available for attachment at the ends of F-actin at the apex of the spine. The principal means of obtaining spine shortening is to activate myosin in the spine, so increasing the retrograde movement of F-actin from the apex of the spine head.

The principal receptors at the plasmalemma that activate signalling cascades that lead to changes in F-actin treadmilling and spine shape are neurotransmitter receptors as well as voltage-dependent calcium-ion channels on the one hand and surface membrane receptors that activate intracellular signalling cascades that control activity of Rho GTPases on the other (Fig. B.2) (Ethell and Pasquale 2005). The main transmitter-activated calcium-permeable channels are the NMDA receptors, which are modulated by ErbB4 receptors, and the main receptors controlling the RhoGTPases are ephrin (Gray and Roth 2007), brain-derived neurotrophic factor (BDNF) (Schubert and Dotti 2007) and plexin (Fig. B.3).

NMDA receptor activation by glutamate leads to the largest calcium influx that any receptor can generate. This calcium can in turn activate calcium-calmodulin kinase (CaMKII), which in turn phosphorylates and activates calmodulin-dependent kinase kinase (CaMKK-CaMK1). These kinases form a multimolecular complex

[2]Reprinted from: Bull Math Biol. Bennett MR, Farnell L, Gibson WG. A model of neuregulin control of NMDA receptors on synaptic spines. Vol. 74(3):717–735. Copyright (2012). With permission from Springer.

Bennett MR. Positive and negative symptoms in schizophrenia: the NMDA receptor hypofunction hypothesis, neuregulin/ErbB4 and synapse regression. Aust N Z J Psychiatry. 43 (8):711–721. Copyright (2009). With permission from Sage.

M. Bennett, J. Lagopoulos, *Stress, Trauma and Synaptic Plasticity*,
https://doi.org/10.1007/978-3-319-91116-8

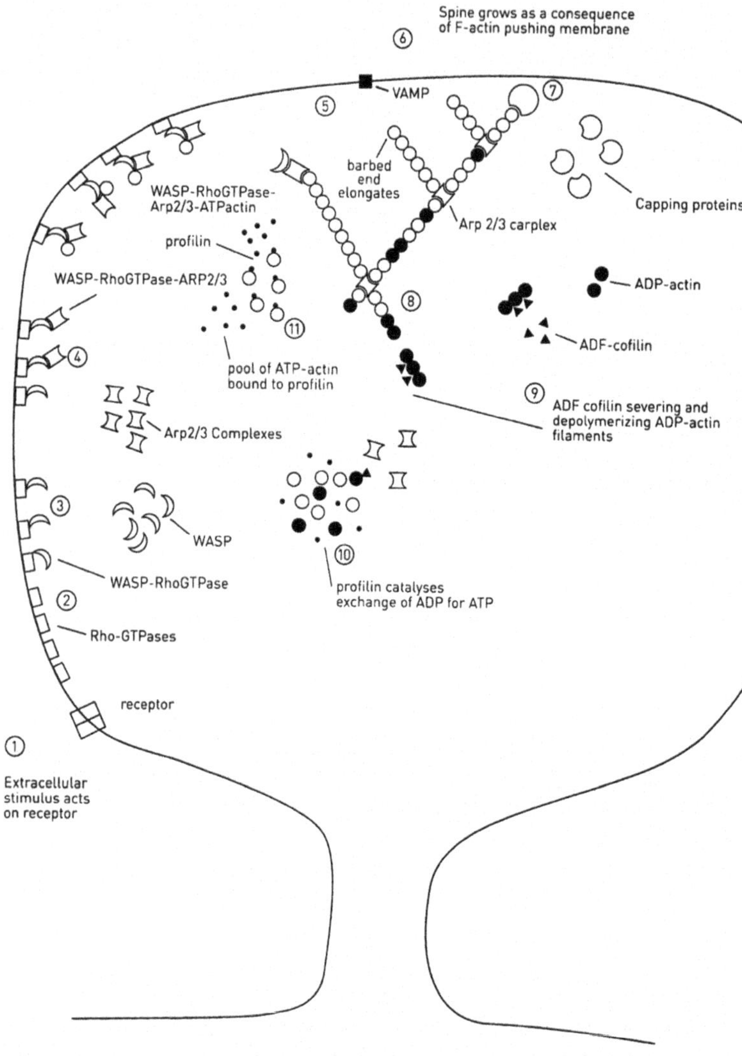

Fig. B.1 The actin nucleation model of growth and regression of synaptic spines. The actin nucleation model of growth and regression of synaptic spines. (1) Receptors activate signalling pathways that lead to Rho GTPases (2). (3) These activate WASP proteins, which, in turn, in (4), lead to activation of the Arp2/3 complex and the subsequent binding of adenosine-5′-triphosphate (ATP)-actin into the complex. (5) This leads to the formation of new actin filaments through attachment of the Wiskott–Aldrich syndrome protein (WASP)-Arp2/3-ATP-actin complex to an existing filament to form a side branch. This new filament grows as a consequence of being supplied from a high concentration of profilin-bound actin (11), with the result that the spine membrane is pushed out (6). (7) Capping proteins bind to the growing barbed ends of the filaments, so terminating elongation or they can be protected from capping by plasmalemma-bound vasodilator-stimulated phosphoprotein (VAMP). Actin-polymerizing factor (ADF)-cofilin severs and depolymerizes the adenosine diphosphate (ADP)-actin in the older regions of the filaments (8, 9). Profilin then promotes dissociation of ADP and binding of ATP to the dissociated subunits (10). (11) ATP-actin then binds profilin, making them available for assembly into actin filaments. (Diagram and description after Fig. 1 in Pollard 2003)

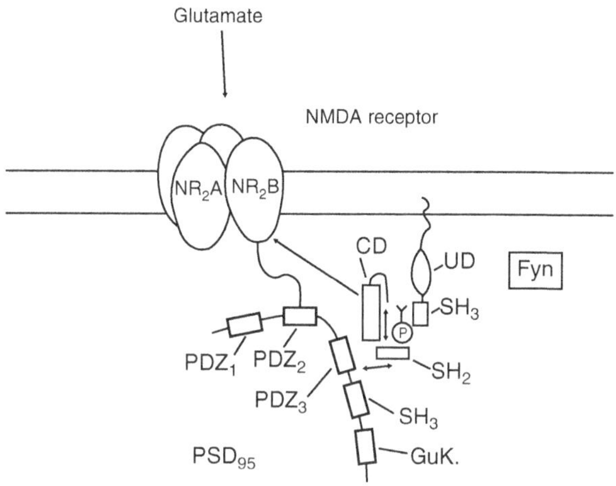

Fig. B.2 Upregulation of *N*-methyl-D-aspartate (NMDA) receptors by Src family kinase Fyn acting on the NR2B receptor subunit. Fyn like Src has domains UD (unique domain), SH3, SH2 and CD (catalytic domain) and also requires phosphorylation (P) in the CD loop in order to undergo a conformational change of an activation loop in order to produce a fully active Fyn kinase. Fyn is associated with postsynaptic density (PSD)-95 through its protein PDZ3 domain, and when the kinase is brought into close proximity with the NR2B subunit of NMDA, it phosphorylates it at Y1472, Y1336 and Y1252 (Fig. B.4), leading to increased efficacy of the NMDA receptor. (After Fig. 1 in Ali and Salter 2001)

with the guanine-nucleotide exchange factor bPIX, as shown in Fig. B.3 (Cingolani and Goda 2008; Saneyoshi et al. 2008). Phosphorylation of this results in activation of Rac1 and Pak1. Pak1-mediated phosphorylation then activates LIM kinase (LIMK), which inactivates actin-depolymerizing protein ADP/cofilin, so promoting stabilization of F-actin and giving spine enlargement (Fig. B.3).

In hippocampal neurons an actin-regulated pathway is directed by Rho small GTPases, such as Rho A (Nakayama and Luo 2000). Rho A recruits and activates its specific Rho-associated kinase (ROCK), which in turn complexes with profilin Iia, so modifying actin stability through a RhoA/ROCK/profiling IIa pathway (Fig. 3) (Da Silva et al. 2003). It is known that RhoA associates with ionotropic glutamate receptors (iGluRs) and metabotropic glutamate receptors (mGluRs) at the plasma membrane of synaptic spines, with activation of the iGluRs leading to detachment of RhoA from these receptors and their recruitment to the mGluRs. This triggers local reduction of RhoA activity at the iGluRs, resulting in inactivation of RhoA-specific kinase ROCK there and so disruption of the ROCK-profilin IIa, producing rapid changes in the underlying actin cytoskeleton and remodelling of spine shape (Schubert et al. 2006).

Activation of the NMDA receptors can also lead to increases in activity of LIMK, and hence inhibition of ADP/cofilin and spine enlargement, by the GTPase RhoA through RhoA-specific kinase (ROCK; Fig. B.3). This pathway can also increase profilin and hence availability of G-ATP-actin for F-actin polymerization. ROCK

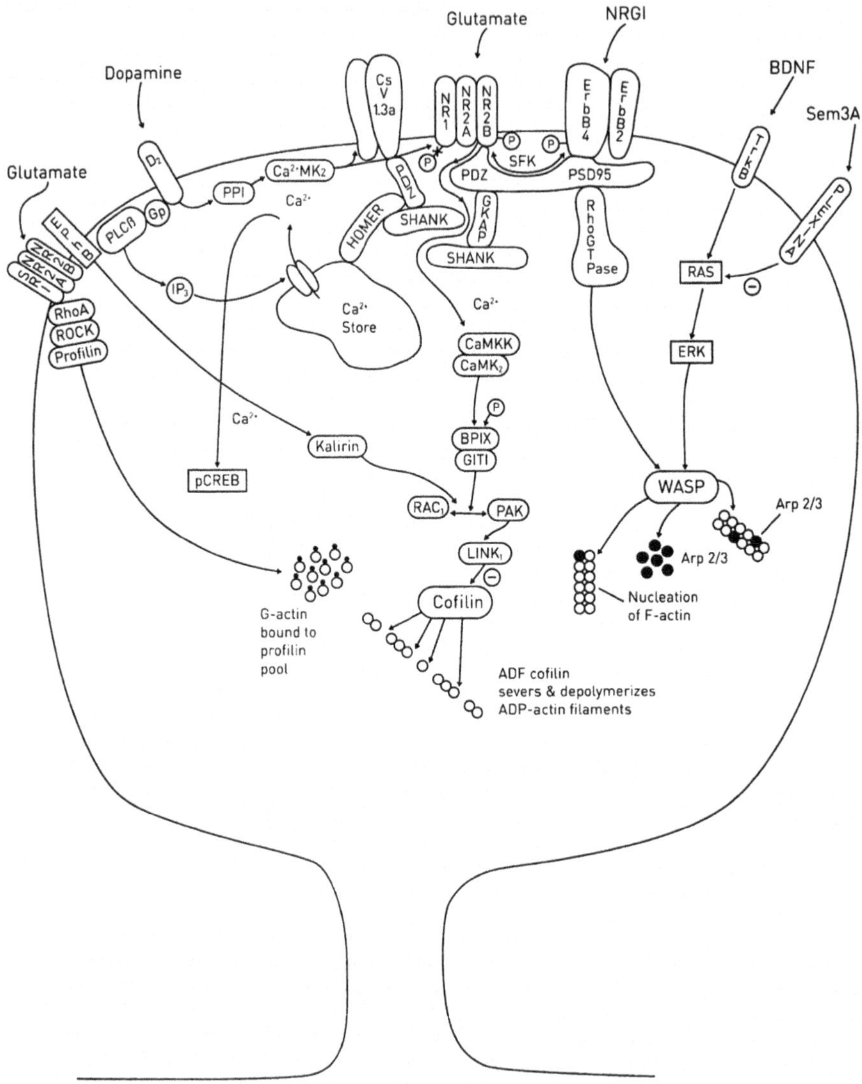

Fig. B.3 Membrane receptors and voltage-gated ionic channel control of synaptic spine actin cytoskeleton and so synapse formation and regression. The following receptor and ion channel pathways are indicated on the spines plasmalemma (read counter-clockwise around the spine). (1) Sem 3A acting on plexin A receptors to decrease RAS, ERK and so WASP nucleation of G-actin around Arp2/3. (2) BDNF acting on TrkB receptors to activate RAS, ERK and so WASP nucleation of G-actin around Arp2/3. (3) NRG-1 on ErbB4/ErbB2 receptors, anchored to the PSD-95 scaffolding protein and RhoGTPase to increase WASP as well as release of SFKs to act on the NR2B subunit of the NMDA receptor to phosphorylate it, so enhancing calcium influx through NMDA. (4) Glutamate acting on NMDA receptors to enhance calcium entry into the cytosol and via CaMKK, CaMK2, BPIX, GIT1, RAC, PAK and LIMK1 to inhibit cofilin depolymerization of ADP-actin in the F-actin filaments. (5) Voltage-dependent calcium channel, CAV1.3a, which is anchored by PDZ and SHANK in the cytosol and through HOMER releases calcium from calcium stores. (6) Dopamine acting on D2 dopamine receptors, which, through PP1, activates calcium calmodulin kinase 2 to modulate CAV1.3a and phosphorylates the NR1 subunit of NMDA.

Fig. B.3 (continued) (7) Dopamine acting on D2 dopamine receptors, which through Gp and PLCb, releases IP3 to activate calcium release from internal stores; this calcium activates pCREB. (8) Glutamate acting on NMDA receptors activates the RhoA, ROCK, profilin pathway to provide G-actin bound to profilin. (9) Ephrin binding to the EphB receptor activates Kalirin, which then acts on RAC1/PAK to excite LIMK1 and so inhibits cofilin depolymerization of F-actin. BDNF, brain-derived neurotrophic factor; BPIX, guanine nucleotide exchange factor for RAC; CaMK2, calcium calmodulin-dependent kinase 2; CaMKK, calcium calmodulin-dependent protein kinase kinase; cofilin severs and depolymerizes ADP-actin; D2, dopamine D2 receptor; EphB, ephrin receptor; ErbB2, receptors for neuregulin; ErbB4, receptors for neuregulin; ERK, extracellular signal-regulated kinases; GKAP, guanylate kinase-associated protein; Gp, G-protein; HOMER, scaffolding protein; IP3, inositol triphosphate; kalirin, Rho guanine nucleotide exchange protein (GEF); LIMK, LIM kinase, phosphorylates ADF/ cofilin; NMDA, *N*-methyl-D-aspartate; NR1, NR2A, NR2B, subunits of the NMDA receptor; NRG-1, neuregulin 1; PAK, downstream effector of RAC (sometimes called P21-activated kinase); pCREB, phosphorylated cyclic AMP response element-binding protein; PDZ, protein domain; PLCb, protein lipase Cb; plexin A, receptor for Sema 3A; PP1, protein phosphatase 1; profilin, actin regulatory molecule; PSD-95, postsynaptic density 95, a scaffolding protein; RAC, Rho-GTPase; RAS, Rho-GTPase; RhoA, Rho-GTPase; Rho-GTPase, Rho-family GTPases, a subgroup of the superfamily of GTPases; ROCK, Rho-associated kinase; sema 3A, semaphorin 3A; SFK, src family kinase; SHANK, scaffolding molecule; TrkB, BDNF receptor; WASP, Wiskott–Aldrich syndrome protein that triggers actin polymerization via Arp 2/3 complex (from Bennett 2009)

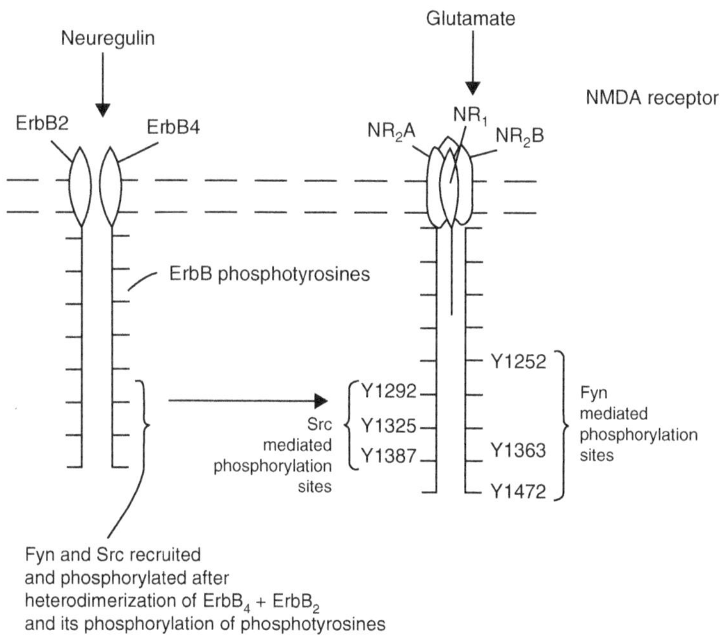

Fig. B.4 ErbB-receptor tyrosine family members ErbB4 and ErbB2, after heterodimerization on activation by neuregulin, can phosphorylate *N*-methyl-D-aspartate (NMDA) receptor subunits NR2A and NR2B through mediation by interaction partners Src and Fyn of the ErBB-receptor tyrosine residues. These upregulate the function of NMDA receptors through Fyn acting at Y1252, Y1363 and Y1353 sites to phosphorylate NR2B and through Src acting at Y1492, Y1353 and Y1387 to phosphorylate NR2A. Src and Fyn are family kinases (from Bennett 2009)

can also activate myosin regulatory light chain (MLC) and simultaneously inhibit myosin light chain phosphatase (MLCP), so leading to myosin motor activity and increased retrograde flow of F-actin (Schmidt et al. 2002) with concomitant spine shortening. Collectively, glutamate stimulation of NMDA receptors can, through both CaKM11 and RhoA, lead to inhibition of ADF/cofilin (with perhaps a small enhancement from the phosphatase calcineurin) and increase profilin, which probably more than offsets in most instances the increased retrograde flow of F-actin due to enhanced myosin motor activity contingent on the increase in ROCK.

B.2 Neuregulin-ErbB4 Receptor Control of NMDA Receptor Efficacy

It is then of special interest to inquire into how the ErbB4 receptor regulates the NMDA receptor leading to its hypofunction with subsequent spine loss (Fig. B.3).

B.3 Neuregulin Transphosporylation of Receptor Kinase ErbB4 and Recruitment of Src Family Kinases

The ErbB receptors are found in the plasmalemma (detergent-soluble) and membrane caveoli (detergent-insoluble), with translocation between these occurring on activation of the receptors with NRG (Zhou and Carpenter 2001). Thus, NRG promotes the translocation of both ErbB4 to the caveoli and its heterodimerization partner ErbB2, which is there tyrosine-phosphorylated to a level fourfold greater than ErbB2 in the non-caveoli regions of the membrane. NRG binds ErbB receptors to induce formation of ErbB4/ErbB2 heterodimers (Fig. B.4). This then activates intrinsic kinases, which leads to the phosphorylation of specific tyrosines located in the ErbB's cytoplasmic region. Such phosphorylated residues are docking sites for non-receptor protein kinase (Src) family signalling molecules, the recruitment of which stimulates the intracellular signalling cascades leading to upregulation of NMDA-receptor function (Holbro and Hynes 2004) (Fig. B.4). The crystal structure of ErbB4 kinase has been described and shown, on activation by NRG, to adopt an asymmetric dimer confirmation (Qiu et al. 2008).

NRGs, encoded by four different genes, namely NRG-1, NRG-2, NRG-3 and NRG-4, are ligands for ErbB4 and promote its autophosphorylation (Zhou and Carpenter 2001). Receptor dimerization occurs following formation of ligand–receptor complexes, so initiating signalling. ErbB4 most frequently forms a heterodimer with ErbB2. Autophosphorylation of ErbB receptors takes place at 18 tyrosine residues within the carboxy-terminal region (Chattopadhyay et al. 1999: Schulze et al. 2005), of which approximately five are potential autophosphorylation sites (Fig. B.4). When phosphorylated, these conserved tyrosine residues can function as docking sites for downstream signalling molecules (Fig. B.4). Different downstream molecules are recruited, depending on the ErbB4 ligand, because this determines the pattern of its phosphorylation (Sweeney et al.

2000) ErbB4 interacts directly with the first two PDZ domains of PSD-95 (Murphy and Bielby-Clarke 2008; Garcia et al. 2000; Huang et al. 2000), through its carboxy-terminal sequence TVV.

B.4 NMDA Receptor Modulation by Phosphorylation Through Src Family Kinases

Src and other Src family kinases, such as Fyn, regulate the activity of NMDA receptors, with Src and Fyn phosphorylating and thereby upregulating the function of NMDA receptors in long-term potentiation (Yu et al. 1997; Kalia et al. 2004), increasing currents through the receptor as well as intracellular calcium (Wang and Salter 1994). The Src adaptor protein NADH dehydrogenase subunit 2 (ND2) is required for Src to be anchored to the NMDA receptor (Fig. B.5) (Ali and Salter 2001; Gingrich et al. 2004).

Activation of Src depends on the focal adhesion non-receptor protein kinase CaKb (Fig. B.5) (Huang et al. 2001). Activation of CaKb is required for inducing long-term potentiation, and such activation is indirectly dependent on calcium, probably through stimulation of protein kinase C and perhaps calcium-calmodulin-dependent kinases protein (CaMKII) (Guo et al. 2004).

Activation of CaKb leads to its autophosphorylation at Tyr-402 on the linker between the catalytic domain (CD) and the band 4.1, Janus Kinase family, ERM protein family and focal adhesion kinase family domain (band 4.1-JEF domain) as well as on Tyr-579/580 in the CD (Fig. 5) (Ali and Salter 2001). Phosphorylation on Tyr-402 creates an SH2 ligand through which CAKb binds to the SH2 domain of Src and activates this kinase by relieving autoinhibition (Salter and Kalia 2004). When this occurs, autophosphorylation of Y416 in the CD results in a conformational change of the activation loop, producing a fully active kinase. The SH2 domain binds to peptide motifs that contain phosphorylated tyrosine. The CD contains phosphorylated tyrosine. The CD contains the tyrosine kinase activity that phosphorylates the NR2A subunit of the NMDA receptor in its carboxy (C)-terminal tail at least at sites Y1292, Y1325 and Y1387 (Fig. 5) (Salter and Kalia 2004). Fyn is associated with PSD-95 that, like the adaptor protein for Src, brings the kinase into close proximity with the NMDA receptors (Fig. 2) (Ali and Salter 2001; Tezuka et al. 1999). This leads to phosphorylation of the NR2B subunit of *N*-methyl-D-aspartate receptor (NMDAR) at Y1472, Y1336 and Y1252 (Cheung and Gurd 2001). During long-term potentiation (LTP) induction, there is increased phosphorylation of NR2b subunit (Rostas et al. 1996) and Fyn mutant mice have impaired LTP that can be rescued by postnatal expression of Fyn (Kojima et al. 1997).

The kinetics of phosphorylation by the SH3 domain of Fyn have been described (Bhaskar et al. 2005), as has the kinetics of phosphorylation by Src of the G-protein-coupled receptor kinase GRK2 (Sarnago et al. 1999) as well as of dynamin (Solomaha et al. 2005).

Src facilitates NMDA receptor function by increasing tyrosine phosphorylation of the NR2A subunit (Lu et al., 1998), whereas Fyn does so by phosphorylating the

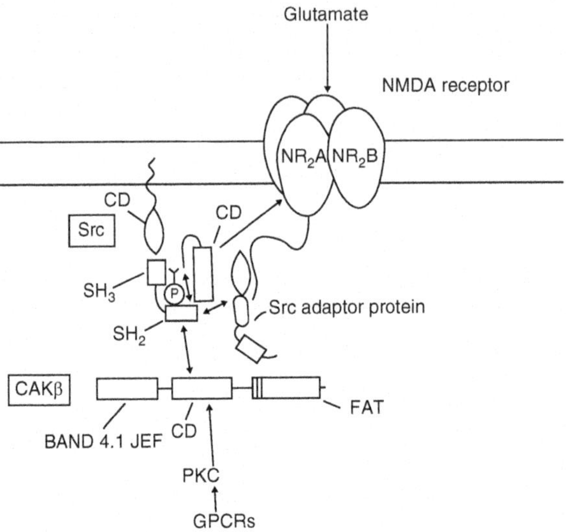

Fig. B.5 Upregulation of *N*-methyl-D-aspartate (NMDA) receptors by Src family kinase Src acting on NR2A receptor subunits. Src has domains UD (unique domain), SH3, SH2 and CD (catalytic domain). Phosphorylation (P) in the CD results in a conformational change of the activation loop, producing a fully active Src kinase. This can now act, via its tyrosine kinase activity, through an adaptor protein, to phosphorylate the NR2A subunit of the NMDA receptor in its carboxy (C)-terminal tail at least at sites Y1292, Y1325 and Y1387 (Fig. B.4). Besides Src becoming a fully active kinase through its interaction with the ErbB receptor, it can also be activated by the non-receptor protein kinase cell adhesion kinase beta (CaKb), which possesses three domains: BAND 4.1JEF (Janus kinase/ERM/FAK), CD (catalytic domain) and FAT (focal-adhesion targeting). CaKb is activated by protein kinase C (PKC) and therefore indirectly by calcium and by G-protein-coupled receptors (GPCRs). (After Fig. 1 in Ali and Salter 2001)

NR2B subunit (Figs. B.2 and B.5) (Salter and Kalia 2004). The Src-dependent regulation of NMDA receptor activity is regulated by an interaction between C-terminal Src kinase (Csk) and the NMDA receptor, probably through direct association of Csk with the Src-phosphorylated NR2 subunits (Xu et al. 2008). In this way, Csk maintains constant the excitatory synaptic transmission at impulse rates that do not evoke long-term potentiation by inhibiting calcium increases of Src kinase-dependent increases

B.5 Kinetics of Erb (EGFR) Transphosporylation and of Subsequent Src Family Kinases

A detailed kinetic model of epidermal growth factor receptor (EGFR)/ErbB signalling pathways, describing the activation, dimerization and tyrosine phosphorylation of EGFR/ErbB, followed by the binding and activation/phosphorylation of an adaptor/target protein such as Src, has been developed by Kholodenko et al. (1999) (Moehren et al. 2002). Two kinetic modules may be considered here, one concerning the dimerization of EGFR/ErbB on activation with its subsequent phosphorylation and the other module the binding and activation/phosphorylation of the target protein in this case Src. Considering first the second module concerning the target protein: this is bound and phosphorylated by EGFR/ErbB through activated EGFR/ErbB kinase and a coupled cycle of phosphorylated and dephosphorylation of the target protein by specific tyrosine phosphatase. Other considerations of these kinetic schemes are given in (Kholodenko et al. 1999; Hatakeyama et al. 2003; Nakakuki et al. 2008; Schoeberl et al. 2002).

B.6 Kinetics of Src Family Kinases (Src, Fyn) Interaction with NR2B Subunit of the NMDA Receptor

There have been no quantitative models of Src and Fyn interaction and subsequent tyrosine phosphorylation of the NMDA receptors, although such sites are identified. Kinetic parameter constants have been determined for calcium/calmodulin-dependent protein kinase (CaMKII) binding to specific residues on NR2B subunit (Mayadevi et al. 2002). An appropriate kinetic model, however, has probably been provided by (Castellani et al. 2005). This gives the following kinetic modules: dimerization transphosphorylation of ErbB to give the activated receptor (Fig. B.5), subsequent phosphorylation (dephosphorylation) of the target/adaptor protein given by Src (or Fyn or other Src family kinase (SFK)) and then binding of Shc (or Fyn or other SFK) to the NR2B subunit of the NMDA receptor (Figs. B.2 and B.5).

B.7 Quantitative Measures of ErbB4 Control of NMDA Receptor Activity

Early studies examined the acute effects of NRG-1 application on NMDA receptors and reached the conclusions that NRG-1 inhibits synaptic plasticity (Huang et al. 2000; Ma et al. 2003; Kwon et al. 2005) and reduces NMDA receptor activity and synaptic transmission. These studies used tonic application of a functional domain, such as the epidermal growth factor (EGF) domain, of particular isoforms of recombinant NRG-1 (Huang et al. 2000: Kwon et al. 2005; Gu et al. 2005; Roysommuti et al. 2003), but the function of endogenous NRG-1 at synapses may be different. The location or duration or physiological concentration of these functional domains is not known (Woo et al. 2007), nor their action on different ErbB receptors. Indeed, NRG-1-deficient mice show reduction in NMDA receptor activity, compared with wild type, rather than the opposite effects that would be expected from the acute studies (Stefansson et al. 2002). Recently, Li et al. (2007) have shown that preventing NRG-1/ErbB4 signalling by using either ErbB4 knockout mice or ErbB4 RNAi leads to loss of NMDA receptor currents in hippocampal neurons and this is accompanied by a decrease in spine size and number.

If Fyn is the SFK under consideration, then the Fyn kinase activity as a consequence of the NRG-1 action on ErbB4 increases to a peak in approximately 7 min, well after the ErbB4 kinase activity reaches a maximum at 1 min. The dose–response curve for NRG-1 increasing kinase activity of ErbB and of Fyn indicates that they have similar ED50s but quite different activation levels (Bjarnadottir et al. 2007).

B.8 Conclusion

The need to remedy NMDA hypofunction where it occurs suggests that agents should be sought that enhance the activity of this receptor. NMDA agonists are not likely to be useful because of their leading, amongst other things, to excitotoxic cell death of the neurons consequent on excess calcium entry. Approaches using the allosteric site on the NMDA receptor complex that bind glycine, a necessary condition for operation of the receptor, are promising. Indeed, chronic treatment of rodents with glycine does not produce excitotoxicity (Shoham et al. 2001). Another possible approach is to prevent reuptake of glycine, so enhancing its extracellular concentration and NMDA receptor activity (Javitt et al. 2005). Of particular interest is the NRG-ErbB4 pathway in NMDA regulation. Rather than directly attempting to upregulate the NMDA receptor, it is suggested that compounds should be sought that prevent hyperphosphorylation of ErbB4 or the subsequent hypophosphorylation of NMDA receptors through changes in the recruitment of the SFKs. NMDA hypofunction, with subsequent loss of spines, emphasizes the need for fast-throughput assays to identify compounds that will restore the functioning of NMDA receptors and hence synapses with the resultant re-establishment of neural circuitry, holding out the possibility of some rehabilitation of those suffering from schizophrenia.

References

Ahmed R, Zha X-M, Green SH, Dailey ME (2006) Synaptic activity and F-actin coordinately regulate CaMKIIalpha localization to dendritic postsynaptic sites in developing hippocampal slices. Mol Cell Neurosci 31(1):37–51. https://doi.org/10.1016/j.mcn.2005.08.020

Ali DW, Salter MW (2001) NMDA receptor regulation by Src kinase signalling in excitatory synaptic transmission and plasticity. Curr Opin Neurobiol 11(3):336–342

Applewhite DA, Barzik M, Kojima S-i, Svitkina TM, Gertler FB, Borisy GG (2007) Ena/VASP proteins have an anti-capping independent function in filopodia formation. Mol Biol Cell 18(7):2579–2591. https://doi.org/10.1091/mbc.E06-11-0990

Asanuma K, Kim K, Oh J, Giardino L, Chabanis S, Faul C et al (2005) Synaptopodin regulates the actin-bundling activity of α-actinin in an isoform-specific manner. J Clin Investig 115(5):1188–1198. https://doi.org/10.1172/JCI200523371

Asrican B, Lisman J, Otmakhov N (2007) Synaptic strength of individual spines correlates with bound Ca2+-calmodulin-dependent kinase II. J Neurosci 27 (51):14007–14011. https://doi.org/10.1523/jneurosci.3587-07.2007

Bennett M (2009) Positive and negative symptoms in schizophrenia: the NMDA receptor hypofunction hypothesis, neuregulin/ErbB4 and synapse regression. Aust N Z J Psychiatry 43(8):711–721. https://doi.org/10.1080/00048670903001943

Bennett MR, Farnell L, Gibson WG (2011) A model of NMDA receptor control of F-actin treadmilling in synaptic spines and their growth. Bull Math Biol 73 (9):2109–2131. https://doi.org/10.1007/s11538-010-9614-4

Bhaskar K, Yen SH, Lee G (2005) Disease-related modifications in tau affect the interaction between Fyn and Tau. J Biol Chem 280(42):35119–35125. https://doi.org/10.1074/jbc.M505895200

Bjarnadottir M, Misner DL, Haverfield-Gross S, Bruun S, Helgason VG, Stefansson H et al (2007) Neuregulin1 (NRG1) signaling through Fyn modulates NMDA receptor phosphorylation: differential synaptic function in NRG1+/- knock-outs compared with wild-type mice. J Neurosci 27(17):4519–4529. https://doi.org/10.1523/jneurosci.4314-06.2007

Bramham CR, Wells DG (2007) Dendritic mRNA: transport, translation and function. Nat Rev Neurosci 8(10):776–789. https://doi.org/10.1038/nrn2150

Buchs PA, Muller D (1996) Induction of long-term potentiation is associated with major ultrastructural changes of activated synapses. Proc Natl Acad Sci USA 93 (15):8040–8045

Cai Y, Biais N, Giannone G, Tanase M, Jiang G, Hofman JM et al (2006) Nonmuscle myosin IIA-dependent force inhibits cell spreading and drives F-actin flow. Biophys J 91(10):3907–3920. https://doi.org/10.1529/biophysj.106.084806

Carlisle HJ, Kennedy MB (2005) Spine architecture and synaptic plasticity. Trends Neurosci 28(4):182–187. https://doi.org/10.1016/j.tins.2005.01.008

Carvalho AL, Caldeira MV, Santos SD, Duarte CB (2008) Role of the brain-derived neurotrophic factor at glutamatergic synapses. Br J Pharmacol 153(Suppl 1): S310–S324. https://doi.org/10.1038/sj.bjp.0707509

Castellani GC, Quinlan EM, Bersani F, Cooper LN, Shouval HZ (2005) A model of bidirectional synaptic plasticity: from signaling network to channel conductance. Learn Mem 12(4):423–432. https://doi.org/10.1101/lm.80705

Chattopadhyay A, Vecchi M, Ji Q, Mernaugh R, Carpenter G (1999) The role of individual SH2 domains in mediating association of phospholipase C-gamma1 with the activated EGF receptor. J Biol Chem 274(37):26091–26097

Cheung HH, Gurd JW (2001) Tyrosine phosphorylation of the N-methyl-D-aspartate receptor by exogenous and postsynaptic density-associated Src-family kinases. J Neurochem 78(3):524–534

Cingolani LA, Goda Y (2008) Actin in action: the interplay between the actin cytoskeleton and synaptic efficacy. Nat Rev Neurosci 9(5):344–356. https://doi.org/10.1038/nrn2373

Da Silva JS, Medina M, Zuliani C, Di Nardo A, Witke W, Dotti CG (2003) RhoA/ROCK regulation of neuritogenesis via profilin IIa–mediated control of actin stability. J Cell Biol 162(7):1267–1279. https://doi.org/10.1083/jcb.200304021

Delorme V, Machacek M, DerMardirossian C, Anderson KL, Wittmann T, Hanein D et al (2007) Cofilin activity downstream of Pak1 regulates cell protrusion efficiency by organizing lamellipodium and lamella actin networks. Dev Cell 13(5):646–662. https://doi.org/10.1016/j.devcel.2007.08.011

Edwards DC, Sanders LC, Bokoch GM, Gill GN (1999) Activation of LIM-kinase by Pak1 couples Rac/Cdc42 GTPase signalling to actin cytoskeletal dynamics. Nat Cell Biol 1(5):253–259. https://doi.org/10.1038/12963

Ethell IM, Pasquale EB (2005) Molecular mechanisms of dendritic spine development and remodeling. Prog Neurobiol 75(3):161–205. https://doi.org/10.1016/j.pneurobio.2005.02.003

Fifkova E, Morales M (1989) Calcium-regulated contractile and cytoskeletal proteins in dendritic spines may control synaptic plasticity. Ann N Y Acad Sci 568:131–137

Garcia RA, Vasudevan K, Buonanno A (2000) The neuregulin receptor ErbB-4 interacts with PDZ-containing proteins at neuronal synapses. Proc Natl Acad Sci USA 97(7):3596–3601. https://doi.org/10.1073/pnas.070042497

Gingrich JR, Pelkey KA, Fam SR, Huang Y, Petralia RS, Wenthold RJ, Salter MW (2004) Unique domain anchoring of Src to synaptic NMDA receptors via the mitochondrial protein NADH dehydrogenase subunit 2. Proc Natl Acad Sci USA 101(16):6237–6242. https://doi.org/10.1073/pnas.0401413101

Gray JA, Roth BL (2007) The pipeline and future of drug development in schizophrenia. Mol Psychiatry 12(10):904–922. https://doi.org/10.1038/sj.mp.4002062

Gu Z, Jiang Q, Fu AK, Ip NY, Yan Z (2005) Regulation of NMDA receptors by neuregulin signaling in prefrontal cortex. J Neurosci 25(20):4974–4984. https://doi.org/10.1523/jneurosci.1086-05.2005

Guo J, Meng F, Fu X, Song B, Yan X, Zhang G (2004) N-methyl-D-aspartate receptor and L-type voltage-gated Ca2+ channel activation mediate proline-rich tyrosine kinase 2 phosphorylation during cerebral ischemia in rats. Neurosci Lett 355(3):177–180

Haeckel A, Ahuja R, Gundelfinger ED, Qualmann B, Kessels MM (2008) The actin-binding protein Abp1 controls dendritic spine morphology and is important for spine head and synapse formation. J Neurosci 28(40):10031–10044. https://doi.org/10.1523/jneurosci.0336-08.2008

Hatakeyama M, Kimura S, Naka T, Kawasaki T, Yumoto N, Ichikawa M et al (2003) A computational model on the modulation of mitogen-activated protein kinase (MAPK) and Akt pathways in heregulin-induced ErbB signalling. Biochem J 373 (Pt 2):451–463. https://doi.org/10.1042/bj20021824

Higgs HN, Pollard TD (2001) Regulation of actin filament network formation through ARP2/3 complex: activation by a diverse array of proteins. Annu Rev Biochem 70:649–676. https://doi.org/10.1146/annurev.biochem.70.1.649

Holbro T, Hynes NE (2004) ErbB receptors: directing key signaling networks throughout life. Annu Rev Pharmacol Toxicol 44:195–217. https://doi.org/10.1146/annurev.pharmtox.44.101802.121440

Honkura N, Matsuzaki M, Noguchi J, Ellis-Davies GC, Kasai H (2008) The subspine organization of actin fibers regulates the structure and plasticity of dendritic spines. Neuron 57(5):719–729. https://doi.org/10.1016/j.neuron.2008.01.013

Hosokawa T, Rusakov DA, Bliss TV, Fine A (1995) Repeated confocal imaging of individual dendritic spines in the living hippocampal slice: evidence for changes in length and orientation associated with chemically induced LTP. J Neurosci 15 (8):5560–5573

Huang YZ, Won S, Ali DW, Wang Q, Tanowitz M, Du QS et al (2000) Regulation of neuregulin signaling by PSD-95 interacting with ErbB4 at CNS synapses. Neuron 26(2):443–455

Huang Y, Lu W, Ali DW, Pelkey KA, Pitcher GM, Lu YM et al (2001) CAKbeta/Pyk2 kinase is a signaling link for induction of long-term potentiation in CA1 hippocampus. Neuron 29(2):485–496

Huber F, Käs J, Stuhrmann B (2008) Growing actin networks form lamellipodium and lamellum by self-assembly. Biophys J 95(12):5508–5523. https://doi.org/10.1529/biophysj.108.134817

Javitt DC, Duncan L, Balla A, Sershen H (2005) Inhibition of system A-mediated glycine transport in cortical synaptosomes by therapeutic concentrations of clozapine: implications for mechanisms of action. Mol Psychiatry 10(3):275–287. https://doi.org/10.1038/sj.mp.4001552

Kalia LV, Gingrich JR, Salter MW (2004) Src in synaptic transmission and plasticity. Oncogene 23(48):8007–8016. https://doi.org/10.1038/sj.onc.1208158

Kholodenko BN, Demin OV, Moehren G, Hoek JB (1999) Quantification of short term signaling by the epidermal growth factor receptor. J Biol Chem 274 (42):30169–30181

Kojima N, Wang J, Mansuy IM, Grant SG, Mayford M, Kandel ER (1997) Rescuing impairment of long-term potentiation in fyn-deficient mice by introducing Fyn transgene. Proc Natl Acad Sci USA 94(9):4761–4765

Kwon OB, Longart M, Vullhorst D, Hoffman DA, Buonanno A (2005) Neuregulin-1 reverses long-term potentiation at CA1 hippocampal synapses. J Neurosci 25 (41):9378–9383. https://doi.org/10.1523/jneurosci.2100-05.2005

Li B, Woo RS, Mei L, Malinow R (2007) The neuregulin-1 receptor erbB4 controls glutamatergic synapse maturation and plasticity. Neuron 54(4):583–597. https://doi.org/10.1016/j.neuron.2007.03.028

Lin CH, Espreafico EM, Mooseker MS, Forscher P (1996) Myosin drives retrograde F-actin flow in neuronal growth cones. Neuron 16(4):769–782

Lu YM, Roder JC, Davidow J, Salter MW (1998) Src activation in the induction of long-term potentiation in CA1 hippocampal neurons. Science 279(5355):1363–1367

Ma L, Huang YZ, Pitcher GM, Valtschanoff JG, Ma YH, Feng LY et al (2003) Ligand-dependent recruitment of the ErbB4 signaling complex into neuronal lipid rafts. J Neurosci 23(8):3164–3175

Mayadevi M, Praseeda M, Kumar KS, Omkumar RV (2002) Sequence determinants on the NR2A and NR2B subunits of NMDA receptor responsible for specificity of phosphorylation by CaMKII. Biochim Biophys Acta 1598(1–2):40–45

Moehren G, Markevich N, Demin O, Kiyatkin A, Goryanin I, Hoek JB, Kholodenko BN (2002) Temperature dependence of the epidermal growth factor receptor signaling network can be accounted for by a kinetic model. Biochemistry 41 (1):306–320

Mogilner A (2006) On the edge: modeling protrusion. Curr Opin Cell Biol 18(1):32–39. https://doi.org/10.1016/j.ceb.2005.11.001

Mogilner A, Edelstein-Keshet L (2002) Regulation of actin dynamics in rapidly moving cells: a quantitative analysis. Biophys J 83(3):1237–1258

Murphy SP, Bielby-Clarke K (2008) Neuregulin signaling in neurons depends on ErbB4 interaction with PSD-95. Brain Res 1207:32–35. https://doi.org/10.1016/j.brainres.2008.02.063

Nakagawa T, Engler JA, Sheng M (2004) The dynamic turnover and functional roles of alpha-actinin in dendritic spines. Neuropharmacology 47(5):734–745. https://doi.org/10.1016/j.neuropharm.2004.07.022

Nakakuki T, Yumoto N, Naka T, Shirouzu M, Yokoyama S, Hatakeyama M (2008) Topological analysis of MAPK cascade for kinetic ErbB signaling. PLoS One 3 (3):e1782. https://doi.org/10.1371/journal.pone.0001782

Nakayama AY, Luo L (2000) Intracellular signaling pathways that regulate dendritic spine morphogenesis. Hippocampus 10(5):582–586. https://doi.org/10.1002/1098-1063(2000)10:5<582::aid-hipo8>3.0.co;2-k

Okamoto K, Nagai T, Miyawaki A, Hayashi Y (2004) Rapid and persistent modulation of actin dynamics regulates postsynaptic reorganization underlying bidirectional plasticity. Nat Neurosci 7(10):1104–1112. https://doi.org/10.1038/nn1311

Okamoto K, Narayanan R, Lee SH, Murata K, Hayashi Y (2007) The role of CaMKII as an F-actin-bundling protein crucial for maintenance of dendritic spine structure. Proc Natl Acad Sci USA 104(15):6418–6423. https://doi.org/10.1073/pnas.0701656104

Okubo-Suzuki R, Okada D, Sekiguchi M, Inokuchi K (2008) Synaptopodin maintains the neural activity-dependent enlargement of dendritic spines in hippocampal neurons. Mol Cell Neurosci 38(2):266–276. https://doi.org/10.1016/j.mcn.2008.03.001

Pollard TD (2003) The cytoskeleton, cellular motility and the reductionist agenda. Nature 422(6933):741–745

Pollard TD, Borisy GG (2003) Cellular motility driven by assembly and disassembly of actin filaments. Cell 112(4):453–465

Qiu C, Tarrant MK, Choi SH, Sathyamurthy A, Bose R, Banjade S et al (2008) Mechanism of activation and inhibition of the HER4/ErbB4 kinase. Structure 16 (3):460–467. https://doi.org/10.1016/j.str.2007.12.016

Racz B, Weinberg RJ (2006) Spatial organization of cofilin in dendritic spines. Neuroscience 138(2):447–456. https://doi.org/10.1016/j.neuroscience.2005.11.025

Rácz B, Weinberg RJ (2008) Organization of the Arp2/3 complex in hippocampal spines. J Neurosci 28(22):5654–5659. https://doi.org/10.1523/JNEUROSCI.0756-08.2008

Rohm B, Rahim B, Kleiber B, Hovatta I, Püschel AW (2000) The semaphorin 3A receptor may directly regulate the activity of small GTPases. FEBS Lett 486(1):68–72. https://doi.org/10.1016/S0014-5793(00)02240-7

Rostaing P, Real E, Siksou L, Lechaire JP, Boudier T, Boeckers TM et al (2006) Analysis of synaptic ultrastructure without fixative using high-pressure freezing and tomography. Eur J Neurosci 24(12):3463–3474. https://doi.org/10.1111/j.1460-9568.2006.05234.x

Rostas JA, Brent VA, Voss K, Errington ML, Bliss TV, Gurd JW (1996) Enhanced tyrosine phosphorylation of the 2B subunit of the N-methyl-D-aspartate receptor in long-term potentiation. Proc Natl Acad Sci USA 93(19):10452–10456

Roysommuti S, Carroll SL, Wyss JM (2003) Neuregulin-1beta modulates in vivo entorhinal-hippocampal synaptic transmission in adult rats. Neuroscience 121(3):779–785. https://doi.org/10.1016/s0306-4522(03)00503-7

Salter MW, Kalia LV (2004) Src kinases: a hub for NMDA receptor regulation. Nat Rev Neurosci 5:317. https://doi.org/10.1038/nrn1368

Saneyoshi T, Wayman G, Fortin D, Davare M, Hoshi N, Nozaki N et al (2008) Activity-dependent synaptogenesis: regulation by a CaM-kinase kinase/CaM-kinase I/betaPIX signaling complex. Neuron 57(1):94–107. https://doi.org/10.1016/j.neuron.2007.11.016

Sarnago S, Elorza A, Mayor F (1999) Agonist-dependent phosphorylation of the G protein-coupled receptor kinase 2 (GRK2) by Src tyrosine kinase. J Biol Chem 274(48):34411–34416. https://doi.org/10.1074/jbc.274.48.34411

Schmidt JT, Morgan P, Dowell N, Leu B (2002) Myosin light chain phosphorylation and growth cone motility. J Neurobiol 52(3):175–188. https://doi.org/10.1002/neu.10083

Schaus TE, Taylor EW, Borisy GG (2007) Self-organization of actin filament orientation in the dendritic-nucleation/array-treadmilling model. Proc Natl Acad Sci USA 104(17):7086–7091. https://doi.org/10.1073/pnas.0701943104

Schoeberl B, Eichler-Jonsson C, Gilles ED, Muller G (2002) Computational modeling of the dynamics of the MAP kinase cascade activated by surface and internalized EGF receptors. Nat Biotechnol 20(4):370–375. https://doi.org/10.1038/nbt0402-370

Schubert V, Dotti CG (2007) Transmitting on actin: synaptic control of dendritic architecture. J Cell Sci 120(Pt 2):205–212. https://doi.org/10.1242/jcs.03337

Schubert V, Da Silva JS, Dotti CG (2006) Localized recruitment and activation of RhoA underlies dendritic spine morphology in a glutamate receptor-dependent manner. J Cell Biol 172(3):453–467. https://doi.org/10.1083/jcb.200506136

Schulze WX, Deng L, Mann M (2005) Phosphotyrosine interactome of the ErbB-receptor kinase family. Mol Syst Biol 1:2005.0008. https://doi.org/10.1038/msb4100012

Sharma K, Fong DK, Craig AM (2006) Postsynaptic protein mobility in dendritic spines: long-term regulation by synaptic NMDA receptor activation. Mol Cell Neurosci 31(4):702–712. https://doi.org/10.1016/j.mcn.2006.01.010

Shen K, Teruel MN, Subramanian K, Meyer T (1998) CaMKIIbeta functions as an F-actin targeting module that localizes CaMKIIalpha/beta heterooligomers to dendritic spines. Neuron 21(3):593–606

Shoham S, Javitt DC, Heresco-Levy U (2001) Chronic high-dose glycine nutrition: effects on rat brain cell morphology. Biol Psychiatry 49(10):876–885

Solomaha E, Szeto FL, Yousef MA, Palfrey HC (2005) Kinetics of Src homology 3 domain association with the proline-rich domain of dynamins: specificity, occlusion, and the effects of phosphorylation. J Biol Chem 280(24):23147–23156. https://doi.org/10.1074/jbc.M501745200

Stefansson H, Sigurdsson E, Steinthorsdottir V, Bjornsdottir S, Sigmundsson T, Ghosh S, Brynjolfsson J, Gunnarsdottir S, Ivarsson O, Chou TT, Hjaltason O, Birgisdottir B, Jonsson H, Gudnadottir VG, Gudmundsdottir E, Bjornsson A, Ingvarsson B, Ingason A, Sigfusson S, Hardardottir H, Harvey RP, Lai D, Zhou M, Brunner D, Mutel V, Gonzalo A, Lemke G, Sainz J, Johannesson G, Andresson T, Gudbjartsson D, Manolescu A, Frigge ML, Gurney ME, Kong A, Gulcher JR, Petursson H, Stefansson K (2002) Neuregulin 1 and susceptibility to schizophrenia. Am J Hum Genet 71(4):877–892

Svitkina TM, Bulanova EA, Chaga OY, Vignjevic DM, Kojima S, Vasiliev JM, Borisy GG (2003) Mechanism of filopodia initiation by reorganization of a dendritic network. J Cell Biol 160(3):409–421. https://doi.org/10.1083/jcb.200210174

Svitkina TM, Verkhovsky AB, McQuade KM, Borisy GG (1997) Analysis of the actin-myosin II system in fish epidermal keratocytes: mechanism of cell body translocation. J Cell Biol 139(2):397–415

Sweeney C, Lai C, Riese DJ 2nd, Diamonti AJ, Cantley LC, Carraway KL 3rd. (2000) Ligand discrimination in signaling through an ErbB4 receptor homodimer. J Biol Chem 275(26):19803–19807. https://doi.org/10.1074/jbc.C901015199

Tezuka T, Umemori H, Akiyama T, Nakanishi S, Yamamoto T (1999) PSD-95 promotes Fyn-mediated tyrosine phosphorylation of the N-methyl-D-aspartate receptor subunit NR2A. Proc Natl Acad Sci USA 96(2):435–440

Ultanir SK, Kim JE, Hall BJ, Deerinck T, Ellisman M, Ghosh A (2007) Regulation of spine morphology and spine density by NMDA receptor signaling in vivo. Proc Natl Acad Sci USA 104(49):19553–19558. https://doi.org/10.1073/pnas.0704031104

Van Troys M, Huyck L, Leyman S, Dhaese S, Vandekerkhove J, Ampe C (2008) Ins and outs of ADF/cofilin activity and regulation. Eur J Cell Biol 87(8–9):649–667. https://doi.org/10.1016/j.ejcb.2008.04.001

Wang YT, Salter MW (1994) Regulation of NMDA receptors by tyrosine kinases and phosphatases. Nature 369(6477):233–235

Woo RS, Li XM, Tao Y, Carpenter-Hyland E, Huang YZ, Weber J et al (2007) Neuregulin-1 enhances depolarization-induced GABA release. Neuron 54 (4):599–610. https://doi.org/10.1016/j.neuron.2007.04.009

Xu J, Weerapura M, Ali MK, Jackson MF, Li H, Lei G et al (2008) Control of excitatory synaptic transmission by C-terminal Src Kinase. J Biol Chem 283 (25):17503–17514. https://doi.org/10.1074/jbc.M800917200

Yu XM, Askalan R, Keil GJ 2nd, Salter MW (1997) NMDA channel regulation by channel-associated protein tyrosine kinase Src. Science 275(5300):674–678

Zhou W, Carpenter G (2001) Heregulin-dependent translocation and hyperphosphorylation of ErbB-2. Oncogene 20(29):3918–3920. https://doi.org/10.1038/sj.onc.1204517

The manufacturer's authorised representative in the EU is Springer Nature Customer Service Centre GmbH, Europaplatz 3, 69115 Heidelberg, Germany. If you have any concerns regarding our products, please contact ProductSafety@springernature.com

Printed and bound by CPI Group (UK) Ltd, Croydon, CR0 4YY
05/10/2026
02228129-0004